AN INTIMATE HISTORY
OF HUMANITY

Theodore Zeldin, educated at Birkbeck College London
and Christ Church Oxford, is senior fellow of St
Antony's College Oxford. He has been awarded the
Wolfson Prize for History, been elected a member of
the European Academy, and figures on the *Magazine
Littéraire*'s list of the hundred most important thinkers
in the world today.

BY THEODORE ZELDIN

A History of French Passions
The French
Happiness

PRAISE FOR *AN INTIMATE HISTORY OF HUMANITY*

'It is a narrative of private lives but it extends much further. It is a
universal history but it has an extraordinary intimacy of tone...In
trying to understand the inflections of human behaviour, he is able to
uncover the secret springs that lie within it'
Peter Ackroyd, *The Times*

'A vivid and encyclopedic look at everyday life and all the little
things we take for granted – a brilliant piece of work'
J. G. Ballard, *The Times*

'Brilliantly illuminating...accessible, audacious and though-provok-
ing...deeply humane and sensitively concerned with finding answers
to the knots in which we tie ourselves'
Alain de Botton, *The Independent*

'Never boring, never obscure, learned but with a smile, caustic when
it is necessary, joyfully massacring the platitudes and clichés which
push us into pessimism, Zeldin...offers us a guide book for a painless
crossing from this century to the next'
Jean David, *V.S.D.*

'A mixture of kindness and lucidity, a book full of discoveries and
surprises, a superb antidote against stereotyped thinking'
L'Express

'A courageous, often profound, and extraordinary attempt by one of
England's best historians to cut through the pessimism and parochial-
ism of the profession and to find the bonds of humanity underlying its
conventional divisions...No short review can do justice to the richness,
humor, humanity, and range of this important book'
Kirkus Reviews

'History at its most supple, teasing and curious'
The Tatler

Theodore Zeldin

AN INTIMATE
HISTORY OF
HUMANITY

VINTAGE

Published by Vintage 1998

23 25 27 29 30 28 26 24 22

Copyright © Theodore Zeldin 1994

First published in Great Britain by
Sinclair-Stevenson in 1994

First published in paback by Minerva in 1995

Vintage
The Random House Group Limited
20 Vauxhall Bridge Road, London SW1V 2SA

Random House Australia (Pty) Limited
20 Alfred Street, Milsons Point, Sydney,
New South Wales 2061, Australia

Random House New Zealand Limited
18 Poland Road, Glenfield,
Auckland 10, New Zealand

Random House (Pty) Limited
Endulini, 5a Jubilee Road, Parktown 2193, South Africa

The Random House Group Limited Reg. No. 954009

www.randomhouse.co.uk

A CIP catalogue record for this book
is available from the British Library

ISBN 0 7493 9623 7

Papers used by Random House are natural,
recyclable products made from wood grown in sustain-
able forests. The manufacturing processes conform to the
environmental regulations of the country of origin.

Printed and bound in Denmark by
Nørhaven Paperback, Viborg

Contents

Preface

Our imaginations are inhabited by ghosts. Here are the results of my investigations of the familiar ghosts which reassure, the lazy ones which make us obstinate and, above all, the frightening ones which discourage. The past haunts us, but from time to time people have changed their minds about the past. I want to show how, today, it is possible for individuals to form a fresh view both of their own personal history and of humanity's whole record of cruelty, misunderstanding and joy. To have a new vision of the future, it has always first been necessary to have a new vision of the past.

Each of my chapters begins with the portrait of a living person who has desires and regrets in which you may perhaps recognise something of yourself, but who is also restrained by attitudes inherited from origins long forgotten. The mind is a refuge for ideas dating from many different centuries, just as the cells of the body are of different ages, renewing themselves or decaying at varying speeds. Instead of explaining the peculiarity of individuals by pointing to their family or childhood, I take a longer view: I show how they pay attention to – or ignore – the experience of previous, more distant generations, and how they are continuing the struggles of many other communities all over the world, whether active or extinct, from the Aztecs and the Babylonians to the Yoruba and the Zoroastrians, among whom they have more soul-mates than they may realise.

You will not find history laid out in these pages as it is in museums, with each empire and each period carefully separated. I am writing about what will not lie still, about the past which is alive in people's minds today. However, before I explain what I want to do with these ghosts, I should like to introduce you to a few of them.

How humans have repeatedly lost hope, and how new encounters, and a new pair of spectacles, revive them

'My life is a failure.' That is Juliette's verdict on herself, though she very rarely makes it public. Could her life have been different? Yes, just as the history of humanity could have been different.

She carries herself with dignity, observing all that happens around her, but keeping her reactions to herself. Only in brief moments, hesitatingly, will she reveal a little of what she thinks, which she does in whispers, as though truth is too brittle to be brought out of its wrappings. A glint in her eye says: You may think I am stupid, but I know I am not.

Juliette is fifty-one, and has been a domestic servant since the age of sixteen. She has so mastered the art of looking after a house, preparing and serving meals, that all overburdened mothers who catch a glimpse of her, and who can afford to, have the same thought: how can they persuade this paragon to work for them? Has she a few hours to spare? But though an ideal family help, she has been unable to cope with her own family. At work she is wholly reliable, taking endless care with every detail; but in her own home those qualities have never been sufficient.

Her mother was also a domestic servant. 'I have nothing to complain of,' says Juliette. 'She brought us up very well, even if she did spank us.' Widowed when Juliette was only seven, she went to work early and returned late: 'We didn't see her much.' So Juliette fooled around rather than doing her lessons: 'I didn't see the point of school.' She met no ally who cared for her specially, no mentor from outside her small world to help her, and she left with no paper qualifications, no entry ticket to anywhere.

At the age of sixteen, 'I did something stupid.' So she married the father of her baby and had eight more children. Babies were pure joy to her; she loves to hug them; but only so long as they are babies. Once they grow up, 'They become difficult.' Her husband was a handsome carpenter doing his military service, and at first he was nice to her: 'I was truly in love.' But very soon things went wrong. When her first daughter was six months old, she discovered from her neighbours that he had a mistress. From then on there was no trust between them. He went out a lot, to visit the mistress, she always suspected; then he took to drink, working less and less, saying work was too tiring. He began beating her: 'I have scars all over my body.' But she told nobody, she was so ashamed. 'When I saw him coming home through the garden, I was terrified.' Why did she not leave him? 'I was too frightened. I was alone in his home town, where I knew nobody; I was cut off from my family after my marriage; I didn't see my sisters for fourteen years; he stopped me going out, and it was the children who did the shopping. He stopped me going even to my brother's funeral. I had no women friends any more. I went out only to work.' And that, of course, meant she could not look after the children, who were farmed out to foster parents by the social services. The humiliation has left Juliette very sensitive. When people want to insult her, they say: 'You couldn't even bring up your own children.' She protests, 'People shouldn't say things like that without knowing the facts.'

'I started hitting my husband back in the end; I should have done so earlier.' It was a long time before she managed to leave him. He died one month after the divorce: 'I wasn't sad; indeed, I laughed. I laugh now, but when we were together, I didn't laugh.' Ever since, she has worked with one purpose only: 'My aim in life was to own my own home.' And recently she paid off the mortgage on her flat. It is the foundation of her pride, making her a stronger person. But she is too frightened to live alone, though she has tried to do so. Now she has a man with her: 'It's for security, not to be totally alone at night.' Sometimes she would prefer to have no one at all, and she is adamant that she does not want to marry this man. 'In that I'm like a young girl of today, for whom marriage is no longer essential.' They get on because he is also a divorcee and 'wants peace'. He does the cooking, and she does the shopping: she loves touring the markets on Sundays, just looking, and enjoying the feel of new cloth, which is like a dream unsullied by reality. Having her own money to spend gives her a great sense of freedom. He has bought his own country cottage, because she has made it clear to him that if they quarrel he will have to move out: she constantly reminds him that the flat is hers, and says defiantly, 'I can go out when I like, I can go and see a friend when I like.'

They do not speak much. When she gets home in the evening, her pleasure is to rest, to lie alone on her bed in the dark. Never reading a book, and hardly ever watching television, she prefers instead to think, with the lights out, about her past life: her mother, her husband, her children, and the terror of unemployment. 'If a time came when there was no work for the children, that wouldn't be pretty.' She is sad that their life will not be better than her own: 'It isn't fair.' Her explanation is that there are too many foreigners in France, who take the work and the housing, which means 'the poor French have nothing. I don't want to criticise Arabs or blacks, but I think it's unfair. It's because of them that my children's life is hard.' One daughter works in a factory, another at the prefecture of police, a third is a domestic servant, as though this family is condemned to the worst-paid jobs for all eternity.

And what does Juliette think about at work? 'Why, nothing. At work I don't think, or I think about my saucepans.' Work is relaxation from home. For though she has organised her home life so that she can have peace, people are prickly porcupines to Juliette, and getting on with them involves constant vigilance. Though she feels less fragile now, she is still very easily hurt by what others say about her. She prefers to work alone, as an independent cleaning woman, because she fears the gossip of offices and factories: 'People repeat things about you, twisting your words, and sometimes that can cost you dearly.' There is nothing she hates more

3

than to be criticised; every hint of disapproval is a bruise upon a bruise. Keeping her head high is a constant effort, and dignity requires that she should not complain. She never told her sisters how her husband treated her. When she visits them now, she is careful not to say what she thinks of their way of managing their lives; and they never speak to her about her past: 'They know they would make me really angry.' For example, her younger sister, whose husband is dead and who now lives with a man with whom she is not really happy, often says to him, 'Pack your bags and clear out.' Juliette is careful not to interfere in their quarrels. 'It's her problem.' And if, despite herself, she does spill a tiny drop of criticism, the sister replies, 'Mind your own onions.' All her sisters, she says, are as careful as she is: they do not show their anger.

'In families with children, there are always quarrels.' Of her own children, perhaps the one who is doing best is the eldest, whose husband is also dead and who lives with a man who obeys her: 'She's the boss, and he's an idiot, because she's too hard on him.' But she adds, 'I don't take any interest in the private lives of my children. If they quarrel in front of me, I won't interfere.'

The most irritating person in Juliette's world, like a mosquito that bites and will not go away, is her man's daughter, aged seventeen, who lives in a hostel because her mother is separated after a second failed marriage. Juliette, for all her wisdom, is a classic stepmother. 'You can't come here for Mother's Day because you're not my daughter. You can come for Father's Day.' This girl, she is convinced, is 'really wicked': she has learnt about Juliette's troubles and is always saying, 'You're a failure.' Juliette becomes furious. 'If she was my daughter, I'd give her a spanking': the girl is spoilt, badly brought up, does not help with the housework: the new generation have it too easy. The girl replies that she will complain to the judge: 'You'll go to prison,' and Juliette is frightened of being involved with the law. Her man does not interfere in these disputes: 'He wants peace.' So when the arguing becomes unbearable, 'I go out for a walk with my cheque book.' It is like a passport that proves that Juliette is an independent woman. She feels she is making progress in the art of being independent by the way she uses it. Only a few years ago, she would recover from an insult by spending wildly: 'I didn't think twice before buying, I didn't compare prices. But now I am more stable. It's my friend who has probably influenced me in this. He's careful; he has made me more balanced. I used to be more nervous than I am.' The consumer society is a giant tranquilliser for raw nerves.

When Juliette was young, she worked thirteen hours a day; fewer now, but she still earns less than most people. It would be possible to find better-paid work, but she likes employers whom she can cope with and understand, who do not disturb her with criticism. To ensure a proper balance, she works for several, distributing her hours like someone on a diet. 'It would be impossible if I had an employer who shouted at me all day, and then I had to go home to a man who shouted at me all evening.' One of the women she cleans for does indeed shout, but 'has a good heart'. Another is the granddaughter of a former president of the French Republic, who lies on a sofa all day doing nothing, suffering from various ailments: 'If she was not so sorry for herself, she could do something with her life'; but her kindness is perfect. A third has problems with his children and with his health: 'Take care of yourself, I say. Yes, doctor, he replies.' A fourth is a doctor, who shows no interest in her when she is ill, in contrast to her fifth client, who is all attention the moment she utters a little cough: she remembers it as a high point in her life that he allowed her to go home an hour early once, saying, 'It is not the factory here.'

Some of these employers, at least, she considers to be her 'friends'. To one of them she said, 'Whatever happens, I won't desert you. I wouldn't allow myself to leave you. I couldn't find anyone as kind.' She has worked for the doctor for twenty-four years, despite his limitations, 'because I know his character. I know how to deal with him. I say nothing when I see he is in a bad temper.' The low points come when they complain about her work. 'The mistress of a house must not insult an employee in front of guests: she should go to the kitchen to do it. Otherwise it's vulgar.' Once at a dinner party Juliette forgot to place the potatoes around the meat, putting them by mistake on a separate plate. Her mistress called her a stupid cow. She burst into tears and said she would leave. 'The doctor apologised, but his wife would not.' Juliette stayed. At another house, she was called skivvy. 'I will not tolerate being called that.' But then anger subsides: 'One must adapt oneself to everyone. Every employer presents problems. There are some who understand the life of a *femme de ménage*, but some do not.' And she consoles herself: 'These people rely on me. I become more cultured with them: they tell me things. One of them – he's an educated man – tells me all about his problems, but he says, "Don't tell anyone." So it's between him and me.'

Perhaps Juliette's life could have been different if the meetings which have decided its course had been less silent, superficial or routine, if more thoughts had been exchanged, if humanity had been more able to show itself in them. But they were restrained by the ghosts which continue to influence what employers and strangers and even people who live together may or may not say to each other. Juliette insists that 'given my abilities' she could have had a better job, that working for old people would have appealed to her, and that it was her lack of paper qualifications that obstructed her. It was even more tragic that none of the influential people she worked for believed it was in their interest to help her start a more satisfying career. Her conclusion is, 'My life is finished.'

There are currently several different ways of interpreting this story. One can say: that is what life is like, and there are many reasons why it is so. Or one can hope that if the knots into which humanity ties itself could be untied, and its crazy institutions made more sensible, then life could be changed, and poverty could be abolished, but it may take decades, perhaps even centuries. Or one can hate life for being so cruel and try to bear it by making fun of it, or parodying it, or delighting in minute descriptions of it, all the while protecting oneself against disappointment by refusing to suggest solutions to problems, and by condemning all such efforts as naive.

My purpose is different. Behind Juliette's misfortunes, I see all those who have lived but thought of themselves as failures, or been treated as such. The worst sense of failure was to realise that one had not really lived at all, not been seen as an independent human being, never been listened to, never been asked for an opinion, regarded as a chattel, the property of another. That was what happened publicly to slaves. We are all of us descended from slaves, or almost slaves. All our autobiographies, if they went back far enough, would begin by explaining how our ancestors came to be more or less enslaved, and to what degree we have become free of this inheritance. Legally, of course, slavery has been abolished (not all that long ago: Saudi Arabia was the last country to end it in 1962), but slavery also has a metaphorical, broader meaning: it is possible to be a slave of the passions, or of one's work, or of one's habits, or of the spouse one cannot for various reasons leave. The world is still full of people who, though they have no recognised slave masters, see themselves as having little freedom, as being at the mercy of uncontrollable, anonymous economic and social forces, or of their circumstances, or of their own stupidity, and whose personal ambitions are permanently blunted thereby. The modern descendant of a slave has even less hope than a sinner, who can repent; the impotent, trapped human being can see no comparable instant cure. Juliette is not a slave: nobody owns her. She is not a serf: nobody has a right to her labour. But to think one's life is finished, or that it is a failure, is to suffer from the same sort of despair which afflicted people in the days when the world believed it could not do without slaves. It is therefore important to understand what legal slavery meant.

Humans became slaves in the past for three main reasons. The first was fear: they did not want to die, however much suffering life caused. They agreed to be despised by kings and knights and other addicts of violence, who believed that death in battle was the highest honour, and for whom to enslave humans, and to domesticate animals, was part of the same search for power and comfort. But slaves also put up with being treated like animals, bought and sold, heads shaven, branded, beaten, called by contemptuous names (Monkey, Downcast, Strumpet, Irritation), because oppression seemed to be an inescapable ingredient of life for most people. In Han China, the word 'slave' derived from the word 'child', or 'wife and child'. Similar unquestioning obedience was also imposed on the majority of humanity in most parts of the world, whether officially slave or not.

Before twelve million Africans were kidnapped to be slaves in the New World, the main victims were the Slavs, who gave their name to slavery. Hunted by Romans, Christians, Muslims, Vikings and Tatars, they were exported all over the world. Slav came to mean foreigner; most religions taught that it was acceptable to enslave foreigners; British children who were exported as slaves – the girls fattened up to fetch a higher price – ended up as Slavs. More recently, when Slavs found themselves ruled by tyrants and saw no hope of escaping, some gloomily concluded that there must be something in the character of Slavs which dooms them to being enslaved. This is false reasoning, pretending that what has happened had to happen. No free person can believe that: it is a reasoning imposed on slaves to make them despair.

Fear has nearly always been more powerful than the desire for freedom: humans are not born free. However, the Emperor Maurice of Byzantium (582–602) discovered an exception. He was amazed by three Slavs he captured, who carried no arms. All they had with them were guitars or zithers, and they wandered around singing about the joys of liberty, of being in the open fields and the fresh breezes. They told him, 'It is normal for people who are strangers to war to devote themselves with fervour to music.' Their songs were about free will, and they were known as the free-will people. In 1700 there were still such people, when Peter the Great decreed that there should be no more of them: everyone must be part of a legal estate, with fixed duties. But 150 years later Tara Sevcenko, a liberated Ukrainian serf, was singing poems in the same tradition, lamenting that 'liberty has been put to sleep by the drunken Tsar', insisting that hope could be found in nature:

> *Listen to what the sea says,*
> *Ask the black mountains.*

There was slavery, first of all, because those who wished to be left alone could not keep out of the way of those who enjoyed violence. The violent have been victorious for most of history because they kindled the fear with which everyone is born.

Secondly, humans became slaves 'voluntarily'. In Aztec Mexico, the majority of slaves chose, if that is the word, to be so, overcome by depression, wanting to be rid of their responsibilities – retired players of patolli, the national ball game, for example, ruined by their addiction to the sport, or women tired of love and preferring to be sure of having something to eat: the basis of the slave contract was that a slave had to be fed, and if he was not, he had to be freed. The Muscovites, when they learnt to resist their predators, and began to enslave each other instead, developed eight different forms of slavery, of which by far the most common was the 'voluntary' one. Their frontier society was established without institutions of charity. The hungry sold themselves as slaves. Between the fifteenth and eighteenth centuries, about one-tenth of Muscovites became slaves, so that there were more slaves than there were townspeople, or soldiers, or priests. An American historian has compared these slaves to the American poor living on welfare.

Slavery in Russia was a kind of pawnshop for people who had nothing but themselves to sell; one-third of slaves were normally absconders on the run, but usually they came back, worn out by freedom, unable to cast off the prisoner's mentality: 'Not every slave dreams of freedom. After a few years of being totally dominated, an independent existence in harsh reality became almost unthinkable,' says their historian Hellie. In America escape was harder: the southern states of the USA probably had one of the harshest slave systems in the world, because slaves were employed intensively to make high profits from agriculture, whereas in Russia and China they were used mainly as domestics. But whatever the precise conditions of treatment, the fact that there were so many kinds of slaves, and that each individual could be subject to slightly different forms of abuse, meant that each could consider he had some privileges, that he was not the lowest of the low; envy blinded him to the common suffering; on American plantations, it was possible to find African slaves to whip other African slaves. In other words, once an institution is set up, even those who suffer from it find ways, however slight, of exploiting it, and willy-nilly they help it to survive.

The third kind of slave was the ancestor of today's ambitious executive and bureaucrat. Owning slaves gave prestige; being a slave meant working. Free men considered it beneath them to work for another; Roman aristocrats refused to be bureaucrats for the emperor. So he started a civil service using slaves, and the aristocrats employed slaves to manage their estates. Slaves had no family, no loyalty to anybody but their master. They made the most reliable officials, soldiers, private secretaries. The Ottoman and Chinese empires were often managed by slaves, sometimes by eunuch slaves, who rose to the highest posts and indeed sometimes ended as grand viziers and emperors; castration made sure that they placed loyalty to the state before family. There are no statistics to say how many people are morally castrated by their employers today.

The Russian word for work – *rabota* – comes from the word slave, *rab*. The origin of the leisure society is the dream of living like a master, with the work done by robots, mechanical slaves. The sting in the tail of this history of slavery is that once free, people often become robots, at least in part of their lives. There has been a great reluctance to abandon all forms of slavish behaviour. 'The height of misery is to depend on another's will,' said Publilius, a Syrian slave who became a popular entertainer and mime in ancient Rome. And yet the fantasies of romantic love are based on dependence. The freed slave often preferred to remain dependent, continuing in the same work; the stain of slavery took several generations to fade. In China and Africa the freed slave often became a sort of poor relation; in Europe a client. To live outside the protection of someone more powerful than oneself was too frightening an adventure.

The quality which was most remarkable about slaves – or at least those who did not get permanently drunk to forget their sorrow – was dignity. Many of them succeeded in asserting their autonomy even while being forced into menial work, pretending to accept their humiliations, playing a role, so that the master could live under the illusion that he was in charge, while they knew that he depended on them. 'Play fool, to catch wise,' was the favourite proverb of the Jamaican slave. And occasionally a slave owner did realise that he was not only being fooled but that he was a slave too: 'We use other people's feet when we go out, we use other people's eyes to recognise things, we use another person's memory to greet people, we use someone else's help to stay alive – the only things we keep for ourselves are our pleasures,' wrote Pliny the Elder in AD 77. This Roman slave owner, the author of an enormous Natural History, died as a result of getting too close to

Mount Vesuvius, whose eruption he wanted to witness: he knew he was a parasite, because observing nature is a good way to learn how to recognise parasites.

The solution for slavery was not its abolition, at least that was not a total solution, because new forms of slavery were invented under another name. The factory workers who toiled in poisonous air from sunrise till sunset and never saw daylight except on Sundays, obeying in silence, probably led even worse lives than many ancient slaves. And today, all those who prefer to do what they are told rather than think for themselves and shoulder the responsibility – one-third of Britons, according to a poll, say that is what they prefer – are the spiritual heirs of the voluntary slaves of Russia. It is important to remember that it is tiring, and trying, being free; and in times of exhaustion affection for freedom has always waned, whatever lip-service might be paid to it.

The conclusion I draw from the history of slavery is that freedom is not just a matter of rights, to be enshrined in law. The right to express yourself still leaves you with the need to decide what to say, to find someone to listen, and to make your words sound beautiful; these are skills which need to be acquired. All that the law says to you is that you can play your guitar, if you can get hold of one. So declarations of human rights provide only a few of the ingredients out of which freedom is made.

Just as important have been encounters with others, with people or with places, which have provided the inspiration and courage to escape from dull routines. There has been a waste of an opportunity every time a meeting has taken place and nothing has happened, as when it did not even occur to any of Juliette's employers to help her into the career she dreamed about. In most meetings, pride or caution still forbids one to say what one feels most deeply. The noise of the world is made out of silences.

So rather than begin by summarising ancient Greek philosophy, as is usual whenever freedom is mentioned, I prefer to use a single example of a person assembling the right mix of people and conditions, though it took him half a lifetime to do so. Domenicos Theotocopoulos, surnamed El Greco (1541–1614), would doubtless have remained an obscure artist of no significance, repetitively painting conventional icons, imprisoned by formalities and habits, had he not formed links with others and learned how to draw out the humanity from those who seemed to have none. Having absorbed all he could from the assorted traditions of his native Crete – ruled by Venetians, divided by Orthodox and Catholic Christianity, anchored to the past by refugees perpetuating the dying art of Byzantium – he added new dimensions to his heritage by travelling abroad. In Italy, he met a minor Croatian painter called Julio Glovio, known as 'the Macedonian', through whose introduction he became a pupil of Titian. He could then easily have put himself back in shackles as a pseudo-Italian minor portraitist doing what was asked of him; but he aspired to more than imitation. So at the age of thirty-five he settled in Toledo. Asked why, he replied, 'I am not obliged to answer that question.' It was dangerous to say publicly that here he felt free, that here there were no rivals to haunt him, that his ambition to paint – as he put it – 'more honestly and decently' than Michelangelo could only be achieved in a frontier city. Toledo reverberated with excitement because it knew what both toleration and persecution meant; Christians, Muslims and Jews had once lived in it side by side; one of its kings had been proud to call himself Emperor of the Three Religions and another to have his epitaph inscribed on his tombstone in Castilian, Arabic and Hebrew; and yet El Greco witnessed over a thousand supposed heretics being brought to trial before the local Inquisition. Here, living in the old Jewish quarter, both solitary and sociable, surrounded both by the spiritual fervour of the Counter-Reformation and by philosophising friends, he was stimulated to try to reconcile the seemingly irreconcilable, to paint the divine and the human intertwined, to have the courage to put his colours directly on canvas without preliminary drawing, as though character was too fluid to have hard boundaries. He saw painting as part of the pursuit of knowledge and the understanding of the individual.

It took a long time for Spaniards to recognise him as one of their own: the 1910 catalogue of the Prado Museum still listed him as a member of the 'Italian School'. It takes a long time for people to recognise their soul-mates when they have too limited an idea of who they are themselves. It took Spaniards a long time to realise that their contribution to the history of the harmonisation of incompatibilities was more important than their contribution to the history of pride, or to appreciate Alonso de Castrillo's saying in 1512 that eventually people grow 'weary of obedience' (just as people may eventually grow weary of freedom if they do not know what to do with it).

Today, the whole of humanity can see something of itself in the paintings of El Greco, who died owning only one spare suit, two shirts and a beloved library of books on every subject. Because of him, everybody can feel, to a certain extent, a citizen of Toledo. He is an example of a person who discovers what humans have in common. I shall go deeper into this question of how links are formed, or revealed, between apparently isolated individuals, even across the centuries; but before I do so, I shall say a little more about my method and my purpose.

What we make of other people, and what we see in the mirror when we look at ourselves, depends on what we know of the world, what we believe to be possible, what memories we have, and whether our loyalties are to the past, the present or the future. Nothing influences our ability to cope with the difficulties of existence so much as the context in which we view them; the more contexts we can choose between, the less do the difficulties appear to be inevitable and insurmountable. The fact that the world has become fuller than ever of complexity of every kind may suggest at first that it is harder to find a way out of our dilemmas, but in reality the more complexities, the more crevices there are through which we can crawl. I am searching for the gaps people have not spotted, for the clues they have missed.

I start with the present and work backwards, just as I start with the personal and move to the universal. Whenever I have come across an impasse in present-day ambitions, as revealed in the case studies of people I have met, I have sought a way out by placing them against the background of all human experience in all centuries, asking how they might have behaved if, instead of relying only on their own memories, they had been able to use those of the whole of humanity.

The world's memories are normally stored in such a way that it is not easy to use them. Each civilisation, each religion, each nation, each family, each profession, each sex and each class has its own history. Humans have so far been interested mainly in their own private roots, and have therefore never claimed the whole of the inheritance into which they are born, the legacy of everybody's past experience. Each generation searches only for what it thinks it lacks, and recognises only what it knows already. I want to make a start on summarising that legacy, not by going chronologically through the deeds of the dead, but in such a way that individuals can make use of those parts of their legacy which affect what they care about most.

When, in the past, people have not known what they wanted, when they have lost their sense of direction, and everything appeared to be falling apart, they have generally found relief by changing the focus of their vision, switching their attention. What once seemed all-important is suddenly hardly noticed any more. Political ideals thus collapse abruptly and are replaced by personal concerns, materialism succeeds idealism, and from time to time religion returns. I want to show how priorities are changing today, and what sort of spectacles are needed to observe them. In the course of history, humans have repeatedly changed the spectacles through which they have looked at the world and themselves.

In 1662 the beginning of a major shift in attention was marked by the establishment of the Royal Society of London. It was needed, said its founders, because people did not know what to look for or how. These scientists, and their successors, opened up huge territories for exploration, making the world look quite different. But scientific discovery is a specialist activity; most people can only watch in awe, and it does not help them to decide how to lead their daily lives.

In the nineteenth century, shifts in attention became more frequent and so more confusing. Alexis de Tocqueville's journey to the USA in 1831 was inspired by the conviction that America could provide a glimpse into the future, and that one could discover there what amazing things could be done with liberty; reforming political institutions to make them more democratic became the goal of almost everyone engaged in the pursuit of happiness; but Tocqueville came back with warnings about the looming tyranny of majorities and there is still no place where minorities are wholly satisfied. In the same year, Darwin's journey into the animal kingdom, which humans till then believed to exist for their benefit, switched attention to the struggle for life, which was increasingly seen as dominating every aspect of existence. But Darwin himself complained that his doctrines made him feel 'like a man who has become colour-blind', who has lost 'the higher aesthetic tastes', and that his mind had become 'a kind of machine for grinding general laws out of large collections of facts', causing a 'loss of happiness' and an 'enfeebling [of] the emotional part of our nature'. Marx's journey into the sufferings of the working class, and his invitation to revolution, tore the world apart for a hundred years, though it soon became obvious that revolutions are incapable of keeping their promises, however honestly made. Then, in the last years of the century, Freud embarked on a journey into the unconscious of the neurotics of Vienna, which changed what people saw inside themselves, what they worried about, and whom they blamed, but the hope that they would forgive once they understood has not been realised.

All these thinkers put the idea of conflict at the centre of their vision. The world continues to be haunted by that idea. Even those who want to abolish conflict use its methods to fight it.

However, the originality of our time is that attention is turning away from conflict to information. The new ambition is to prevent disasters, illnesses and crimes before they occur and to treat the globe as a single whole; women's entry into the public sphere is reinforcing the challenge to the tradition that conquest is the supreme goal of existence; more attention is being given to understanding other people's emotions than to making and unmaking institutions.

Yet much of what people do, despite these new yearnings, is governed by old ways of thinking. Both politics and economics have been powerless in the face of the obstinacy of entrenched mentalities. Mentalities cannot be changed by decree, because they are based on memories, which are almost impossible to kill. But it is possible to expand one's memories by expanding one's horizons, and when that happens, there is less chance that one will go on playing the same old tunes for ever and repeating the same mistakes.

Five hundred years ago, Europe underwent a Renaissance, as a result of four new encounters, imbibing four new stimulants, expanding its horizons. First, it revived forgotten memories of freedom and beauty. But it limited itself to the Greeks and the Romans. In this book, I have tried to open up the memories of the whole of humanity, and to use them to place the dilemmas of the present in a perspective which is not dominated by the idea of perpetual conflict. Secondly, in the Renaissance, Europe and America were betrothed, with the help of new technology; but it was more a geographical discovery of a continent than a discovery by humans of each other as persons; there is still silence and deafness among the inhabitants of the globe, though the technology exists to enable them to speak to anybody anywhere. I have investigated why ears remained blocked and how they can be unblocked. The Renaissance, thirdly, was based on a new idea of the importance of the individual. But this was a fragile foundation, because individuals depended on constant applause and admiration to sustain them. There is a shortage of applause in the world, and there is not enough respect to go round. I have searched for methods of increasing the supply. Finally, the Renaissance involved a new idea of what religion should mean. The ultimate purpose of all religions is to draw people together, but so far they have also separated them. Their history is unfinished. I have searched beyond their disagreements for the spiritual values that they share, not only among themselves, but with unbelievers too.

Enough is known, enough has been written, about what divides people; my purpose is to investigate what they have in common. So I have focused particularly on how they meet. The search for new and old types of relationships, both close and distant, has, in my view, been the most important human preoccupation throughout history, though it has been disguised under many names, taking many different paths. Meeting God has been the supreme aim for all those to whom the soul is a divine spark. Enchantment with a hero or guru has been at the heart of growing up. Personal life has been increasingly dominated by the hunt for the 'other half'. Parents have striven, more and more, to be on the same wavelength as their children. A good part of culture has been the process by which artists have been recognised as expressing the feelings of people they have never met. Most thoughts have been flirtations with the thoughts of others, dead or alive. Money and power, however obsessive, have ultimately been a means to a more intimate purpose. I investigate how humanity has become confused about its purpose, and how it is possible to acquire a new sense of direction.

When individuals have looked beyond their familiar surroundings, when they have learned to read and travel, they have discovered that many strangers share their emotions and interests. But fruitful contact between them has been rare. Very few of those who could be mutually sympathetic or stimulating, or who could join together in adventures which they could not undertake alone, have yet met. Now that for the first time better communication has become one of humanity's main priorities, no life can be considered to be fully lived if it has not benefited from all the encounters of which it is capable. Today, hope is sustained above all by the prospect of meeting new people.

All scientific discovery, indeed, is inspired by a similar search, and the meeting of ideas which have never come together before. So too is the art of making life meaningful and beautiful, which involves finding connections between what seems to have no connection, linking people and places, desires and memories, through details whose implications have gone unnoticed. The search for a soul-mate who matches one perfectly occupies only one part of intimate life; individuals are becoming increasingly many-sided; so it is about soul-mates of a less total kind that I write, who have some elements of character or attitude which can combine with elements in another to produce more than either could alone. Just as the science of materials has invented many new comforts by discovering how the same molecules are to be found in objects that are apparently totally different, how these molecules can be rearranged, how seemingly incompatible ones can become receptive to one another, and united by gentle, multiple liaisons, so discovering unrecognised affinities between humans holds out the prospect of reconciliations and adventures which have so far seemed impossible. But merely to wait for mutual recognition is inadequate. The dream which cosmopolitans had, that antagonisms would evaporate naturally, was too simple, and ceased to be credible because it underestimated how different and vulnerable every person – and every group – is. I have investigated how relationships of varying closeness may be established across frontiers without abandoning one's loyalties or uniqueness.

It may seem sheer temerity even to conceive of the possibility of another renaissance, but hope has always emerged from its hibernation, however battered, however long it takes to wake up. Of course, it cannot return with its faith in utopias, for they have caused too many catastrophes. To find a new sense of direction, it will need to incorporate the certainty of failure, to a lesser or greater extent; but if failure is expected, and studied, it need not destroy courage.

Instead of packing my information into conventional categories, which would only confirm that the habitual economic, political and social factors weigh heavily on all that humans do, I have rearranged it, finding new points of contact between the ordinary and the exotic, between the past and the present, so as to be able to address the questions which matter most to the present generation.

I have dealt with only a limited range of people, places and topics, because I am trying to suggest a method and an approach, not to fit all facts into pigeon-holes, and because even several lifetimes would not have been sufficient to remedy my ignorance, or to cope with all the information that is available. What to do with too much information is the great riddle of our time. My solution is to look at the facts through two lenses simultaneously, both through a microscope, choosing details that illuminate life in those aspects that touch people most closely, and through a telescope, surveying large problems from a great distance. I hope I say enough to show that humans have many more options before them than they currently believe.

The gallery of portraits I place at the heart of my book are of individuals and not a statistically representative sample: they are there to stimulate reflection, not to point to facile generalisations. I have chosen to write about women, because I am not one myself, and because I have always preferred to write about subjects which do not tempt me to be so arrogant as to believe that I can ever fully understand them, but above all because many women seem to me to be looking at life with fresh eyes, and their autobiographies, in various forms, are the most original part of contemporary literature. Their clash with old mentalities is the impasse which dwarfs all other impasses; and thinking about how it might be resolved first started me on this book. My conclusion was that I had to write about both sexes at the same time.

Freud wrote about humanity on the basis of meetings with patients largely from one country, even if he did drape an oriental carpet over his couch. I had long conversations with people of eighteen different nationalities in the course of my research, and could have started each chapter with witnesses from a different part of the world, but I did not want to suggest that any particular country might somehow be more liable to a particular worry or weakness. The majority of the living characters in this book therefore also come from a single country. It is a rich one (though it knows poverty too), a free country (but one struggling against many subtle restrictions), a favourite for tourists because of its devotion to the good life, attracting as many each year as it has inhabitants, but nevertheless not finding living all that simple, and it is probably disliked by quite as many foreigners as admire it. I am thus able to ask what there is left for humans to do, once they have won their basic comforts and their freedom, or at least some freedoms.

I met most of these women in France, a country which has been like a laboratory for me for all my adult life, a constant source of inspiration. All my books on France have been attempts to understand the art of life, by the light of the fireworks which that country sends into the sky in its effort to understand itself. What I value particularly is its tradition of thinking about its own problems in universal terms, and, self-absorbed though it is, of going beyond the self-preoccupation which all nations feel. The Declaration of the Rights of Man was made on behalf of the entire world. It seems to me that any new vision of the future must, more than ever before, include the whole of humanity, and that it why I have written my book this way.

Suggestive reading: at the end of each chapter, I have placed a sample of my sources, to suggest directions in which the reader's imagination can travel, according to its own tastes, in the same way as a few drinks might be served after a meal, to allow the conversation to expand. I have given preference to recent books, because I want to give some hint of the extraordinary richness of recent research, and the intellectual liveliness of our times, and of our universities, battered though they are. This is a very incomplete record of my deep indebtedness to innumerable scholars, both professional and lay, from whose labours I have benefited, all the more so because I have saved space by omitting many well-known works, which already line the bookshelves in the minds of the present generation, and because I have mentioned only a small fraction of the examples and arguments I have gleaned in my reading, for otherwise this book would have been ten times as long.

Orlando Patterson, *Freedom in the Making of Western Culture*, Basic Books, NY, 1991; Orlando Patterson, *Slavery and Social Death*, Harvard UP, 1976; Richard Hellie, *Slavery in Russia 1450–1725*, Chicago UP, 1982; David Brion Davis, *Slavery and Human Progress*, Oxford UP, 1984; Philip Mason, *Patterns of Dominance*, Oxford UP, 1970; G. Boulvent, *Domestique et fonctionnaire sous le haut empire*, Belles Lettres, 1974; Robin Lane Fox, *Pagans and Christians*, Penguin, 1986; Jean-Paul Roux, *Les Barbares*, Bordas, 1982; A. M. Duff, *Freedmen in the Early Roman Empire*, 1928; James L. Watson, *Asian and African Systems of Slavery*, Blackwell, 1980; G. Freyre, *The Masters and the Slaves*, 2nd edn., California UP, 1986; M. L. Kilson and R. L. Rotberg, *The African Diaspora*, Harvard UP, 1976; M. I. Finlay, *Slavery in Classical Antiquity*, Heffer, Cambridge, 1960; Jonathan Derrick, *Africa's Slaves Today*, Allen and Unwin, 1975; Gail Saunders, *Slavery in the Bahamas 1648–1838*, Nassau, 1965; Paul A. David, *Reckoning with Slavery*, Oxford UP, 1976; Kenneth M. Stampp, *The Peculiar Institution: Slavery in the Antebellum South*, Knopf, NY, 1956; Lydia Maria Child, *Incidents in the Life of a Slave Girl*, written by Herself, published for the author 1861; T. Mitamura, *Chinese Eunuchs*, Tuttle, Tokyo, 1970.

On changing the way one looks at the world

Eugene W. Nester et al., *The Microbial Perspective*, Sanders, Philadelphia, 1982; Edward R. Leadbetter and J. S. Poindexter, *Bacteria in*

Nature, Plenum, 1985; A. G. Morton, *History of Botanical Science*, Academic Press, 1981; T. S. Kuhn, *The Structure of Scientific Revolutions*, 2nd edn., Chicago UP, 1970; M. Teich and R. Young, *Changing Perspectives in the History of Science*, Heinemann, 1973; Robert Doisneau, *A l'imparfait de l'objectif*, 1989; B. Brodzki, *Life Lines: Theorising Women's Autobiography*, Cornell UP, 1988; M. Ignatieff, *The Needs of Strangers*, Chatto, 1984; Henley Centre, *Leisure Futures*, 1992 (statistics about people who prefer to be told what to do); Reuven Feuerstein, *Don't Accept Me as I Am*, Plenum, NY, 1988 (encouraging research on 'retarded' children); Jérôme Clément, *Un Homme en quête de vertu*, Grasset, 1992.

How men and women have slowly learned to have interesting conversations

Cognac (population 22,000) is worth visiting not just because it makes a famous liqueur, which loosens tongues, nor because it is the birthplace of Jean Monnet, the founder of Europe, which has replaced war by debate, nor because it has an old château, but because it has become conscious of its silences. Here it is possible to observe not just the old traditions of talk, but also the *nouvelle cuisine* of conversation.

At the police station, the job of Corporal Lydie Rosier, aged twenty-seven, is to listen to confessions, but after I had finished talking to her, she left the room all blushing, saying that my questions had been very difficult, that she was not used to answering questions, only to asking them. She has never had much occasion to talk about herself: 'Our profession does not allow us to be introspective. We are taught to be discreet. We can have opinions but we cannot express them.'

Everybody's style of speaking is a mixture of echoes dating from different epochs of the past: hers recalls the prudent and modest civil servants of the last century who were proud that they represented the state and who took care not to compromise themselves by saying the wrong thing. There is no point in her indulging in idle talk: nothing but concentration on her duties, and the ageing process, can win her a third chevron. Her private life is in the freezer. The chit-chat of the factory she once sweated in, where most of the workers were women, is an uncomfortable memory: personal relations there were difficult, she says, because 'women are more secretive and more aggressive towards one another than men.' Does she worry that she cannot speak freely? No, because she reads a lot. She has just finished a book about . . . the secret police. However, she has also recently read the biography of Marie Curie: 'I would have liked to have been her. She had great force of character, great will power.'

Will power is what Lydie believes in, not words: most misfortunes are the result of weakness of will. Though her training did not include any discussion of the causes of crime, she is certain that she does not accept the verbiage of modern psychology. 'There are plenty of people living in awful conditions and they haven't become criminals. No one is forced to take the wrong road, even if strength of character is needed to avoid it.' But where does one get strength of character? 'It's a matter of ambition. You must take hold of your life.'

The talks she likes to have with offenders are ones where they understand this. For example, 'A lad of fourteen stole a car. It was his first offence. He said, "What I did was stupid. I need help. I cannot manage alone." We said OK. Now he wants to join the army and has stopped committing crimes. He has an ambition.' Television should not show people carrying guns as though it was a normal thing to do. She of course carries a gun herself and 'I might have to take it out and shoot someone.' They talk about that in the police station. She might get shot herself. Death alone frightens her, but she does not talk or think about it. If it comes, too bad. Nor does she bother her head with the future: 'I live from day to day.'

'There are always solutions to problems': that is her motto. Her father is a post-office clerk; one sister works for the police, another at the town hall, a third in a school. The universe of junior officials, rotating on its own axis, is large enough to keep conversation within the family. For Lydie, entering the public service has been an escape from the commonplace: 'I always wanted to do something out of the ordinary.' And now her ambition is to prove by her conduct that women can do security work as well as men, though not with the aim of taking their place, nor of challenging them, because the 'force side of the profession must be preserved and women do not express force'. She does not argue with misogynists, whether policemen or not. Sexual crimes give her a chance to demonstrate her value: 'Victims find it easier to speak to a woman.'

Lydie looks very modern in trousers with her revolver on her hip; but she has decided she will be more comfortable in Réunion, in the Indian Ocean, which is her next posting, having already had a taste of New Caledonia: there, she says, 'They live as people lived a hundred years ago.' Yet old ways survive in Cognac, too, where in order to speak to me she had to get the permission of her captain, who asked the colonel, who asked the general . . .

Other old ways are to be found in the surrounding countryside. The growers of Cognac's grapes are also careful about what they say. The Bellenguez family, which has sixteen and a half hectares of vines and thirty of arable land, remains faithful to the tradition of not crossing those in authority: 'We never discuss politics. We're on good terms with everyone. We vote but we don't say how we vote.'

However, each member of the family has an individual relationship with words. The grandmother is its intellectual. Aged sixty-five, a dressmaker by trade, she is famous for her wide reading, for her love of new gadgets, for helping children to do their homework and for making notes when she watches television, whose programmes she loves to discuss. They explain her brilliance by saying that she comes from the north, from the Pas de Calais. One of her granddaughters is thought of as following in her footsteps, because she plunges into a book when she gets home.

The head of the family, who inherited the farm from his father, continues to have soup, pâté and sausages for breakfast, as his forebears always have: he knows everybody in the region, goes shooting, reads the newspaper and is a subscriber to the conservative weekly, *L'Express*. But there are subjects of conversation which remain too delicate. He has no sons: does he regret it? His wife replies, 'He has never said.'

The wife talks readily and fluently, but talk is only an accompaniment to work; she is uncomfortable if she is not constantly busy and on the move, looking after the animals and the house, and taking two hours to cook supper. She never reads, not even the newspaper, and she never helped the children with their homework: 'We trusted them, we are not like other parents who interfere. The children must make their own way, there's no point in forcing them to get better marks – each according to his abilities. I brought them up to be thrifty, and to manage for themselves.'

One daughter, who is studying to be a nurse, hoping to become a midwife, says that in the old days there was more sociability, at least at harvest time, when they used to employ fifteen workers, of all nationalities: the atmosphere was festive, and on Saturday nights there would be a big meal and a dance. She likes 'an ambience of warmth', which means having lots of people around her. Last Sunday, the family had no guests: 'How odd that we are alone,' they said. 'The house seems empty.' But again, there are things which they do not talk about: 'What we don't say to each other, we write in letters.' These are to pen-friends abroad, in Africa, Peru, Korea. The nurse sends three or four letters a week. Strangers are needed to whom one can explain what one is doing and feeling.

In the very centre of Cognac, there is also a shortage of people to talk to. Annette Martineau and her husband have a fruiterer's and greengrocer's which is so bustling that they can allow themselves only one week's holiday a year: she has developed a speciality of artistic, ornate fruit baskets, which people come to buy from fifty miles around. 'I would like,' she says, 'to be like Troisgros [the three-star chef], to be recognised professionally. I didn't know I had this in me. But the tragedy of commerce is that people think we're cash registers. It's true we live off their money, but we want other things too. When one meets people, and hears them converse, they become interesting. I left school at fourteen; I have difficulty in spelling; in correspondence I use a dictionary. But now everything interests me. I feel an insatiable hunger for everything. I'm not a literary person but I like to go to a bookshop sometimes and leaf through a book or magazine. I'm willing to try before saying no. Culture has developed so much in France: it means doing everything, speaking about everything; everything has become cultural because one learns from everything. Television brings many new ideas, and interesting biographies, and one wants to talk about them.' She has no inferiority complex about her lack of education: 'There are people who are learned but stupid.'

As a child, she was taught not to speak at table. 'My parents barely talked to each other. My friends say their husbands don't talk either. It's often like that. Husbands didn't say much in the past, because everything was taboo, and because they had nothing to say. In our dinner parties, the conversation is either non-existent or aggressive.

'My husband gets up at 3 a.m. to do the buying. He is completely immersed in his work and is not a man to waste words. I warned my daughter: "You will enjoy life more if you're with a man you can talk to." I recently bought a book which said that women are more interested in conversation than sex. Friendship begins with talking about nothing, just for amusement, but later one talks about life, and it becomes sharing. A real friend is very rare; she won't repeat what I say, nor dirty what I say. I like many people, but friendship is a stronger word.

'I've taught my daughters to fight for themselves, on their own. I've spoken to the elder one about women's matters. I speak freely, but we're not buddies. I feel I am her mother. I tell them to work, love, respect, learn. I would like them to be real women, that is someone who is respected and loved, who knows how to love and make herself respected; I tell them to live freely, better than me, to be less ignorant.

'A woman needs a man. In the past men gave women no support. My father gave my mother no support, and life used to be more passive. She didn't ask for much, and had many children. But I need support, assurance, affection from a man. Why? I don't know. One needs to lean on someone.

'An example of a real man was Yves Montand. I would not have wanted to marry him, but great men have shoulders – a woman can feel their strength, be understood by them, be supported by them. But despite sexual liberation, friendship between men and women is still difficult, there is always an *arrière pensée*. Women can talk about everything that men do, and they can broaden the conversation; they will pause, reflect more; they are not less intelligent. Relationships are changing, but remain difficult. As men get older, they often still need a mother, or want to return to being eighteen; they want to go on proving that they are men. Whereas a woman lives each stage of her life, she has several lives. Men refuse to. People say Montand was made what he was by women, and it's true.

'In my shop, I am a shopkeeper. With you, I am me.'

Madame Martineau's daughter, aged sixteen, finds her own conversation limited in different ways. In the past, she says, girls confided only in girls, but now it is possible to make friends with a boy, without sex, 'like a brother'. 'There is no difference between boys and girls, one can talk to either.' However, whereas 'girls are willing to experiment and search, boys have fixed ideas, absorbed by money and success.' There is a new confidence in girls, which the boys cannot quite satisfy: she admires her mother who, while liking her work, 'would change it if she could; she's interested in more things, and if she doesn't know, she goes and finds out.' So she dreams of leaving Cognac because 'it is a place only for people over thirty-five' and because the children of the poor have to stick to their own class. But her confidence is mixed with diffidence: 'I don't like serving customers in the shop, because of the risk of not pleasing them: they might be dissatisfied.' What goes on in other people's heads is becoming ever more puzzling.

I ask a woman of forty-four, 'With whom do you have your best discussions?' She replies, 'With my dog. He understands me.' She belongs to the generation of 1968, which believed that once taboos were abolished, and people became frank with each other and spoke freely about what they thought deep down in themselves, a new age would dawn. Lisa has been trying to apply that formula for over twenty years and it has not quite worked. She runs a renal dialysis clinic and is proud that she can provide treatment at a quarter of the hospital's cost. But her colleagues do not speak to her the way she would like them to, so she is giving up her work. When she began, young doctors and nurses treated each other as equals: they were a team and the doctors swore that when they got to the top they would never behave like the tyrannical old specialists, who thought they were God. Now the young doctors have become middle-aged and powerful; they go off on their own to international congresses and have lost interest in the nurses. She complains that they assume experienced nurses like herself will just go on doing their duty, earning little more than young nurses straight out of school: she wants recognition of her experience, not necessarily in money, but in respect. The lack of it has turned everything sour. But doctors also complain that they are not respected by their patients, who value the television repairer more highly; nobody foresaw the world shortage of respect.

Patients come to Lisa from all over the country; she spends four or five hours with each of them, three times a week, and close bonds have been formed. Nevertheless she has decided to rebel. In almost the same breath she says, 'I've had enough of the patients,' and 'I'll miss them very much.' She intends to become an administrator specialising in hospital hygiene. The world does not know how to reward people for what they are, rather than for the grade they occupy in the hierarchy.

How have the doctors reacted to her demand for respect? 'I haven't ever said any of this to them.' Silence again. Pride stands in the way. She feels that they treat her with contempt. But she cannot beg them for encouragement. Once upon a time she met a marvellous professor of pharmacology, in whose laboratory she spent some time. He was world-famous and had worked in the USA, but had not a trace of arrogance, never spoke of *his* discoveries but always of his 'team', where only first names were used. That attitude is her ideal. She forgets that it is easy for world authorities to be friendly, whereas mediocrities have to show how important they are, or nobody would guess. However, Lisa is searching for 'a new motivation'.

She had put her hope above all in her career. She married a doctor who busied himself with his career too. 'We are equals. There are no male and female tasks at home. I am completely liberated. I can do what I want, and won't be criticised if I come home late. We have separate bank accounts. I look after myself. We each have different hobbies. His are tennis and bridge. Mine are squash and physical fitness. But I do all the cooking on Sundays, for the whole week, because cooking is my passion; I express my creativity in it: I don't want his help.' Busy with their work, they did not have children. Now she regrets it. Does he? 'I don't know. We never speak about this. He didn't really want children.'

That, however, was not the only subject they did not mention. Does he love her? She never asks; he never tells her. Does she tell him? 'It is easier for me to say, "You get on my nerves." ' There were times when they hardly saw each other, fully involved as they were with their patients. About five years ago, she decided they must agree to dine together, alone, every Saturday at 8.30; she prepares a meal as though they had guests, and they talk about medicine and people, whose psychologies he is very good at dissecting – 'He taught me how to do it.' He loves good food, but says she spends too much time cooking; and when they do have guests, he says the fuss she makes means the meal is less convivial than she thinks, because she puts the guests in a position of inferiority. She does indeed think about these big dinners a whole month ahead.

'We never speak about ourselves. I don't know why. I think about myself, but don't say what I think. If I were really liberated I would speak about myself with my husband; but I would need a husband who was liberated from his shyness, to help me out of my shyness.' There is no solution for this shyness, beyond going out and proving to herself that she is admirable, which is never quite satisfying. 'I think my husband knows I am exceptional, whereas he is not exceptional; but he agrees with me that I am exceptional only to avoid trouble.' She has accepted an invitation to teach at a business school one morning a week, wearing her smartest clothes: 'I do all I can to look good in front of these adolescents. They notice what I wear. I put forward an image that pleases them, so that they listen to me. I don't want to grow old; I don't want to behave like a woman of forty-four; I find people of my own age older than me. Because I am a sportswoman, I can talk with the young.' Her activity in the sports world means she meets many people: 'I'm not cowed by them; they treat me as an equal, which is important to me.'

There is also a man friend who likes to pour out his problems to her, and sometimes they go out together, even to night-clubs; but her husband refuses to get jealous. His pleasure is the tennis club and his men friends. 'Why can't we enjoy ourselves separately?' he asks. 'Why not together?' she replies. 'If he is enjoying himself, I would like to be there also. I like enjoying myself, but we have a different idea of what enjoying oneself means.' She did persuade him to go to a fancy-dress ball, at which she made herself quite unrecognisable and dressed him up as a clown, and he did enjoy himself: 'It was one of our best evenings.' For all her independence, she wants to be close friends with him, but time seems to have separated them, as the doctors got separated from the nurses. It might be different if people could be unrecognisable more often, could surprise each other.

Perhaps one man cannot provide enough conversation. Sometimes she thinks the ideal would be to have two part-time men. 'I don't say it would never happen. But I don't believe I could do this. I'm not an adventuress. I've no taste for risk.' She says to him, 'You don't know me. You don't realise how far I could go.' But it is only to frighten him: 'He exasperates me, but I couldn't live with an ordinary stereotype of a husband; and I'm not so easy to live with. Perhaps, after all, he is the best husband for me.'

I heard a lot of complaints about husbands in Cognac. Going up the social ladder, the demands on husbands increase, and exasperation grows at forty-five-year-old men desperate to prove their sexual prowess; at that age, 'It is very important to let them think that they are the boss.' However, not all wives are content with this ancient strategy: when they are educated and have some leisure and like to use their minds, business talk bores them. Once upon a time they distracted themselves with charity work. Now there are about a dozen groups of women which meet to discuss literature and ideas, religion and Europe. The religious meetings include men, but the literary meetings are for women only: in these, they read a book each month and talk about it over a meal. They have opened a private library, which buys the latest publications, and lends them out at five francs, or three francs for older books; this has become a permanent meeting place for the discussion of literature, in which all classes of women have participated.

Then these women got together for their supremely bold intellectual adventure: they established a European literary festival, to which internationally known writers, critics and artists are invited; an audience of several hundred attends; and, partly as a result, this small town has four bookshops. I think a new kind of university might one day grow from this seed. Oxford, after all, was a city with only 950 houses when its university was founded. It met the need for educated priests, lawyers and teachers, but now that professional training is no longer enough to satisfy intelligent people, there is room for a new sort of university that is not a ghetto for the young, but a place where all generations can exchange experience, culture and hope.

'My husband is sent upstairs during our women's meetings, and sometimes he eavesdrops and asks me questions afterwards,' says one wife. 'My husband is only interested in science and mechanics,' complains another. 'When he found out I would be away all day at the festival,' says a third, 'he was cross.' 'Everybody has a secret garden,' says a fourth. 'We each appear as an image. If I revealed myself as I am, no one would believe me. So I keep my thoughts to myself. I don't want to give my key away.' 'The men,' concludes yet another, 'earn our living. We think for them.'

Conversation between men and women has barely begun.

Is it inevitable that so many conversations should be fruitless? Why, after centuries of experience, are humans still so awkward, rude, inattentive in conversation, with even 40 per cent of Americans – brought up to regard silence as unfriendly – complaining that they are too shy to speak freely? The answer is that conversation is still in its infancy.

The world's memory has been stuffed with the names of generals rather than of conversationalists, perhaps because in the past people spoke much less than they do now. 'A man who is given much to speech, however wise he may be, is numbered among the foolish,' said the Persian Prince Kai Ka'us of Gurgan, and for most of history the world agreed. Homer's ideal hero, 'the speaker of words' as much as 'the doer of deeds', was a rarity. The Hindu goddess of speech, Sarasvati, dwelt only 'in the tongues of poets', and when ordinary humans spoke, she made them realise they were attempting to be divinely creative. In 1787, an English traveller remarked on the taciturnity of French peasants, in a country whose élite was famed for elegant verbosity.

That ancient peasant silence can still be heard in parts of Finland, which is reputed to be the least talkative country on earth. 'One word,' says a Finnish proverb, 'is enough to make a lot of trouble.' The province of Hame in Finland is the most silent, where they take pride in the story of the farmer who visits his neighbour and sits quietly for a long time, saying nothing, before his host asks him why he has come. He finally brings himself to reveal that his home is on fire. These Finns used to live in isolated farmhouses, not villages, and it was no burden for them to tolerate silence. Anthropologists report that there are places in Central Africa where people 'do not feel under any obligation to talk in a social situation, because speech, not silence, is what gets a person into trouble'. Others have analysed how in Madagascar it is important to be careful about what one says, because information is a scarce commodity to be hoarded, for it gives prestige, and because making a pronouncement which proves to be inaccurate will cause serious loss of face. But this is not peculiar to any part of the world, it is characteristic of many professions and many formal situations elsewhere: there are many reasons for not speaking, above all the fear of making a fool of oneself. A few miles from Oxford an old lady I knew, the widow of a farm labourer, used to receive visits from other ladies who just 'sat with her', saying virtually nothing for a whole hour. What is instructive about Madagascar is that the men are so worried about losing face, or offending other men, that they leave it to the women to do the talking. When they wish to criticise, they ask the women to do it for them, which the women do in French, not Malagasy; men only use harsh words when giving orders to their cows, but only in French. Then the men criticise the women for having long tongues.

Freedom of speech was an empty right until people freed themselves of the feeling that they did not know how to express themselves properly; nor was it enough for them to congregate in cities to learn to converse. They needed first to overcome the old, deeply ingrained dislike of being interrupted, which seemed a sort of mutilation. Then they had to be spurred to talk by the need to discuss what they were uncertain about, and by not knowing what to believe. Tongues were loosened only when scientists and philosophers began saying (as they did in ancient Greece, and as they are now repeating) that it is impossible to know the truth, that everything is constantly changing, and multiple, and very complicated, and only the sceptic knows wisdom. The invention of democracy, too, required that people should say what they thought, and express themselves in public assemblies. Syracuse in Sicily, a city of Greek immigrants – foreshadowing the settlers of New England – was the first democracy to have a teacher of the art of speaking, Corax by name. Soon rhetoric became the supreme skill in the Hellenic world, and the most important part of education. Though some believed that to be an impressive speaker it was necessary to be familiar with all branches of knowledge, most were too impatient, and so a shorter cut to success was invented: the syllabus was reduced to mere training in disputation, the technique of talking on any subject even if one knew nothing about it. The ability to speak persuasively became the new excitement, the new intellectual game, turning politics and the law courts into entertainments, with orators competing against each other like athletes, but with greater, magic power to stir the emotions. The most famous teacher of rhetoric, Gorgias, originally an ambassador to Athens from Syracuse, thought of himself as a magician, rhyming his phrases as though they were incantations.

But this was not conversation. The first known conversationalist was Socrates, who replaced this war of words by dialogue. Perhaps he did not invent dialogue, which was originally a Sicilian mime or puppet play, but he introduced the idea that individuals could not be intelligent on their own, that they needed someone else to stimulate them. Before him, the model for all speech was the monologue: the wise man or the god spoke and the rest listened. But Socrates had been through the trauma of studying science and had been left with the feeling that he would never know what to believe. His brilliant idea was that if two unsure individuals were put together, they could achieve what they could not do separately: they could discover the truth, their own truth, for themselves. By questioning each other and examining their prejudices, dividing each one of these into many parts, finding the flaws, never attacking or insulting, but always seeking what they could agree between them, moving in small steps from one agreement to another, they would gradually learn what the purpose of life was. Wandering through Athens, through the markets and meeting places, Socrates demonstrated how dialogue worked, accosting artisans and politicians and people of all callings, questioning them about their work and opinions. Whatever they happened to be doing at the moment, they must have a reason, they must think it was right, or just, or beautiful; and so he led the discussion to what those words meant. He argued that it was inadequate simply to repeat what others said, to borrow beliefs. One had to work them out for oneself. He was a teacher such as had never existed before, who refused to teach, who refused to be paid, insisting that he was as ignorant as the pupil, and that the way to find a reason for living was to engage in a conversation.

Socrates was exceptionally ugly, almost grotesque in appearance, but he showed how two individuals could become beautiful to each other by the way they talked. 'Those who frequent my company at first appear, some of them, quite unintelligent, but as we go further with our discussion, all those who are favoured by heaven may progress at a rate which seems surprising to others as to themselves, although it is clear that they have never learned anything from me. The many admirable truths they bring to birth have been discovered by themselves from within themselves. But the delivery is heaven's work, and mine.' Socrates' mother had been a midwife, and that is how he saw himself too. For ideas to be born, a midwife is needed. That was one of the greatest of all discoveries.

However, some found Socrates too strange, irritating, subversive, 'a sting ray'. That an opinion was shared by everyone did not impress him, he still questioned it. His irony was disconcerting, because he seemed to be meaning two opposite things at the same time. He scoffed at democracy, too, and chose to be condemned to death to prove that it could be unjust. A life that does not question itself, he told his persecutors, is not worth living.

But conversation is not made just of questions: Socrates invented only half of conversation. Another rebellion was still needed, and it came with the Renaissance. This time it was a rebellion by women.

So long as success in life depended on military strength, or noble birth, or having a patron to protect one, 'to converse' was understood to mean 'to live with, to frequent, to belong to the circle of someone powerful', with no need for speech beyond proclaiming one's obedience and loyalty. Etiquette books for courtiers advised them to concentrate on defending their reputation, with military metaphors to guide them in fortifying their pride: form alliances, use words as weapons and insults as ammunition against your rivals, show your strength by your readiness to accept confrontations, to start a quarrel, to employ bluff. The language of courtiers for long remained coarse, their demeanour ostentatious, their model strutting cocks. But then the ladies of the courts grew tired of this routine, and first in Italy, then in France and England, and finally throughout Europe and beyond, a new ideal of how a human being should behave was invented, demanding the opposite – politeness, gentleness, tact and culture. The model whom everyone ultimately copied was Madame de Rambouillet (née Pisani, she was half Italian). Just as Marilyn Monroe taught a whole generation what it meant to be sexy, Madame de Rambouillet showed what it meant to be sociable in the most refined way, so that it no longer mattered how rich, or how well born, or how physically beautiful one was, provided one knew how to take part in a conversation.

She organised conversation in an entirely new way. A salon was the opposite of the large, royal or baronial hall; its characteristic was intimacy, a dozen people perhaps, two dozen at most; sometimes called an alcove, it was presided over by a lady with a talent for drawing out the best from talented people, whom she invited not on the basis of their status, but because they had interesting things to say, and because in the company she created, they said them even better. Socrates invented the conversational duet. Madame Rambouillet did not try to create a chamber orchestra of talk, because each individual spoke his own words; rather she provided a theatre in which each could judge the effect of those words and receive a reaction. People of all classes and nationalities met in her salon – and in the many other salons which imitated hers – for conversations which looked at life with the same distance as Socrates had favoured, but instead of torturing themselves with self-questioning, they concentrated on expressing their thoughts with elegance.

The salons did for conversation what Garrick's acting did for Shakespeare. They were mediators, as Madame Necker put it, helping 'feelings to pass into the souls of others'. Horace Walpole, who had a horror of the sort of people who went to salons ('freethinkers, savants, the hypocrite Rousseau, the scoffer Voltaire . . . all are to me impostors in various ways'), became a devoted member of Madame Geoffrin's salon nonetheless, discovering that however much men might dislike the pretentiousness of other men, the presence of intelligent women whom they wished to please transformed normally uncomfortable meetings into exhilarating encounters. 'I never saw anybody in my days,' he wrote of his hostess, 'that catches one's faults and convinces one so easily. I never liked to be set right before . . . I make her both my confessor and director. The next time I see her, I believe I shall say, Oh Common-sense, sit down; I have been thinking so-and-so; is it not absurd?'

The mixing of clever women and clever men brought sex and intellect into a different relationship. 'Warm, deep, occasionally passionate friendships were formed, but they were nearly always platonic rather than domestic in expression.' Men and women learned to value one another for their character, rather than their appearance, profiting from their differences to try to understand themselves and each other. Their meetings gave birth to epigrams, verse, maxims, portraits, eulogies, music, games, which were discussed with extraordinary thoroughness but without spite, for the rule was that the participants had to be agreeable. There was a deliberate effort to keep up with all that was new in literature, science, art, politics and manners, but the women who ran these salons were not specialists in any of these subjects. Their achievement was to purge men of the boorish academic legacy which weighed on them, in which the purpose of discussion was to crush others with the weight of one's learning. In so doing, they imbued the prose of the eighteenth century with clarity, elegance, universality, 'filtrating ideas through other minds', encouraging seriousness to be also light-hearted, reason to remember emotion, politeness to join with sincerity. Mrs Katherine Philips, who started a salon in London and of whom more might have been heard had she not died in 1664 at the age of thirty-four, described hers as 'a Society of Friendship to which male and female members were admitted, and in which poetry, religion and the human heart were to form the subjects of discussion'.

However, small groups often limit the individuality of their participants and diminish their capacity to venture outside. The taste which the salons cultivated easily became tyrannical, so that they could not tolerate any other. Though they tried to teach themselves to 'enjoy contact with others' and to value what Montaigne called 'the diversity and discordance of nature', they often ended worshipping their own brilliance, or their imitation of brilliance, and lapsing into conversations which were in effect counterfeit. When the salon became as boring as a royal court, the solution was to withdraw into a tête-à-tête. As the yearning for more intimate conversation grew, and the obsession with sincerity became more absolute, only letters seemed an adequate refuge for the pondered exchange of private thought.

The desire to converse is not enough to keep a conversation alive. Eighteenth-century Spain, for example, developed the art of whispering (*chichisveo*), in which a woman granted a man, not her husband, the privilege of talking to her alone. Medieval knights performed great deeds for their ladies; now men were given a chance to show their skill at words. Husbands did not object because this was supposed to be a platonic relationship, but also because the admirer's duty was to play a comedy of enslavement, of being devoted to a woman whom he could not possess, and indeed he attended her almost like a servant, appearing at nine in the morning to offer her chocolate in bed, give his opinion on what she should wear, escort her on her walks, send her flowers and bonnets. But when neither he nor she had very much to say, the conversation contained no more than gossip and complaints about the servants. 'A lady who can talk about hats, cabriolets, harnesses and horseshoes thinks that she has reached the peak of wisdom and that she can set the tone of a conversation. And men to please women learn the same dictionary and become ridiculous.' This false start occurred in Italy, too, and doubtless elsewhere: 'We Genoese husbands,' wrote one of them in 1753, 'are too busy, while our wives are not busy enough to be satisfied to get along unaccompanied. They need a gallant, a dog or a monkey.'

There was a crucial factor missing: education. Maria de Zayas y Sotomayor had denounced the ignorance in which most women were kept as early as 1637, but it was not easy for them to rebel while they regarded men only as potential suitors. The Church fulminated against the very idea of women talking to men, as for example in Gabriel Quijano's essay on *The Evil of Social Gatherings: The Excesses and Harm of Conversation, otherwise known as cortejo* (Madrid, 1784). The *chichisveo* or *cortejo* could have been the beginning of something new (which I shall explore in chapter 18), but it deteriorated into a series of 'attentions, gallantries and courtesies so rigid and obligatory that they lost their initial tinge of passion and became fixed in a code as tedious and stiff as marriage'.

The difficulty of talking was revealed with particular force in England. Dr Johnson is England's king of conversation, and will remain so until a better biographer than Boswell overthrows him by providing an alternative. But talk was a dust-cloud which he blew around him to conceal the terrors, evils and gloom which constantly haunted him, and which he considered to be the very essence of life, to the point that he got angry with anyone who denied that life was necessarily unhappy. Combating such thoughts was pointless, he insisted, one could only divert them, by directing one's mind to other topics; so that he envied women who knitted and knotted, and indeed tried to learn knotting and music in vain. Talk was his supreme pleasure because it provided him with relief, but his talk was not genuine conversation, it was not exchange. His talent was for uttering opinions perfectly formed, in immaculate prose, on any subject. Disagreement did not interest him, because he believed that it needed to be ended by one of the parties emerging victorious and he battled ferociously to ensure that he always won himself. He never discovered the value of being contradicted. People admired him because he could sum up a problem in an epigram, but the effect was to end a conversation, not to start it. His sententious judgements – for example, that 'when a man is tired of London he is tired of life, for there is in London all that life can afford', or that he was 'willing to love all mankind, except an American', or that the 'French are gross, ill-bred, untaught people' – belie his more interesting statement that 'I look upon every day to be lost in which I do not make a new acquaintance.' Dr Johnson, for all his many admirable qualities, represents a dead end. He has been imitated by equally sad, equally brilliant, Oxford dons I have known, and their talk did not diminish their sadness.

The salons kindled conversation between great minds, but they could not teach how to converse with strangers, or those who had no pretensions. The history of English conversation shows how obsession with class distinctions was perpetuated by fine talk, leaving different sections of the population glorying in their mutual incomprehensibility. A doctor wrote in 1908 that she doubted whether 'any real conversation between members of two classes is possible. All conversations with my patients and their friends have been of an exceedingly one-sided character... in some cases I talked, and in some cases they did, but we never took anything like equal parts. A question, a shade of surprise, the faintest dissent from their views, the lack of constant approbation, would generally be enough to silence them, and in many instances to cause them to veer round suddenly, and bring forward opinions in direct opposition to those they had already expressed.'

Enormous effort has been devoted to obstructing the development of a common language between people. 'All those who have had close intercourse with the labouring classes,' wrote this same doctor, 'well know with what difficulty they comprehend words not of a Saxon origin; and frequently addresses to them are unintelligible from the continual use of terms of a Latin or Greek derivation.' That was precisely how the newly educated middle class tried to distinguish itself, by 'talking like a book', as polysyllabically as possible – a style which survives today in caricatures of the official language of the police. And the upper classes adopt their own slang, distinguishing U from non-U, to prove their superiority. The purpose of snobbery is to limit conversation. Disraeli described how it was done by the use of fashionable clichés: 'English is an expressive language but not difficult to master. Its range is limited. It consists, as far as I can observe, of four words: nice, jolly, charming and bore, and some grammarians add, fond.'

The United States does not seem to have escaped similar obstacles to conversation, exacerbated by differences in ethnic and national origin. But worst of all, it seems to have lost hope that women and men will ever be able to speak the same language. Deborah Tannen, after a lifetime of research, concludes that they cannot understand each other, that they mean quite different things when they speak, that women want comfort from those they converse with, while men seek solutions to problems. Women, she claims, complain in order to increase their sense of community, they gossip because, like children, they believe that telling secrets is the way to make friends; so they willingly listen to each other's troubles, because their overriding purpose is not to feel alone. Men, however, do not like to listen, because 'it makes them feel subordinate'; they are always supposed to be jockeying for supremacy and have no time to show sympathy. Telling people to change their behaviour, say Tannen, just does not work. Her solution is that they should study sociolinguistics, which will convince them that the sexes 'play different games', that their dissatisfactions are not due to personal defects but to 'gender differences'. The two sexes are brought up in 'different cultures' and have to realise that they are like foreigners who will never communicate properly; they must accept that they speak different languages. She implies that they do not even try to converse, citing the sad statistic that American married couples, on average, spend only half an hour a week 'talking to each other'. I do not believe the myth that the United States is the land of the lonely crowd, but many Americans have convinced themselves that it is true, because they dream that conversations should be more wonderful than they normally are.

Has nothing changed in two thousand years? Han Fei Tzu saw what the problem was in the third century BC. He could not get people to listen to him. He was always misunderstood: if he tried to be witty, he was accused of being flippant; if he was compliant, he appeared insincere; if he spoke out of turn, he was punished; different people found him alternately awkward, conceited, boastful, a coward and a flatterer. Was it any wonder, he said, that he was diffident about speaking, and worried when he did? And yet he liked to talk and give his opinion, which eventually earned him a death sentence. Han Fei left a book of essays about 'Solitary Indignation' and 'Difficulties in the Way of Persuasion', showing that he knew what he ought to have done, but could not: the obstacle to conversation was 'not knowing the heart' of the person one spoke to, 'thereby to fit my wording into it'. He saw that the problem was that humans are a mystery.

Which is, of course, why they are interesting, why they are worth talking to. If they were predictable, there would be no point in conversation, which draws its inspiration from the differences between humans. Conversation is quite unlike confession or its secularised variants, quite unlike the practice of pouring out one's troubles to anybody who will listen, paying them to listen if necessary. The healer who listens aims to put an end to the confession, by providing an explanation of it, often on the basis of a childhood or a sexual experience, or a scapegoat of one sort or another. Conversation, by contrast, demands equality between participants. Indeed, it is one of the most important ways of establishing equality.

Its enemies are rhetoric, disputation, jargon and private languages, or despair at not being listened to and not being understood. To flourish, it needs the help of midwives, of either sex: women have generally shown more skill at this task, but there were times in the history of feminism when some gave up on conversation, and staked all on persuasion. Only when people learn to converse will they begin to be equal.

Kai Ka'us Ibn Iskandar, Prince of Gurgan, *A Mirror for Princes: The Qabus Nama*, tr. Reuben Levy, Cresset Press, 1951; David Kinsley, *Hindu Goddesses: Visions of the Divine Feminine in the Hindu Religious Tradition*, California UP, 1986; *The Complete Works of Han Fei Tzu*, tr.

W. K. Liao, Arthur Probsthain, London, 1939; Etienne Balazs, *Chinese Civilisation and Bureaucracy*, Yale UP, 1964; J. V. Neustupny, *Communicating with the Japanese*, Japan Times, Tokyo, 1987; Tullio Maranhao, *Therapeutic Discourse and Socratic Dialogue*, Wisconsin UP, 1986; George A. Kennedy, *Classical Rhetoric and its Christian and Secular Tradition from Ancient to Modern Times*, Carolina UP and Croom Helm, 1980; Nicolas P. Gross, *Amatory Persuasion in Antiquity*, Delaware UP, Newark, 1985; Georges Vlastos, *Socrates, Ironist and Moral Philosopher*, Cambridge UP, 1991; M. Magendie, *La Politesse mondaine et les théories de l'honnêteté en France au dix-septième siècle*, Félix Alcan, n.d.; Ian Maclean, *Woman Triumphant: Feminism in French Literature 1610–52*, Oxford UP, 1977; Elizabeth C. Goldsmith, *Exclusive Conversations: The Art of Interaction in Seventeenth-Century France*, Pennsylvania UP, 1988; Carmen Martin Gaite, *Love Customs in Eighteenth-Century Spain*, California UP, 1991; W. Jackson Bate, *Samuel Johnson*, Chatto, 1978; C. B. Tinker, *The Salon in English Letters*, Macmillan, NY, 1915; K. C. Phillipps, *Language and Class in Victorian England*, Blackwell/Deutsch, 1984; John S. Nelson, *The Rhetoric of the Human Sciences*, Wisconsin UP, 1987; William Carey, *Dialogues Intended to Facilitate the Acquiring of the Bengalese Language*, 3rd edn., Mission Press, Fort William, 1818; N. H. Itagi, *Communication Potential in the Tribal Populations of Assam and Madhya Pradesh*, Central Institute of Indian Language, Mysore, 1986; K. S. Rajyashree, *An Ethno-Linguistic Survey of Dharavi, a Slum of Bombay*, Central Institute of Indian Language, Mysore, 1986; Raymonde Carroll, *Cultural Misunderstandings*, Chicago UP, 1988; Lila Abu Lughod, *Veiled Sentiments: Honor and Poetry in a Bedouin Society*, California UP, 1986; Deborah Tannen, *You Just Don't Understand: Men and Women in Conversation*, Morrow, NY, 1990 and Virago, 1991; Deborah Tannen, *Perspectives in Silence*, Ablex, Norwood, NJ, 1985; Jan Bremmer and Herman Rooden, *A Cultural History of Gesture*, Polity, Cambridge, 1991; Elinor Ochs Keenan, 'The Universality of Conversational Postulates', in *Language in Society*, vol. 5, 1976, pp. 67–80.

How people searching for their roots
are only beginning to look
far and deep enough

A greying man and a young woman are dancing. He is English, and he says to her, 'You have such a lovely face; it's a pity about your accent.'

Her face is half Japanese and half European, perhaps German or Slav or French . . . she has to explain it. Her accent is American. She does not think that he is right to be sorry for her; on the contrary, she feels she was born in the right century because she belongs entirely neither to the East nor to the West.

Educated in a girls' school in Tokyo, she obediently copied what the teachers wrote on the blackboard; by the age of fifteen, 'I had learned how to be Japanese.' Not completely, for when she was disobedient her schoolmates shunned her and called her American. So she went to the USA to discover her father's civilisation, and to study music, where origins do not matter. However, she was not convinced that she was talented enough to be a full member of the frontierless nation of artists. In Washington DC, she acquired instead a university degree in international relations and became an expert on China. Two years in Taiwan made her fall in love with an Englishman. She thinks of herself as a citizen of the world, though she sees no reason to shout it out loud. She assumes that she will never be a great artist or a great person, but she has found an alternative role.

In London she discovered that citizens of the world are not necessarily welcome. Her mother-in-law-to-be disliked not just her accent but, even more, her aloofness, and could not see how the family could absorb her. After six years, her boyfriend still could not make up his mind whether they were suited. What is a modern woman who does not fit neatly into an obvious slot supposed to do? What modern woman does fit neatly into an obvious slot?

Of all the cities in the world, Maya has now chosen to live in Paris, where 'whatever I do is all right; in France anything goes'. But of course that is because she has not tried to become French. 'I like to consider myself a gypsy . . . I feel I am made of different colours, I cannot say which part is me – the American or the Japanese: I am the sum, the total.' She is now a presenter for Japanese television. When she interviews a Westerner in English, she is completely Western; but when she puts on her make-up and presents the interview in Japanese on NHK, her eyes light up in a different way, the language transforms her facial expressions and she appears a different person. 'I am a chameleon,' she says. The advantage of that is that she can feel comfortable with a far wider range of characters. She has interviewed the pretender kings of Europe, without strain, 'because they have all suffered', and she has walked alone in the most violent parts of Washington, 'because I am one of a minority too'.

Maya lives alone, as do half the inhabitants of Paris, the city of solitary thoughts. 'I am a distant person,' she says. 'When I have problems, I close the door and just think; I don't discuss them.' Her path to independence has been simple: she puts the blame for misadventures on herself; and she finds that a relaxing sensation, because it means there is hope – it is only herself she has to change. Changing other people is too hard. 'I have this idea that I am polishing myself and that I have a basis to work on.' Her mother taught her that if she could pity those who upset her, she would be half way there. Following that rule, she believes, makes her free. She seeks freedom, too, in trying not to accumulate too many objects around her; she says she is not attached to material things. When someone broke her favourite teapot, she did feel anger for two minutes, but then she said to herself, 'Everything has a life, everything has an end.' What she wants is not to possess her surroundings, but to be stimulated by them, and that means they have to change all the time.

The philosophy, currently popular in the USA, that an individual's feelings matter more than anything else, are sacred and inviolable and need to be protected, does not appeal to her. It used to depress her as a child that she could not form a firm opinion, that she could see both sides. She cannot emulate the wisdom of her Buddhist grandmother, who seems to know straight away what is the right thing to do. The Christian Sunday school she attended as a child failed to give answers to her questions. Being able to express her opinions is not her top priority, first because her opinions may not be right and secondly because she might change her mind. Bigots make her feel that they are trapping her, compelling her to say, 'You are wrong.' 'I hate to say I don't like these people.' To step on others, intentionally, to be deliberately mean, is for her 'the worst crime'. So her ambition is not to seek power, nor to force others to do her bidding. 'I prefer to put myself aside and do what I am good at, which is bringing people together.'

She is an intermediary, and intermediaries 'cannot be great'. But 'being on neither side, I can stand between people who have different opinions and persuade them to talk.' Her sister, brought up in an identical way, had none of her problems; after obtaining a master's degree in literature in the USA, she went back to Japan and works for a bank. Maya at first also believed that she was more Japanese than American, because Japan was where she had discovered life, but then she decided she could no longer live there, because 'one part of me would have to die. In Japan I am too tall, too outspoken, I feel boxed in; the houses are too small.' But every country is perhaps too small for her.

Sometimes her solitary existence is painful. When she is sick she misses having no one to take care of her; sometimes she regrets that she has no one from whom to ask advice; sometimes she wants to share her joys. Though she delights in taking walks alone in familiar surroundings, distant travels are less easy: she longs to walk in the Middle East, to add another continent to her homeland. If only she were a man, she would develop her adventurous side to its limits. Conquering her fears is a constant aim. So she has learnt to sail, because the sea always frightened her, from the time when as a child she lived on a cliff and had nightmares about waves overwhelming her house. She dreams of taking flying lessons, she promises herself she will get up earlier in the morning and fit more adventure into her day.

For all her success, she feels that there is something missing in her life (as anyone who has ideals must feel). If only she had an obsession, a passion, such as artists have (who give the false impression that they know where they are going, and that they have no doubts). If only there were not such gaps in her knowledge of herself. Why did her boyfriend's mother dislike her so much? But she will not settle for any compromise which does not offer the hope of combining her ideals in both love and work. 'I am very greedy.' Her ideal man, as she describes him, is not of this world: he must find joy in the same things as herself, be totally relaxed, and show it in the way he eats. Eating is the sign by which you can recognise the beauty inside a person. She hates it when the Japanese make slurping noises eating spaghetti, but likes them to slurp their own soup. She has met only a handful of people who know how to eat beautifully, who treat it as an art. She likes those who treat life as an art. And the question is, though one can admire the art of another, does each person have an art of his own?

When people tell the story of their life, the way they begin reveals at once how free they consider themselves to be and how much of the world they feel at home in. Until recently, who one's father was mattered more than anything else. The ideal human was like an oak tree, with roots set firmly in a birthplace. Living on the same piece of land as one's ancestors won respect and gave prestige, however personally odious one was, so that aristocrats, having more roots than anybody else, claimed that the present as well as the past belonged to them. But there is no longer any need to imitate aristocrats. There is another way of finding one's place in the general history of humanity.

What are the roots of one's pleasures and emotions? These are quite different, deeper sorts of roots, extending further back than the genealogy of one's own family, and one can only find them by searching across continents through all the centuries. The link with the days when humans were explorers setting out from the forests of Africa and Asia is a reminder that they have been on the move as often as they have settled down. Today, more and more people have a Chinese eye, which looks at nature as having its own life, most beautiful when irregular and untamed; the first person to have had that vision, and to be called an artist, was Ko Shou, the sister of the Emperor Shun, 2,000 years before Christ. More and more have an Arab and Persian heart, for it was from the Middle East that romantic love emerged. Europeans have chosen to forget not only that their language originates in India, but that it was there that the most modern view of sexual pleasures was conceived. More and more Westerners are discovering common emotions through African music and dance. As constant travel and escape from urban smog become indispensable to their sense of freedom, their imaginations register echoes in the fantasies of the Mongolian and Scythian nomads who once mocked the dwellers of cramped cities. One may feel isolated in one's own town, but one has forebears all over the world.

However, the history taught in schools does not emphasize such links, nor is it designed to reveal what memories matter most to the present. If a film were to be made compressing into a couple of hours all that, according to the textbooks, is ever supposed to have happened, with half a century flashing by every minute, the world would look like the moon, grey and desolate, remarkable only because there are a few craters on it. The craters are civilisations – thirty-four major ones so far – each exploding and then dying out, having briefly lit up parts of the globe, but never all of it; some last for a few hundred years, others for a couple of thousand. Meanwhile, on every side around the craters, dunes of grey dust stretch as far as one can see: they are the people not mentioned in history books, for whom civilisations never did very much, whose lives were largely meaningless suffering. Some volcanoes are erupting still, but there is no suspense about what will happen next: they will fall silent sooner or later: all civilisations, so far, have decayed and died, however magnificent they have been in their glory, however difficult it is to believe that they can vanish and be replaced by desert or jungle.

Maya cannot expect to be a heroine in a film like this, or to be considered suitable for any role other than that of a piece of drift-wood floating uncertainly between two civilisations, a problem child, someone abnormal. But more and more people are becoming abnormal, and are not fitting neatly into a single civilisation. The film suggests that normal people should be proud of the civilisation into which they are born, because they need roots and self-esteem. Nevertheless, the dramatic action arises from the disappointment of those who have not personally tasted the full delights of their civilisation, who see no chance of influencing its evolution, whose family, as far back as they know, have been excluded from most of what civilisation offers, because of their poverty – financial or cultural or spiritual – and who complain that admiring the great figures of their civilisation does not do very much to make them feel fulfilled. They may learn what their roots are, but not how to grow into an ornament of the landscape, nor how to branch out, nor how to flower. Maya, as an educated woman, would probably be uncomfortable even if she had perfectly ordinary roots.

However wonderful civilisations have been, there is always sad-ness in their story: their happiness is so brief. I am not writing this book to spread gloom, nor to encourage criticism and scepticism that leads nowhere, nor to tell histories of decline and fall. I reject the obsession with death, and with memory that is only a filling for the gaps between gravestones. Another scenario is possible, which does not make Maya look like either driftwood or dust. The film could be composed of different pictures.

If the past is replayed too fast, life seems futile, and humanity resembles water flowing from a tap, straight down the drain. A film of history for today needs to be in slow motion, showing every person who has ever lived as a star, though dimly visible in a night sky, a mystery still unexplored. The focus would move into close-up to reveal the amount of fear there is in the eyes of each one, and how much of the world each can confront without terror. They each cast their light on as much space as they feel at home in, revealing the real limits of their personal, private civilisation. In such a setting, what they have in common depends less on when or where they were born than on the attitude they have towards their fellow beings. You belong with those people with whom you can sympathise, in whatever century they lived, in no matter what civilisation. Such a film would create its surprises by placing side by side people who thought they were strangers, but were not.

There are interesting encounters to be made. For example, the peasants of the Bigouden in Brittany, in the year 1920, do not know that there is a sea only ten kilometres away; the world is black beyond their own villages, filled with devils and danger; their lives are faint, flickering candles. In 1950 the 3,000 people living on a hill overlooking the Argenton valley, in Bressuire, can still wrap their horizons around themselves like a warm blanket: a quarter of those under thirty have never visited the local capital town Niort, fifty-five kilometres away, where they could encounter the majesty of French civilisation in the shape of the secondary school, the barracks, the jail, the tax-collector and the pharmacist who is the local intellectual.

By contrast a thirteenth-century Venetian merchant, who can speak Persian and Mongolian, writes his memoirs in French to describe his travels in India and Japan. He is so appreciated by the Kublai Khan that he is appointed to be the vice-governor of a city in China. Marco Polo has such curiosity that he forgets fear. The diversity of human beings amuses rather than terrifies him, even though he does not get rid of all the darkness – no one ever has, there is always the puzzle that cannot be understood, the shapeless spectre that automatically makes hair stand on end. The Venice from which he comes is Europe's largest city, with 160,000 inhabitants, but Hangchow in China at the same time has a population of one million; city dwellers of today have more in common with the latter than with the former.

In every life there is an element of victory over fear, which needs to be searched for, though it may be a false victory. Again and again, apparently intelligent people ooze contempt to protect themselves from what they cannot understand, as animals defend their territory with foul smells. Gains in liberty are regularly lost. Or else people become so broad-minded that they do not know where they are going.

That is what the modern search is for: where to go next. There are already enough books praising the dead, or congratulating the living on knowing more than their ancestors, or telling them why they are what they are. Now the stones of history need to be re-used to construct roads which lead to where one wants to go. That means giving up the illusion that humans can be understood simply as examples of their civilisation, or nation, or family.

It is no longer enough to have roots of the sort with which they have contented themselves so far. To know who one's ancestors were, and what they were proud of, can no longer suffice for people who think of themselves as different from their parents, as being unique with opinions of their own, and who feel uncomfortable with traditions embedded in violence. People who want to be free need to dig over a much wider area, and deeper, to understand their personal emotions and ambitions. Looking at one's most obvious roots does not automatically equip one to choose one's friends, one's partner, one's life's work, nor to cope with anger, loneliness and other inadequacies. To discover in what direction one wishes to go, one needs to acquire memories with a new shape, memories which point into the future, and which have direct relevance to one's present preoccupations.

What humans have thought they could do in their own lives has been coloured by what they saw happening in nature around them. Their ideas about how the universe came into being and functions have limited their imagination when they considered how much freedom they personally had. Their ideas about roots are a good example. In modern botany, roots are not merely anchors through which all nourishment is drawn: they produce hormones too. So if humans want to compare themselves to plants, on the basis of what is known about plants today, they should not assume that roots give nothing but stability: they could say that roots also create moods. Everything that is remembered about the past is washed, and often drowned, in nostalgia, pride, illusions and passions of all kinds. Indeed, few people can extract solutions to their problems from their roots. The past no longer speaks with a clear voice; nobody seems to be agreed any more about what the lessons of history are. The old-fashioned sort of root could feed humanity only so long as differing opinions were treated like weeds, gouged out or poisoned. A new way of looking is needed in a world which believes that every individual has a right to flower, within limits, in his or her own way.

Until the eighteenth century, how plants functioned was a mystery. Then leaves, rather than roots, were gradually revealed to be responsible for the most important bio-physical process in the world, supplying most of its food and energy, and making it possible for plants to be adaptable, capable of surviving in cold and dry terrain. The 250,000 different kinds of leaves are alternative ways of dealing with the surprises of the environment. So far, only photosynthesis (the word dates from 1893) has produced living matter out of elements which are not alive. Only the present generation, looking through electron microscopes (1965), has been able to watch the pores of leaves open to daylight each morning, however grim yesterday was.

It is as odd to say that humans need roots as to say they need foliage. But a generation which values adaptability as much as tradition, which seeks energy and creativity and openness of mind, must like the idea of drinking in the light of the sun, from whatever direction it shines. The smell of the air becomes sweeter when roots produce leaves, which make it possible for other forms of life to exist. Applied to humans, this means that it is not just where they come from that matters, but where they are going, what kind of curiosity or imagination they have, and how they use it, both by day and by night.

Maya is already free in one respect: she is free of the illusion that she has no illusions. But in her desire to be an intermediary – an ambition that will recur in this book, because quite a few of my characters share it – is she a victim of another illusion when she imagines that it is possible to establish more intimate communication between the myriad stars of which humanity is composed, and which at present barely tolerate each other? To get beyond despair at humanity's inability ever to agree requires new ways of thinking, and in particular new images. A start can be made with ideas about loneliness, to which I now turn.

Nomads

C. Nelson, *The Desert and the Sown: Nomads in the Wider Society*, California UP, 1973; Centre G. Pompidou: Centre de création industrielle, *Errants, nomades, voyageurs*,1980; C. Bataillon, *Nomades et nomadisme au Sahara*, UNESCO, 1963; Sevyan Vainshtein, *Nomads of South Siberia*, Cambridge UP, 1980; J. P. Digard, *Techniques des nomades baxtyari d'Iran*, MSH and Cambridge UP, 1981; Elizabeth Atwood Lawrence, *Hoofbeats and Society: Studies in Human–Horse Interactions*, Indiana UP, Bloomington, 1985 (on relations with nomadism and much else); Elizabeth Atwood Lawrence, *Rodeo: An Anthropologist Looks at the Wild and the Tame*, Tennessee UP, Knoxville, 1982; Paul Vigneron, *Le Cheval dans l'antiquité gréco-romaine*, Nancy, 1968.

Migrations

Chen Yuan, *Western and Central Asians in China under the Mongols: Their Transformation into Chinese*, California UP, 1966; Werner Sollors, *Beyond Ethnicity: Consent and Dissent in American Culture*, Oxford UP, NY, 1986; J. L. Amselle, *Les Migrations africaines*, Maspero, 1976; Robert Mirak, *Torn Between Two Lands: Armenians in America, 1880–1914*, Harvard UP, 1983; Hasia R. Diner, *Erin's Daughters in America: Irish Immigrant Women in the Nineteenth Century*, Johns Hopkins UP, Baltimore, 1983; Jean I. Martin, *The Migrant Presence: Australian Responses 1947–77*, Allen and Unwin, Sydney, 1978; O. Lattimore, 'La Civilisation, mère de la barbarie', in *Annales*, 1962; Michael Todd, *The Everyday Life of Barbarians, Goths, Franks and Vandals*, Batsford, 1972; B. F. Manz, *The Rise and Rule of Tamerlane*, Cambridge UP, 1989; Albert Chan, *The Glory and Fall of the Ming Dynasty*, Oklahoma UP, Norman, 1982; Françoise Ascher, *Tourisme, sociétés transnationales et identités culturelles*, UNESCO, 1984 (international travel increased tenfold in the 1980s); Yves Lequin, *La France mosaique*, Larousse, 1988; Clive Roots, *Animal Invaders*, David & Charles, 1976 (on animal migrations).

Encounters of civilisations

Joseph A. Tainter, *The Collapse of Complex Societies*, Cambridge UP, 1988; Arnold Toynbee, *A Study of History*, revised edn., Oxford UP, 1972; Arnold Toynbee, *Mankind and Mother Earth*, Oxford UP, 1976; Jean Casemajou and J. P. Martin, *La Crise du Melting Pot*, Aubier, 1983; Alberto Asor Rosa, *En Marge: l'Occident et ses autres*, Aubier, 1978; Michael Grant, *From Alexander to Cleopatra*, Weidenfeld, 1982; Roland

Martin, *L'Urbanisme dans la Grèce antique*, Picard, n.d.; Stephen L. Collins, *From Divine Cosmos to Sovereign State*, Oxford UP, 1989; Shusako Endo, *Foreign Students*, Sceptre, 1965 (Japanese reactions to Europe); Donald M. Topping, *Thinking Across Cultures*, Lawrence Erlbaum, Hillsdale, NJ, 1989; Centre Aixois de recherches anglaises, no. 9, *L'Étranger dans la littérature et la pensée anglaises*, Aix, 1989; Raphael Samuel, *Patriotism: The Making and Unmaking of British National Identity*, Routledge, 1989; Walter Goffart, *Barbarians and Romans A.D. 418–584: The Techniques of Accommodation*, Princeton UP, 1980; Barry Cunliffe, *Greeks, Romans and Barbarians: Spheres of Interaction*, Batsford, 1988; Michael Grant, *The Fall of the Roman Empire*, Weidenfeld, 1986; J. P. V. D. Balsdon, *Roman Women*, Bodley, 1960; R. D. Milns, *Alexander the Great*, Hale, 1968 (on how he adopted Persian clothes and fell in love with Persian absolutism); Naphtali Lewis, *Life in Egypt under Roman Rule*, Oxford UP, 1983; *Nationalisme et Cosmopolitanisme dans les littératures ibériques au 19e siècle*, Lille, 1973 (BN 8. Z.45156 (1973)); M. Fogarty, *Irish Values and Attitudes*, Dublin Dominican, 1984 (at that date 2 per cent believed they belonged to Europe first of all and 6 per cent to the world as a whole); Yasuoka Shotaro, *A View by the Sea*, Columbia UP, 1984; Toshiaki Kozakai, *Les Japonais sont-ils des occidentaux?*, L'Harmattan, 1991; Françoise Barret-Ducrocq, *Traduire Europe*, Payot, 1992; Sudhir Kamar, *The Inner World: A Psychoanalytic Study of Childhood and Society in India*, Oxford UP, Delhi, 1978; Ninian Smart, *The World's Religions*, Cambridge UP, 1989; Pierre Riche, *Education and Culture in the Barbarian West*, South Carolina UP, 1976; Centre de Recherches archéologiques, Valbonne, *Le Phénomène des grandes 'invasions'*, CNRS, 1983.

Intermarriage

Susan McRae, *Cross Class Families*, Oxford UP, 1986; Kenneth Little, *African Women in Towns*, Cambridge UP, 1973; Geneviève Vinsonneau, *Les Relations de couple mixte entre noirs africains et françaises*, thèse de 3e cycle Paris V, 1978 (in Marguerite Durand Women's Library); Augustus Barbara, *Mariages sans frontières*, Centurion, 1985; Gabrielle Varro, *La Femme transplantée: Étude du mariage franco-americain en France et le bilinguisme des enfants*, Lille UP, 1984; Martine Muller, *Couscous pommes frites: Le couple franco-maghrébin d'hier à aujourd'hui*, Ramsay, 1987; Julius Drachsler, *Intermarriage in New York City*, NY, 1921 (three times as common in the second as in the first generation).

The science of affinities

Robert P. Multhauf, *The Origins of Chemistry*, Oldbourne, 1966 (chapter 14 on 'affinities'); Colin T. Burrows, *Processes of Vegetation Change*, Unwin, 1990; James D. Mauseth, *Plant Anatomy*, Cummings, Menlo, California, 1988; D. S. Koltun, *The Quantum Mechanics of Many Degrees of Freedom*, Oxford UP, 1988; P. D. Medawar, *The Uniqueness of the Individual*, Dover, NY, 1981; Lorraine Dennerstein, *Hormones and Behaviour*, Excerpta Medica, Amsterdam, 1986.

4

How some people have acquired an immunity to loneliness

The inspector of taxes is aware that she frightens people. But what frightens the inspector of taxes? Belonging to what she calls 'an oppressive branch of government, neither appreciated nor loved' is not a serious problem. She simply avoids revealing what she does. Otherwise people ask for her help with their finances, 'particularly doctors anxious to earn as much as possible and pay as little as possible'.

At the beginning of her career Colette used to be frightened. Her parents were poor; they could not afford to give her a university education. The department of taxes, unable to find recruits because it 'had a bad image', was offering scholarships. Being enticed unwillingly into its gloomy corridors was almost as traumatic as entering a nunnery. 'I understood what had happened. The bars closed around me. It was the worst period of my life. I realised that a great road ahead of me was being shut off.' She had longed to travel and to learn languages; now she had to put aside all dreams of adventure. Today she says she would prefer to be doing a totally different job. What would her ideal be, if she were free to choose? There is a long silence. She says she has never thought about it. After almost twenty years collecting taxes, she is a firmly realistic person. The fear of living in a world made out of numbers has gone.

All ambition, instead, has become a plant needing to be kept neatly pruned, lest it grow wild: ambitious people, she insists, are never satisfied. She was not raised to be ambitious. Her grandfather, a mason, fulfilled his ambitions simply by surviving, which he did by emigrating from Sicily to France. Her father thought a minor civil service job, a wife and children were enough for him. When Colette dared to become a financially independent woman, her rebellion revealed much more ambition than her parents considered reasonable; they had not imagined that girls needed liberty. Very occasionally, wondering whether she could do better, she says to herself, 'You are a nobody.' But she answers, 'To be a somebody, you have to begin very early. A daughter of poor parents can't afford to become a somebody.'

Being an inspector of taxes is not frightening any more because she has learnt to find interesting nooks in the corridors. A master's degree in jurisprudence and a diploma in labour law have opened doors and windows. For six years she was the most intimidating of all tax officials, an investigative inspector, who combs through taxpayers' accounts as though searching for lice. But the experience was nauseating. Fascinating though such detective work was, she felt sorry for her victims. The intolerable part of the job was the 'sometimes brutal' confrontation with the private tax advisers who looked after the books of small businesses, and in whom she repeatedly spotted dishonesty, both towards the state and to their clients. It was because they relied on crooks that helpless merchants found themselves at the mercy of the tax authorities. Battered by greed and deceit and the panic cries of the trapped, she became increasingly proud that she was in the service of the public. Money was not what mattered most to her. The last thing she wanted was to yield to tempting offers to go into private practice.

Another sort of fear struck when she was investigating the affairs of a large patisserie in the second arrondissement of Paris. The owner was very ill. There were many things wrong with his accounts. She went to her superior and said, 'We must stop the audit or we will kill him. He's bound to die in a few months' time in any case. Let's suspend proceedings.' Her superior asked his superiors for instructions, who, isolated from the human realities, were not moved to compassion. So the patissier was sent a huge tax demand. It came back marked 'deceased'. Colette gave up being an inquisitor in search of fraud.

She has never been afraid of getting her priorities wrong, as between public and private life. Her next assignment gave her a greater sense of being useful to her country, helping local councils to manage their financial affairs, getting immersed in local life, showing mayors how to pay for the services they needed. But though that was interesting and rewarding, she gave it up when her husband was promoted and transferred to another city.

There are always new corridors to explore: she trained in computing and became the tax authorities' computer troubleshooter. Officials baffled by the complexity of their system appeal to her, and over the telephone she and her ten assistants sort out their problem, like air traffic controllers bringing a plane down to land in fog. There is a lot of prestige in this job, fortnightly trips to Paris, many training courses: she is one of only six such experts in the whole country. 'I am privileged' is her conclusion on her career. 'I have succeeded in doing things differently from others. I was in the right place to get interesting tasks.' However, she does not aim to climb any higher. In the grade above her, the jobs go almost exclusively to men.

This worries her only in principle. She is not afraid that she will be unfulfilled in her work. 'As soon as one recognises that one is not a genius, one must organise oneself. I never tried to become a minister, only to do as little harm as possible.' The great advantage of her job is that she can give herself entirely to her family when she gets home. 'I have had more contact with my daughter than I had with my mother, who did not work. It is not how many hours you devote to your children that matters, but the quality of your attention.' She has always used almost half her pay to get help at home; she makes no apology for always having worked; financial independence is essential for a woman, and she is certain that her children have lost nothing by it. 'I help them with their homework, play with them, answer their questions, discuss, devote as much time to them as they need.' She is proud of her success. The elder boy is 'brilliant, warm-hearted, amusing' and is preparing for the highest posts in the land, his sights on the top colleges. Her daughter is 'a tornado, convivial, she will not become a submissive wife, she wants to be a somebody, and does not believe anything will stop her'.

It is a classic case of rising from the bottom to the top of the social scale in three generations. But the process was not automatic or natural. The family is an institution, she argues, which requires 'constant effort to sustain it. It's important not to want everything to be perfect. Couples divorce because they expect too much from each other, refuse to make concessions, can't wait patiently for better times.' She describes herself as a 'Mediterranean', because her mother is Corsican, but even that does not guarantee that a family remains close. When her husband first met her family, he thought they were always quarrelling: 'That was a false impression. Now he has come to appreciate it, because the old Mediterranean family gives the man a prominent role.' Colette works hard to please her husband, who is much more ambitious than she is and very absorbed by his work as a hospital administrator. She made a conscious decision to subordinate her career to his. He needs to succeed in life, she says. 'Everyone is vulnerable, but men have more need to be told they are admired, that they are important; they need compliments every day.' Is that not true of her too? 'No. I know what I am. I need attention. In return I give attention to others.' Climbing the professional ladder any further might damage the family.

So when her husband comes home and pours out the daily story of everything he has done, and who said what to whom, she is indulgent: he has no colleagues he can really confide in; he needs to be listened to; she lets him talk until he gets it all out of his system; only after that does he ask about the children and her. She puts up with a lot because he is an excellent father and husband, always agreeable at home. 'I am more selective in the anecdotes I tell about my office. Every day there is work to be done, interesting oneself in one's husband's problems, participating in his professional life, knowing what he thinks. One must make an effort, one must give affection, remain agreeable with the people around one if one does not want to lose them. Otherwise it's like being on the edge of a cliff.'

Colette is not afraid of her trade union either. When it organised a strike, she did not join it. They want a wage increase for everybody, but she is against officials being paid the same however well or badly they do their work. 'That is not the way to win the respect of the public.' She is willing to fight for her subordinates, whose salaries are ridiculously low, but unwilling to exploit the power of civil servants, who cannot be sacked unless they 'rob the till or murder their boss'. To strike for a salary rise for herself seems wrong while there are still so many inequalities.

No, the only thing the inspector of taxes is really afraid of is loneliness. 'Loneliness is the worst kind of suffering. I cannot live alone. If my husband and children are away and I am alone, I lose my bearings. I cannot make use of the opportunity to go out alone. Loneliness really scares me. I've never suffered from it, but I've always thought about it. I have two sisters; I married very young; I have always had people around me. The worst punishment is to find yourself alone.' That belief is the foundation of her life. It is not by chance that she lives in the city centre, so that friends and her children's friends can drop in all the time, so that she is never alone.

Colette's achievement is all the more remarkable because virtually none of her friends has managed to have a stable family life like her own. One has been abandoned by her husband. Another had a child by a foreigner who lives in another country. A third was wonderfully successful in her work, but was plagued by loneliness, persuaded a man to marry her and give her a child, and then called the police to throw him out because he became violent: 'Her life was ruined by the fear of loneliness.'

The fear of loneliness has been like a ball and chain restraining ambition, as much of an obstacle to a full life as persecution, discrimination or poverty. Until the chain is broken, freedom, for many, will remain a nightmare.

But the history of loneliness is not one of simple tyranny, and it shows that loneliness is not an inevitable flaw in the human condition. Whether a person is afraid of being physically alone – with no hand to hold, hearing no voice but the wind's – or of being socially alone – in a crowd but unnoticed, ignored, unloved – or of being spiritually alone – conversing normally, but finding no one who understands – the difficulties caused by loneliness can be traced to a myth.

The story we are usually told is this: in the beginning everybody lived cosily in a family or tribe, people did not originally even know what loneliness was, never conceiving of themselves as separate individuals. Then suddenly, quite recently, togetherness crumbled. Now, not only is an epidemic of loneliness sweeping the world, hand in hand with prosperity, but the more successful you are, the more likely you are to suffer from it; and money cannot buy you out.

Feminists were the latest group to be thwarted by it. Simone de Beauvoir's idea that work would be a protection, a better one than the family, proved mistaken. Even she, who claimed 'I am sufficient to myself', found she needed someone who could 'make me pleased with myself'; even she was 'made stupid by falling in love'; even she felt lonely when Sartre, in his last years, was no longer himself. All movements for freedom come to a halt at the wall of loneliness.

One-quarter of Americans (26 per cent) classify themselves as chronically lonely. In France, the same percentage say they often feel lonely, while 54 per cent say they have suffered from loneliness at some time. The French have questioned themselves with particular thoroughness on the subject. It is not living alone that troubles them most, for among those who do live alone there are just as many who love it as hate it. There is almost as much loneliness among the married as the unmarried. Fifty-nine per cent of those who say they feel lonely are women, and 41 per cent men, but it is impossible to be sure what reticence conceals: in Britain only 14 per cent of adults admit to being lonely even once a month. Those who move house most often are apparently most lonely; those who have most friends are hungriest for even more. Doctors add that the lonely are twice as likely to fall ill in other ways.

But it is not true that loneliness is a modern ailment. The Hindus in one of their oldest myths say that the world was created because the Original Being was lonely. Even when all humanity was religious, there were sufferers from loneliness, as the prophet Job, in the fourth century BC, testified in one of the most poignant of all poems about it:

'My kinsfolk have failed, and my familiar friends have forgotten me. They that dwell in mine house, and my maids, count me for a stranger: I am an alien in their sight. . . . My breath is strange to my wife, though I intreated for the children's sake of mine own body. Yea, young children despised me; I arose, and they spake against me. All my inward friends abhorred me: and they whom I loved are turned against me.'

However, loneliness is not incurable, any more than smallpox is. Its history shows that some people have developed more or less immunity from it by four methods. What these methods have in common is that they have followed the principle on which vaccination works, using loneliness itself, in calculated doses, to avoid being destroyed by it.

The pioneers were the hermits. They were men and women who felt out of place in the world, who did not like its greed, cruelty and compromises, or who believed they were misunderstood; as one of them, Narcissus, Bishop of Jerusalem, said in the year 212, 'Weary of the world's slanders against him, he retired to the wilderness.' Instead of feeling alienated in society, they left it to become professional aliens, aiming deliberately to be 'strangers' or 'exiles', and making that condition a noble one. The reward they sought was internal peace. Some subjected themselves to painful mortifications, almost starving, or tying themselves up with heavy chains, or living in graves, in order to have spiritual illuminations; some became deranged; but the famous ones were those who triumphed and emerged with a sense of having discovered the realities that mattered; and they radiated an internal peace which was immensely impressive: admirers flocked to get their blessing.

Almost every civilisation has tried this method. Probably invented by the Hindus, it travelled eastwards under the influence of Buddha, the prince who became a hermit, and westwards through the Middle East to Europe, attracting perfectionists. Among the early Christians, St Anthony became a hero – an illiterate Egyptian who at the age of thirty-five retired to live alone in the desert, and stayed for twenty years to battle with 'demons', who were the great scourge of those days – that was the ancient name for worries, doubts, fears and guilt. He beat the demons; a biography made him famous; large numbers followed his example, to get rid of their own demons. Being alone seemed to be a cure.

Being a hermit did not necessarily mean being totally alone. In Syria, Simeon the Stylite (390–459) sat on top of a tall pillar to escape the crowds hoping to get rid of their troubles without making the sacrifices he made, simply by obtaining his prayers. There were certainly some hermits who wanted nothing to do with humanity; but many came back to the world, finding that flight from it gave them a sense of responsibility for it, a desire to help. The best-known Western hermit of modern times, Thomas Merton of Kentucky, said, 'Solitude is not separation'; though he was a Trappist monk who had taken a vow of silence, he insisted that he was 'open to the whole world', passing on his vision of true values through popular books. Medieval Irish hermits provided another model: they withdrew twice over by going into exile abroad, and by living among the heathen.

The effects of being a hermit and of being a convict sentenced to solitary confinement were often similar. Some hermits became incapable of tolerating human company, like Pachomius (290–346), who after seven years of fasting reacted angrily to every slight disagreement. But others emerged as Dostoevsky did from Siberia, where he suffered first solitary confinement and then imprisonment with never a moment alone, emotionally isolated among criminals: he went into this experience in a state of despair about humanity, on the point of nervous breakdown, and came out exulting, with a faith in its goodness.

The pure hermit became, with time, an exaggerated oddity, but he (and she, for there were quite a few women hermits among the early Christians) impressed ordinary people enough to encourage them to try a spell of withdrawal from time to time. Martin of Tours (316–77) popularised this idea in France by punctuating his normal living with retreats into solitude. How valued brief retreats could be was shown by the Argentinian Maria Antonia de San José de la Paz (1730–89), who organised them for more than 100,000 people in her lifetime. In Burma, boys leaving school spend some months in a Buddhist monastery to prepare themselves for adult life. The pre-Columbian Amerindians sent their children into a period of isolation and fasting, in order to make contact with a spirit which would guide them through life, after which they were regarded as repositories of a sacred force: the Canadian Athapascans did this at the age of five, the Algonquin at twelve.

The retreat often turned the original Hindu idea back to front, making it a preparation for life rather than for death. As life has got longer, more and more old people have preferred to live alone, but near their relatives. The Hindus had pioneered something similar, but for a poor society, with neither pensions nor health service, so isolation was aimed at enabling the old to triumph mentally over poverty and illness. Ideally, Hindus divide their time on earth into four stages, two of them sociable – as students, then raising a family – followed by two of withdrawal: when their hair turns white and they have seen the birth of their children's children, they should become hermits in the forest, or in a shed at the bottom of the garden, learning to forget about material things, imposing more and more hardships on themselves, living in the open in the rainy season, wearing wet clothes in winter, until, in a final stage, having succeeded in breaking all earthly ties, they become homeless wanderers, owning only a begging bowl and the rags they wear. What Western modernity does is to urge people to mix up these four stages, tempting them to alternate between independence and responsibility.

The second form of immunisation against loneliness involved not moving out of society, nor searching for God, but turning inwards, with the aim of reinforcing one's powers of resistance, by introspection, by understanding oneself, by emphasising one's uniqueness, even though at first that might increase the loneliness.

The Mediterranean family is reputed as a home of togetherness, but it also gave birth to explorers in search of an independent, individual style of life. That required courage. The celebrities of the Italian Renaissance began with self-confidence obtained from mastery of their craft. Until then artists were willing to be told what to do, and to follow tradition. The idea of being different from others was as terrifying to them as to everyone else. People closed their eyes to their own originality, feeling their way through life by imitation, guided by the memory of established models of excellence, believing that obedience was the way to win both divine and human favour.

A mutation occurred and a few artists began to be dissatisfied with what they were doing. However, expressing ideas of their own was such a bold adventure that they needed the support of constant praise. The Renaissance artists were fanatically competitive, desperate for the first prize, addicted to applause as though to alcohol – once they did not do what they were expected to, there could be no certainty that they were doing the right thing. They were troubled by a kind of desire which was insatiable because it had no clear object, a restlessness they could not understand.

Just how different individuals are from each other was demonstrated by a physician, Girolamo Cardano (1501–76), who wrote both an autobiography to show it in his own case, and large scientific works to apply the idea to the whole of nature. Studying himself as though he was his own patient, he examined all his peculiarities, physical and mental, in minute detail. He recorded all his ailments, his haemorrhoids, the exact amount of urine he produced each day, his troubles with his genitals, 'so that from my twenty-first to my thirty-first year I was unable to lie with women'. 'I have fourteen good teeth and one which is rather weak.' He explained what his precise tastes were at each meal, 'my manner of walking and thinking: because I think as I walk, my gait is uneven; how I walk depends on how I feel'. He described the odd shape of his feet, 'so that I can scarcely find shoes'. He was careful to include a chapter on 'Things in which I have failed'. Minute details, for him, made all the difference in life and he thought they ought to be 'analysed into their infinitely minute components'. 'Intuition . . . an intuitive flash of direct knowledge, certainly the most highly perfected faculty which man may cultivate' gave all these details meaning; in other words, he could not explain how he made sense of them. But he tried to look at humans impartially, in the same way as he looked at animals, of whom he said that they had not been created to serve man, but existed for their own sakes. The big books which made his reputation were entitled *The*

Variety of Things and *The Subtlety of Things*. Instead of telling people to conform, he concluded, 'It makes for happiness to be what you can, when you cannot be what you would.' That was the first assertion of the difficulty that individuals have in putting on a conformist exterior. The new human beings of the Renaissance were mere beginners in the art of being lonely. They wanted both to be different and to be admired despite that. However, being an exception requires a new attitude to other people.

The awareness that one was alone, unable to fit into the normal human pattern, sometimes stimulated worry that one was not consistent, not a coherent person. Petrarch, having been crowned poet laureate in Rome (1341), was unhappy, even though he had reached the top. He went off to hide in Provence, where he wrote *The Secret Conflict of My Cares*, complaining that he was victim of a 'terrible plague of the soul, melancholy', the worst part of which was that 'I feed upon my tears and sufferings with a morbid attraction'. In other words, he hated parts of himself, but did not want to give them up: 'I see the better course and I cling to the worse.' Seeing oneself as a double person has been the beginning of creativity ever since; it makes possible a new kind of relationship with others, without sacrificing one's own integrity. But Petrarch did not realise this: he kept asking himself how he could 'be himself'. Avoid following the example of the masses, was his answer; get away from the world which might 'falsify' him. But the desire for fame was also part of him: he did not wish to cut himself off from the world. He travelled abroad, he bought a remote country house in the Vaucluse, but there was no escape. Born in exile, he turned himself into a professional exile, uncomfortable, alone with his fame.

Benvenuto Cellini (1500–71) tried to avoid these pitfalls. Having become a world-famous jeweller and sculptor, he believed he had found a method for avoiding self-doubt. He advised everybody to write their autobiography, not in order to understand themselves, but, more simply, to assert their individuality. He went so far as to kill people who he thought were obstructing the flowering of his genius, and boasted of the beautifully made daggers he used. He conversed with God, who told him, 'Have no fears.' Obtaining the Pope's pardon for 'all the homicides I had committed, or ever should commit, in the service of the Church Apostolic', which meant himself, he was convinced that he must continue to be what nature made him: 'A man must do what he must.' Cellini confused individuality with selfishness and megalomania, insisting that his gifts entitled him to be a law unto himself, eliminating loneliness by blinding himself to others, one of many wrong turnings in the history of loneliness.

The idea of the independent individual does not derive just from Italy. There were exceptional individuals meditating on loneliness in most countries. Some Germans, for example, tempted by loneliness, extended its appeal beyond the artist. The Romantics claimed that each individual combines human attributes in a unique way, and that one should aim at expressing one's uniqueness in one's manner of living, just like an artist expressing himself in his creative act. To sympathise with another person was not enough. 'The truly spiritual man feels something higher than sympathy': he feels the individuality of other people, and he considers that individuality sacred, not because of how important or powerful its possessor is, but because it is individuality. Such opinions expanded the dreams of the Renaissance by demanding that one should like a person because he is different. My quotation is from A. W. von Schlegel (1767–1845), who put that idea into practice by translating both Shakespeare and the *Bhagavad-Gita*.

In the vast emptiness of America, loneliness was from the beginning an enemy to be conquered; nowhere was war waged on it more determinedly, and yet the Americans were too astute to reject it completely. The cowboy who rode off into the sunset alone never made it his purpose to overcome his loneliness: he was not a member of a team, and he bore the inconveniences of independence with stoicism. Sometimes he converted his sadness into music, sometimes into sport, sometimes into taming horses, sometimes into imagining he could tame women in the same way; but though wildness and unpredictability were his scourge, when he won victory over them, he at once became restless again. No country has taken loneliness more seriously than America, none has created so many specialist organisations to combat it, but at the same time nowhere has examination of one's inner self become such a widespread passion.

'To be unhappy means to be alone, unhappiness comes from solitude.' Bishop Bossuet (1627–1704) was expressing conventional wisdom when he said that, particularly in a country like France which has made good conversation one of its main passions. But even the French have sometimes been exhausted by too much conversation, by the effort of having to think up witty epigrams all the time and be polite to people they do not like. Escaping from social life for part of the year, a month in the country, became a national institution, necessary to restore the appetite for company. But there was a hitch: boredom.

'Solitude is my greatest dread,' wrote Jean-Jacques Rousseau, 'I fear the boredom of being alone with myself.' When boredom became one of the great powers in the world, it took possession of loneliness. There is a limit to how far you can get on self-knowledge.

The third way of tolerating loneliness was by an injection of the absurd. British eccentrics combined loneliness with humour and extracted courage from the mixture. Eccentrics have sadly tended to be left out of history books, which have a misplaced idea of what it means to be serious. Eccentrics are monuments, unafraid of being alone. John Stuart Mill argued that since humanity is imperfect, different varieties of character should be given free scope, as experiments in the art of living; he regretted not so much that people 'choose what is customary in preference to what suits their inclinations, [but that] it does not occur to them to have any inclination, except what is customary'.

The first victory of eccentrics was to be considered amusing rather than dangerous. That was due, perhaps, to the powerful aristocrats who became eccentric because they did not have to worry about what people thought. The fourteenth Baron Berners, a diplomat, used eccentricity to mock those who were frightened by it: he obtained a compartment all to himself when travelling by train, by wearing a black skull cap and dark glasses, and beckoning to people to sit with him; if anybody did venture to, he would soon get rid of him by bringing out a large thermometer and taking his temperature every five minutes, to the accompaniment of heavy sighs. The fifth Duke of Portland, a maniac for privacy, refused to admit even his doctor into his bedroom, requiring him to make his diagnosis standing outside, questioning and taking his temperature through the medium of a valet. But he dreamed of a convivial world, regarding privacy as a preparation for something else, building a ballroom in his home big enough for 2,000 guests, a lift to carry twenty people, a library with twelve billiard tables, none of which were ever used, employing 15,000 construction workers, but disguising himself so as to remain unrecognised, which was his idea of freedom.

John Christie, who built himself an opera house at Glyndebourne at which formal evening dress was required, often wore old tennis shoes with his; he treated his customers as guests, but liked to introduce them to each other by the wrong names; he died before he was able to build a cafeteria for accompanying dogs. Eccentricity for him was a way of shuffling the cards of life. It was eccentric to treat animals as equals – it could lead to the shuffling of cards not just between species, but between every hierarchy.

Women have been among the boldest eccentrics in history, when they have turned clothes into theatre, breaking the rules of both fantasy and decorum. The couturier Worth deserves to be remembered as the person who tried to make eccentricity universal. Fashion, as he conceived it, offered to give each woman individual clothes, like nobody else's. Had he succeeded, had fashion not become degraded into imitation, loneliness might not have turned into fear so often.

The final form of immunisation has been achieved by thinking that the world is not just a vast, frightening wilderness, that some kind of order is discernible in it, and that the individual, however insignificant, contains echoes of that coherence. People who believe in some supernatural power have their loneliness mitigated by the sense that, despite all the misfortunes that overwhelm them, there is some minute divine spark inside them: that is how they are immunised. Those who have no such faith can develop a sense of being useful to others, and can recognise a link of generosity between themselves and others, rational and emotional connections which mean that they are part of a wider whole, even though they may be unable to decipher fully its enigmas and cruelties.

Much of what is called progress has been the result of solitary individuals saved from feeling totally alone, even when persecuted, by the conviction that they have grasped a truth, a fragment of a much wider one too large to capture. But getting beyond loneliness in this way does not eliminate all forms of loneliness, any more than one vaccination will protect against all forms of disease.

Thinking one's private thoughts has often had bizarre results: the Russian zoologist Metchnikoff (1845–1916) believed that eating yoghurt was the way to live to an enormous age, and he twice attempted suicide by ludicrously incompetent methods, but nevertheless he did set the science of immunology on its feet, putting loneliness in a completely new perspective. The discovery of how the immune system works has shown that every individual is ceaselessly building up resistance to the unfriendly outside world, and that each has got to do it both independently and in concert with others. Human bodies are not identical, all equally menaced by the germs around them. Medicine cannot protect them simply by finding a culprit for each disease and waging war on it. Instead of conquest, the emphasis has moved to the understanding of compatibilities and incompatibilities between bodies, and the flexible boundaries of tolerance and rejection. Once life is seen as being maintained by an immune system, each individual appears as unique, a peculiar combination of multiple characteristics which are found in others, but which very rarely occur in exactly the same combination in two individuals. The discovery of allergies has emphasised how individuals differ in their sensitivity; that of blood groups has divided them up in ways which cut across nations, religions and colour; the recognition of the influence of stress on health has made clear that reactions can never be wholly predicted; and Aids has dramatically revealed how serious are the results of the immune system becoming confused in its efforts to distinguish between itself and others. Medicine can no longer treat the individual as a machine that infallibly obeys strict rules. God has been renamed the Generator of Diversity.

It is now clear, therefore, that everyone needs small doses of foreign bodies, that in order to survive side by side with others, it is necessary to absorb a minute part of them. It is impossible to cut oneself off, or to destroy one's enemies for ever. Curiosity about others can no longer be thought of as a luxury or a distraction: it is indispensable to one's very existence.

None of these four methods is a guarantee against loneliness. Their effect is not to abolish loneliness, but to diminish the fear of being alone: only then can one relate to others on terms of mutual respect.

The Chinese word for alone is *Tu*. That is a nice coincidence. *Tu* is sometimes used in Chinese to indicate a weakness, when applied to a person acting waywardly, in isolation from sound principles, but it is also used in praise of the Taoist sage, and of his right to act independently, to be alone, because he knows what he is doing: 'Only one who is clear about men and spirits is able to conduct himself alone.'

But the stimulus of other people is necessary for clear thoughts and for knowing where one wants to go; only knowledge of humanity's previous experience can save one from suffering disillusionment. Having won the right to be alone, to be an exception to generalisations (which can be even more dangerous to freedom than generals), having freed oneself from the generalisation that humans are condemned to suffer from loneliness, one can stand it on its head: turn being alone upside down and it becomes adventure. How to find companions for one's adventures is the subject of the next chapter, and several more.

Loneliness

Robert Sayre, *Solitude in Society* Harvard UP, 1978; M. Hojat, *Loneliness: Theory, Research and Applications*, Sage, 1989; Michel Hannoun, *Nos Solitudes*, Seuil, 1991; Anthony Storr, *Solitude*, Flamingo, 1988; M. D. S. Ainsworth, *Patterns of Attachment*, Erlbaum, NY, 1978; Vladimir Shlapentokh, *Public and Private Life of the Soviet People*, Oxford UP, NY, 1989; Jean Heuchlin, *Aux origines monastiques de la Gaule du Nord: Ermites et reclus du Ve au XIe siècle*, Lille UP, 1988; Jean Pierre Vernant, *L'Individu, la mort, l'amour: Soi-même et l'autre en Grèce ancienne*, Gallimard, 1989; Ch. A. Fracchia, *Living Together Alone: The New American Monasticism*, San Francisco, 1979; Margaret Mary Wood, *Paths of Loneliness*, Columbia UP, 1953; Richard Kiekhefer and G. D. Bond, *Sainthood: Its Manifestations in World Religions*, California

UP, 1988; K. J. Weintraub, *The Value of the Individual: Self and Circumstance in Autobiography*, Chicago UP, 1978; Charles Taylor, *Sources of the Self: The Making of the Modern Identity*, Cambridge UP, 1989; R. Kuhn, *The Demon of Noontide: Ennui in Western Literature*, Princeton UP, 1978; Giti Amirami, 'Depression in Medieval Islam', unpublished post-graduate essay, Oxford, 1992; O. John Rogge, *Why Men Confess*, Da Capo, NY, 1959/1975; David Schweitzer and R. F. Geyer, *Alienation Theories and De-Alienation Strategies*, Science Reviews, Northwood, 1989; Markus Fierz, *Girolamo Cardano 1501–76*, Birkhauser, Boston, 1983; Jack Stillinger, *Multiple Authorship and the Myth of the Solitary Genius*, Oxford UP, 1991.

Immunology

Elie Metchnikoff, *Essais optimistes*, Maloine, 1914; Olga Metchnikoff, *Vie d'Elie Metchnikoff*, Hachette, 1920; Anne Marie Moulin, *Le Dernier langage de la médecine: Histoire de l'immunologie de Pasteur au Sida*, PUF, 1991; Niels Mygind, *Essential Allergy*, Blackwell, 1986; William F. Jackson, *Colour Atlas of Allergy*, Wolfe Medical, 1988; M. H. Lessof, *Allergy: Immunological and Clinical Aspects*, Wiley, Chichester, 1984; Thomas J. Kindt, *The Antibody Enigma*, Plenum, NY, 1984; W. R. Clark, *The Experimental Foundations of Modern Immunology*, Plenum, NY, 1984; Debre J. Bibel, *Milestones in Immunology*, Springer, Madison, 1988; J. J. Merchalonis, *Antibody as a Tool*, Wiley, 1972.

Eccentrics

Catherine Caulfield, *The Emperor of the USA and Other Magnificent British Eccentrics*, Routledge, 1981; Enid Welsford, *The Fool: His Social and Literary History*, Faber, 1935; C. R. Snyder and H. L. Fromkin, 'Abnormality as a Positive Characteristic: The Development and Validation of a Scale Measuring Need for Uniqueness', *Journal of Abnormal Psychology*, 86, 1977, pp. 518–27; A. J. Berry, *Henry Cavendish*, Hutchinson, 1960.

5

How new forms of love
have been invented

The world's first children's revolution broke out in France in 1990. Simply by demonstrating in the streets, about 100,000 teenagers were able to force a government to surrender four and a half billion francs. No group of adults had ever achieved a triumph so quickly, so completely. Schoolteachers have been protesting for decades, in vain. Nurses who went on strike, even though they held the power of life and death, were not able to achieve as much. But the children were neither grateful nor impressed.

Mandarine Martinon was regarded as the leader of the schoolchildren of Lyon. The newspapers were amazed that such a frail blonde girl of sixteen could have been so intimidating to the authorities, but for her part, she could see that the government was frightened of her and her friends. The politicians who attempted to flatter the children by inviting them to the ministry, and even the presidential palace, seemed contemptible: their speeches, she says, were 'very cunning, giving the impression that they understood everything, but we were not taken in'. Alone, each child might have felt overwhelmed; together they dismissed the rhetoric as demagogic. The politicians are not stupid, thinks Mandarine, but they are schemers, intent on doing underhand deals. There is not a single one of them who 'reflects my ideals. I will find it difficult to vote when I get a vote.' Above all, the children are determined not to be used by the politicians, nor even by their elders in the universities, with whom they carefully avoided making common cause. They want their world to remain independent, because the adult alternative is no longer attractive.

As a provincial, Mandarine is suspicious of Paris, resenting the assumption of the Parisian children that they can speak for the rest of France. As a resident of the working-class suburb of Villeurbane, she resents the pretentiousness of Lyon's wealthy bourgeoisie. Under no circumstances can she tolerate any clique imposing its ideas on her movement. She repeats that she is a spokeswoman, not a leader offering a solution; her only purpose is to enable each school to decide for itself how it wants its own affairs to be run.

For these children are, in spirit, not children. Their backs are bent, like those of old sages, by the memory of all the failures of their parents, and of all past generations. So much knowledge has been fed into them that there is very little they can admire with enthusiasm. All previous revolutions – the history books say so – have ended in disaster of one kind or another. The risings in Eastern Europe have revealed tragic problems: 'We know too much about the world. We have lost our ideologies; we know they cannot be made to work in real life.' There are no foreign models any more, as China or Russia once were. Mandarine has been to England, but found it 'too respectful'; the USA is 'even worse'.

Personally she does have some ideals: equality is what matters most, democracy too, resistance to state oppression, but keeping the state's role in culture and television; she is for demilitarisation but without abolishing the army; she wants to help the poor; she is in favour of innovation and change, but also afraid of change, and she doubts whether others want change. So she does not attempt to convert or persuade. Her school is setting up committees to discuss very modestly how it wants its individual budget altered and what rights each group within it should have; its only ambition is to transform the school from an examination factory into a 'place to live in'. The school has become the children's other home; they accept it as they accept their own home; all they want is to try to make the best of it.

However, when Mandarine reveals her deeper thoughts, it emerges that it is not just general disillusionment that has made her so different from her utopian parents. She is on very friendly terms with her father, who was a Maoist in 1968 and who does the cooking at home; when she made the posters for her demonstrations, he said to her, 'That is not the way to make posters.' He had produced marvellous ones in his youth; and they laughed a lot redoing them. But the crucial difference is that Mandarine has no confidence in herself. 'I am not creative. I do not feel I am a capable person. Perhaps I lack ambition.' And why? 'Because there is less risk in having limited ambitions.' She plans to work in 'communications', organising cultural events for a municipality, or helping in a theatre, but not putting on plays herself: 'That would involve too much stress. I am an anxious person. I feel stress during examinations in particular, because each one involves my whole existence being judged.' It is not the competition among young people for the best jobs that worries her, but the war within herself, to prove her competence to herself: 'Perhaps I will always be stressed.'

It does not help much that girls are being so successful at school, that more girls than boys are elected class representatives. During the demonstrations, she was fascinated to observe how easily the girls could take the lead, even when there was a majority of boys: 'We could use charm on them.' Mandarine started the demonstrations in Lyon with one other girl: 'We never thought of ourselves as girls.' The woman question never arises, is never discussed. That is all over. Inequality between the sexes is a problem, but they will only meet it when they start working. They will deal with it then. Meanwhile, they have problems different from those which tormented their parents, obsessed by the thrill of the pill.

What Mandarine discusses at great length with her friends is love. She has had boyfriends since the age of twelve, but has still not found love, only friendships. 'We discuss whether friendship is possible between men and women. We decide it is, but that it is difficult, because the desires of the body complicate relationships.' Free sex is no longer the miracle solution; sex is no longer even a way to friendship; instead, it is a problem within friendship.

They are shocked by the crude way the boys talk about it, 'but we think they are trying to conceal the hold that girls have on them, to pretend they are strong, to hide their emotions'. Unable to reach a conclusion about how to behave, the girls divide into two groups. Some change partners frequently, so that they can remain uncommitted. But the fashion is increasingly to find a partner while still very young and to remain loyal throughout the school years. The aim is security. 'We feel stronger when there are two of us.' The children of divorced parents either try to create a stable relationship, or reject the very idea of a couple.

Mandarine and her friends approve highly of love, but cannot find it. 'We see it in American films, but not at home. We experience it only through pictures.' All the discussion of Aids has removed the mystique from sex. But she is struggling to keep alive the romantic idea of love, which means deciding for oneself, individually, what love is. 'We do not want love to be made ordinary, banal. It should be something personal. I don't want to make the details of my private life public. If we are forced to talk about our private lives, they will become a matter of duty, and will lose their interest.'

From which it can be seen that love is one of the last refuges where a person can feel that he or she is able to achieve something noble, and receive the approval of another person: one of the few forms of success that can hold its own against self-doubt.

Love is no longer what it was. There are two types of women in the world today of whom there were very few in the past: the educated and the divorced. Every time new sorts of people emerge, they give a new direction to the passions. They may appear to continue to believe that love is mysterious, talking about falling in love uncontrollably, as though love never changes. However, in the past they have often taken apart the different elements of which love is composed and recombined them to suit themselves, twisting, adding, suppressing. Humans have been far less helpless in the face of passion than legend makes out. They have been able to introduce new meanings into it again and again, as surprisingly as they have transformed grain into bread and dumplings and millefeuille cakes.

Everybody knows that passionate love was given a new form by Germany's Romantic poets, and before them by the knights and troubadours of France, who themselves transformed echoes of emotions refined by the Arab conquerors of Spain. But these changes do not point in the same direction; the history of love is not a sweeping movement towards greater freedom, but an ebb and flow, a whirlpool, and long periods of calm. The contracepted Westerners of today have many alternatives before them. It is surprising, now that love is more highly valued than ever, that schools do not teach its history, its battles, the rise and fall of its dominions, its diplomatic methods and rhetoric, and the hypocrisy of its economics. Perhaps sex education will prove to be the first lesson of a much longer syllabus.

The process of kneading passionate love into different shapes can be observed particularly well in the moonlight of *The Thousand and One Nights*, because the medieval Arabs were once the world's most sophisticated lovers.

In the Arabian desert, the Bedouin nomads, leading the simplest of lives, had no use for passionate love. In their songs of the sixth century, they saw it as sorcery, the work of djinns, with an effect like wine, a challenge to custom; and they mocked the husband who loved his wife too much. This is an attitude that has prevailed in most countries at one time or another; it is normal, because it is based on fear, and fear is normal. However, among the Bedouins, easy familiarity between the sexes was allowed, joking relationships were part of etiquette, men and women could say almost anything to each other. It was through this playfulness that the extraordinary idea developed that two people could love each other to the point of abandoning everything else. Occasionally joking between a local girl and a visiting stranger (to whom the Bedouin's tradition of hospitality allowed liberties which they denied themselves) challenged tribal loyalty. Humour, which was the safety valve of custom, went out of control, and the excitement of breaking rules, of taking risks, of an adventure into the unknown, of thinking one was right against the opinion of the whole world, preferring the mysterious to the familiar face, became a conspiracy of passion. 'What between us two brought love, in the valley of Bagid?' asks a Bedouin song, and it answers that it was the joking insults the pair exchanged, the repartee which gradually broke down apparently impregnable truths. The foreigner could make custom look ridiculous. Attraction became explosive when ignited by fun. Ibn Hazm, the most famous Arab authority on love, said, 'Of love the first part is jesting and the last part right earnestness.'

It was in Mecca and Medina, in the first century of the Muslim era (beginning in AD 622), that women introduced another ingredient into their feelings, creating new moods with music. That this happened at a time of upheaval, when people were losing their taste for old ways and were agitated by the new ones offered to them, was no accident. The cities were wealthy, devoted to pleasure and feasting, frantic to forget the dangers around them. Singers were 'omnipotent', like the pop stars of today. Rich women were expanding their traditional liberties, imposing conditions on their suitors before marriage, rejecting any suggestion that a wife should be like a slave. Sukayna, the granddaughter of Ali the cousin of the Prophet, was one such free spirit, who neither wore the veil, nor obeyed her husband, but organised literary and musical salons (the Arabic word was *majlis*, meaning a gathering of notables). Rich young men flocked to them, to drink forbidden wine or the less objectionable fermented juice *nabid*, and to listen to poets and singers. Sukayna got the most famous singer, Umar b. Ali Rabia, to meet her and her friends at night in the desert, and they would talk till dawn, about feelings. His songs were above all about feelings. 'Keep women far away from song, because it is a summons to adultery,' said a proverb. Umar gave his heart to several women at the same time, never sighing for the absent ones: 'Ah, how many girl-friends I have had, whom I left without ever hating them, always valuing them.' This was not exactly the passion these women had intended to create, and Umar's songs are full of their complaints; their attempt to make life more interesting also made it sadder. But courage invariably leads to unexpected results; that is what defines it, the willingness to meet the unexpected.

When only the music of the tambourine was approved of in

religious feasts, these singers of Mecca and Medina now brought
in from Persia the lute, ancestor of the guitar, and despite protests
that it was a lascivious instrument, the husbands, busy with their
own pleasures, did not interfere. It worked like a charm. The
singers were 'youths of outstanding beauty', wearing their hair
down to their shoulders, a sign that they drank; often freed slaves,
with no tyrannical family loyalties, suspected of being the illegitimate children of noblemen, they were always in trouble with the
authorities, but were regularly saved from punishment by their
women admirers. Whereas the old songs were about war, these
singers sang only about love, the women demanding lyrics in
which they could recognise their own feelings. Exploring alternative loyalties between individuals, beyond those of tribe and
family, was again accomplished with foreign aid. The importation
of strange melodies protected this bold behaviour by wrapping it
up in mystery. One famous singer, Ibn Muhriz, travelled to Persia
to study its music and its tradition of refined love, of 'voluptuous
contemplation', of tales which said that a sovereign is unworthy to
reign if he does not know how to love; then he went to Syria
to study Greek music, returning with sounds no one had ever
heard before. There was enthusiasm for novelty in the cities, provided it was mixed with old Bedouin strains, like the irregular beat
of the camel song. Foreign music was the second ingredient, after
humour, in the restyling of passionate love, as it was to be many
times again, as African and American music have been in our own
day. Arab love was transmitted to the French troubadours not by
philosophers, but through music. The musicians on either side
of the Pyrenees understood each other, because a mood is more
infectious than an idea. The word troubadour may come from the
Arabic *tarab*, meaning music.

Islamic piety eventually silenced the self-questioning of these
women who had devoted themselves to elevating the relationship
of the sexes. So then passion was reshaped into a different form. In
the next century, the bustling city of Basra experienced a similar
epoch of 'giddy uncertainty': values were in a whirl, 'free-thinking,
morally lax addiction to light-headed pleasures produced
emotional exaltation', and 'mystical experiences were observed
among women'. Another sort of heretic now used love to cope

with doubt. The most famous, Bashar b. Burd, was an angry young man who became even angrier in old age: a gifted poet of Persian origin, uncomfortable among Arabs, he felt he was not appreciated, and indeed he eventually died after being sentenced to flogging for annoying the calif. Despising the world and its ways, recognising no morals or religions, rejecting authority, torn between materialism and a search for absolute redemption, he revenged himself in unrestrained lampoons. He was not afraid to write poems telling his rivals what he thought of them:

> *You son of a rutting animal, you are*
> *A pustulous, foul, filthy bum.*

Around him a multitude of heresies questioned almost everything. The mutazilites, for example (the word meant 'to keep one's distance'), claiming that no one was all right or all wrong, argued against dogmatism for 'a position between two positions', in favour of free will. But the one thing they were certain about was that passion was an inescapable physical or cosmic force. It was in this atmosphere of disillusionment that love's passions were exalted as the supreme value. Bashar sang not of his love for any particular woman, but in praise of perseverance in love in general, against all the obstacles it encounters. It was the fight that appealed most to him. Then he mocked his own insincerity: 'I have lied enough to have the right to tell the truth.' Passionate love was made into a flag of rebellion.

Al Abbas b. Al Ahnef, another famous Basra poet, proclaimed, 'There is no good in those who do not feel love's passion', as though lovers were a secret society, dedicated to self-torture. He became wildly popular after emigrating to Baghdad to the court of Haroun al Rashid, where his poems were set to music. The calif, miserably in love with slave girls from whom he could theoretically demand anything he pleased, felt something was still beyond his all-powerful reach. Al Abbas explained passion as a yearning for the unattainable. The partner at whom it was directed was not its real object. Love was a sign of a weakness within oneself, inevitably hopeless. Al Abbas sang of unhappy, chaste, paradoxical love, protesting that he was happiest when his loved one was beyond reach. Love and sex parted company: 'He who loves with unrequited love and remains chaste dies a martyr.' The idealisation of women did not of course improve the way real women were treated; on the contrary, it represented despair with real women.

Men of the world who had no taste for metaphysical discussions about what love meant, and who simply wished to enjoy pleasure without pain, clearly needed more practical clothing for their desires. They longed for a happy ending, not the inconveniences, enslavement or extremes of rebellious passion. Their affections had to be compatible with loyalty to the community which gave them their sense of status. It suited them not to enquire too closely into the individuality of their loved one. *The Book of Stories* by Abul Farag al Isfahani (died 967) put their attitude into words. He was a precursor of the jet set, famous for his good humour, fond of joyous company of all nationalities, on the fringes of high society, a constant traveller, always accompanied by a young man. He showed how love could avoid theatricality or pathos or tragedy or worry, with women who always played only a subordinate role; how love could be tamed and rendered harmless, to be boasted about afterwards, like another brand of nostalgia.

But for men of the world who were not attracted by such easy and superficial relationships, an alternative was offered by Ibn Hazm of Cordova (994–1064). His treatise on love, *The Ring of the Dove*, was the climax of a life in which everything had gone wrong for him; he complains of having been exiled, betrayed, oppressed, robbed, driven to despair, 'waiting for the further blows of destiny'. Muslim Spain, he believed, was 'self-destructing'. One of his solutions was a new attitude to love. He did not write about love just to praise it: having fallen in love three times over the years, he analysed his intimate experiences in his book, even though he knew that people would consider it unseemly for a public figure like him, who had been a minister to the calif and a scholar. He did say that love renewed life, that it made greedy people generous, boorish ones gracious, stupid ones wise, magically transmuting defects into qualities, so that everybody could hope to be loved: there was no greater joy in the world than two people in love. He exalted sexual intercourse as a necessary part of love, 'completing the circuit and allowing the current of love to flow freely into the soul'.

But what was most original about Ibn Hazm was his conviction that love mattered above all because it could be much more than a tranquilliser or a private consolation. He wanted to make love the central experience of life in a new way. Normally, he said, people used the words of love as masks, to pretend to be what they were not, whereas it could be the opposite, a set of mirrors which revealed to them what they were really like. His book was aimed at helping them to know what the words and acts of love meant. And he wanted the meaning to be clear, protesting against the tradition of reading mysterious significance into statements which were essentially simple. Though an expert himself, he mistrusted experts. Clear communication was his panacea. Truth, he said, should be understood as simply as possible, ignoring the explanations of over-subtle learned theologians. Humans were wrong to tie themselves up in knots with their cleverness. Though in his sentimental moments he called love 'a most delightful malady, a most desirable sickness', he saw it as, or wanted it to be, part of the search for self-understanding. He said he had been brought up by women, and he was interested by what women thought. Though he could only love blondes, because his first love had been a blonde slave and he had never quite got over her, he could only become interested in another one 'after long association and long converse'. To want to understand the soul of a woman was to open a door most men preferred to keep shut.

He belonged to a sect called the zahirites, who tried to make Islam simpler to understand. The chief poet of courtly love in Baghdad, Ibn Daud (died 909), author of *The Book of the Flower*, an erotic anthology, was the son of the founder of this sect, Daud of Isfahan. Another zahirite, Ibn Arabi (1165–1240), born in Spain, one of the most influential of all Muslim thinkers, said that he had himself been much influenced by several remarkable women; he revealed, in *The Interpretation of Longings*, what this understanding gained through love could lead to:

> *My heart is open to all the winds:*
> *It is a pasture for gazelles*

And a home for Christian monks,
A temple for idols,
The Black Stone of the Mecca pilgrim,
The table of the Torah,
And the book of the Koran.
Mine is the religion of love.
Wherever God's caravans turn,
The religion of love
Shall be my religion
And my faith.

So here are five quite distinct kinds of passionate love developed in a comparatively short time in just one part of the world. Sexual desire, emotion, fantasy and everything that is called instinct can doubtless be combined in a thousand other ways. And when love is rewarded, however momentarily, by that even more elusive sensation, self-confidence, it becomes not just a private secret, but a public force.

Thus the playfulness which started it all off leads further than to mere laughter. Playing means giving oneself temporary freedom from duty and necessity, voluntarily taking risks and being excited because one does not know the outcome; 'pretending' is self-conscious delight in alternative possibilities, and appreciation of the fact that no victory is final. Is it an accident that the verb to win derives from the Indo-European root *wen*, to desire, and the verb to lose from the root *los*, to set free? Can playing at winning and losing be an apprenticeship in freedom? The Spanish for to win, *ganar*, derives from the Gothic *ganan*, to covet, while *perder* (to lose) comes from the Latin *perdere*, which originally meant to give completely. The courtly lover who did not want to possess his ideal, who played to lose, discovered that whereas business and war were prosaically about possession, in love it was the play that mattered most. Being willing to play is one of the conditions of creativity. Love, far from being a distraction from creativity, is a branch of it.

All the five stories have a stranger somewhere in them, which is not surprising, because love is always for what is strange, for the unique, for the person who is like no other, but then it turns the frightening into the familiar. In the past what lovers have probably feared most of all has been loneliness; but now imprisonment in a static relationship has become even more worrying. The thirst for new experiences, for the unknown, for strangers, is greater than ever. So uniting two exiles to make a safe, self-sufficient family has ceased to be enough. A wider creativity is the modern temptation. The fascination with what is strange is, like play, a step to creativity.

For most of history, love has been considered a threat to the stability of the individual and of society, because stability was usually valued more highly than freedom. In the 1950s still, only a quarter of engaged American couples said they were head over heels in love, and in France less than a third of all women claimed to have experienced a *grand amour*. Forty years later, half of Frenchwomen were complaining that the men in their lives were not romantic enough, and wishing they would at least say 'I love you' more often. Their general view is that modern life makes amorous passions more difficult than in the past, but the golden age never existed. In despair, many of them say they feel more passionately about animals and sport than about people. In Russia, at the dawn of glasnost, even among newly-weds, love came only fifth in the list of eighteen reasons for marrying. Which means that passion is an art that people have yet to master, that love is an unfinished revolution.

For about ten centuries, Europe has echoed mainly two of the strains in Arab love – the idealisation of women and the fusion of the lovers' souls – neither of which can satisfy the longings of those whose ambition is to understand their partners as they are, and to continue to exist as a more or less independent being. Idealisation once seemed to be a chivalrous answer to the impermanence of affection, and fusion offered a romantic solution for loneliness; in both cases love was used as a remedy, because the world was passing through a hypochondriac phase of history, dominated by the sense of sin, or guilt, or shame, and unceasing complaints that humans were inadequate, unable to reach divine perfection. Love worked no worse and no better than other folkloric medicines in making them feel better. After all these experiments with passion over the centuries, it remains as ephemeral as ever, while loneliness continues to expand its empire.

The choice for Mandarine, in seeking an alternative to her parents' idea of love, is not limited to a return to the 'romantic'. The sociologists who have interrogated educated girls like her can think of no other word to express more fully their longing to escape from cynicism, their desire for more than mere contentment, more than a balanced life occasionally seasoned with a piquant, erotic sauce. However, when these girls say, 'We want to make life more beautiful' and when they rank love as one of the arts, they are not interested in reproduction art: to replay the past is not enough, when one knows how grim the realities of the past have been. They would like to invent a new art of love, and there are many precedents which show that it is possible to do so.

However, every invention needs new ingredients, or the jettisoning of at least some old ones. The most time-worn belief, ready for the dustbin, is that couples have no one to rely on except themselves, which is as unfounded as the belief that modern society condemns individuals to loneliness. Now that boys and girls are being educated together and forming friendships at school of a kind that have not existed before between the sexes, love can assume other shapes. What these other shapes might be will become clear as the possibilities of other passions are examined one by one.

J. Ashtiany, *Abbasid Belles Lettres*, Cambridge UP, 1990; Jean Claude Vadet, *L'Esprit courtois en Orient dans les cinq premiers siècles de l'hégire*, Maisonneuve, 1968; A. G. Chejne, *Ibn Hazm*, Kazi, Chicago, 1982; Ibn Hazm, *The Ring of the Dove*, Luzac, 1953; 'Ishk', article in *New Encyclopedia of Islam*, 1978; Émile Dermengheim, 'Les grands thèmes de la poésie amoureuse chez les Arabes précurseurs de poètes d'Oc', in *Les Cahiers du Sud*, special issue on Le Génie d'Oc, Marseilles, 1943; J. M. Ferrante, *Conflict of Love and Honour*, Mouton, 1975; Lois Anita Giffen, *Theory of Profane Love among the Arabs*, New York UP and London UP, 1971; Clinton Bailey, *Bedouin Poetry*, Oxford UP, 1991; L. F. Compton, *Andalusian Lyrical Poetry and Old Spanish Love Songs*, New York UP, 1976; W. Dols, *Majnun: The Madman in Medieval Islamic*

Society, Oxford UP, 1992 (chapter 11 on 'the romantic fool'); Alan H. Gardiner and K. Sethem, *Egyptian Letters to the Dead*, Egyptian Exploration Society, 1928 ('What I have done to thee', by a widower to his dead wife 'in the beautiful west'); H. R. P. Dickson, *The Arab in the Desert: Badawi Life in Kuwait and Saudi Arabia*, Allen and Unwin, 1949; Richard Boase, *The Origin and Meaning of Courtly Love*, Manchester UP, 1977; Claude Tapia, *Jeunesse 1986: au delà du sexe: Psychosociologie de la vie affective de la jeunesse*, Harmattan, 1987; F. Dubet, *Les Lycéens*, Seuil, 1991; Love statistics in Armelle Oger, *Enquête sur la vie très privée des Français*, Laffont, 1991, pp. 317–23; P. Bancroft, *Human Sexuality*, Churchill Livingstone, 1983; V. Shlapentokh, *Public and Private Life of the Soviet People*, Oxford UP, NY, 1989, p. 178; Jérôme Duhamel, *Vous les Français*, Albin Michel, 1989, p. 334; E. Burgess and P. Wallis, *Engagement and Marriage*, Lippincott, Philadelphia, 1953; Reuben Fine, *The Meaning of Love in Human Experience*, Wiley, NY, 1985; Irving Singer, *The Nature of Love*, 3 vols., Chicago UP, 1984–7; 'Évolution historique du sentiment amoureux', in *Futuribles*, July 1990.

6

Why there has been more progress
in cooking than in sex

What happens to a Spanish girl who is sent to a school run by French nuns, who has mademoiselles at home to teach her manners, and goes to French families for holidays to polish her accent? She becomes someone impossible not to notice in the street. Alicia R. Ivars always wears amazing clothes. But they are in her own peculiar style, never the latest Paris fashions.

How did it affect her life that, as a young woman, she should have got drunk on Franco-Germanic philosophical jargon, and become a vitalist, voluntarist, gestaltic rupturist existentialist devotee of Bergson, Bachelard, Fischer, Chiva, Calvo . . . ? She has ended up as a world authority on olive oil. Nothing is inevitable or predictable.

Beneath the dramatic clothes, individually designed, behind the combination of shyness and exhibitionism, Alicia is trying to be a geisha. A highly intelligent woman, she prefers to cultivate sensual pleasure more than any other side of her personality. Despite her long years of theoretical training, she concentrates on making herself pleasant to other people, avoiding getting lost in questions like Who am I? A geisha is of course the opposite of a bunny girl, admired not for youthful beauty but for a skill older than that of the oldest profession, closer to that of a priestess, performing rituals to reconcile men to the fact that they cannot entirely have what they want.

What is taught by teachers is one thing; what goes on in the soul is another. The legacy of Alicia's childhood is her shyness; until the day when, in her twentieth year, a man suddenly declared his love for her, she had not even noticed his interest. Not only had she been to a girls' school, but her family consisted, apart from her father, exclusively of women: three sisters, a grandmother, two maids, a mother more enamoured of nature than of men. Men had not entered her world, which she' had learnt to enjoy without them. Her reply to this first man in her life (He wrote, 'I have been trying to express my feelings and you take no notice') was to fall in love in return, 'out of gratitude' for something unknown. Ever since she has been fighting shyness, deliberately pretending to be an extrovert, almost becoming one in the process. The life of a shy person was not as interesting, she decided, as that of someone who allows curiosity to run wild, who is 'half wild, half well-mannered'.

It does not matter all that much with whom you are in love, she believes; it is a mistake to want to be loved by the person you love. No, there is a 'cosmic justice' in the distribution of love, which overrides our personal amorous disappointments. Affection given is always a good investment; you may not get it back from the people you offer it to, but you will get it from somebody, and the more you give the more you will get. There is a deficit in the balance of affection because people hesitate to give it, because they do not want affection from just anybody, so they limit their opportunities, developing a narrow image of themselves, as being no more than a certain sort of person. As a result, it becomes harder for the unexpected lover to find them and to make surprising discoveries in them.

So Alicia tries to have as 'fluid' a personality as possible. Her religion is the 'cult of daily life'. 'I believe it is possible to modify your previous or expected roles, not to mention manias or all sort of psychological burdens and self-determined limitations.' Her knowledge of English may not be perfect, but that is in keeping with her principles. She prefers to spread herself, to get by in four languages rather than be faultless and impregnable in a single one.

You can avoid being a neurotic if you stop obsessively analysing what you imagine to be your character: never mind your faults, stop moaning about your complexes, do not pour out confessions about what you can and cannot do, like or desire. Treat each meeting with a person as an independent event. 'A geisha is always ready to produce pleasure without considering her own needs.' Put aside personal ambitions and the expectations you have of yourself. Learn to be a geisha by being, first of all, a geisha to your own body; look after it, cook meals for yourself when you are alone as though giving yourself a treat; look after your mind by feeding it with poetry and music. Avoid creating an excessively rigid idea of your desires. Look upon yourself as an amoeba, floating through life, dividing: do not be afraid of losing your identity. Or look upon yourself as a collection of electric light bulbs: do not put all the electricity into a single bulb, or it will explode; allow your energy to circulate freely through the many sides of yourself. The looser, the more open and limitless your identity, the better. Treat your emotions as a garden needing to be kept tidy. Be generous, and that will stimulate new resources within yourself, new ideas. Follow the 'laws of nature'. It is up to you.

With such beliefs, it was not enough to earn a living as a professor, even as an eccentric one. Alicia opened a restaurant. First she cooked three days a week and lectured at the university for three days. Then she resigned from academia and put all her energies into The Garden of Delights. This restaurant was her theatre; every day the doors were opened and the customers had to be surprised. 'I was so grateful people came dressed up, I always tried to wear something new.' Her inventiveness in dress is extraordinary; bizarre, surrealistic little touches mean she is always a performance. When the diners demanded that she should come out of the kitchen, she always changed before appearing. But it was more than just putting on two performances a day. People never know quite what they want. Her pleasure was to reveal their desires to them, to offer herself as an expert in fantasy, 'a culinary interpreter', translating vague longings into amazing meals, wrapped in heavy symbolism. 'The geisha cook is attentive, silent at times, but capable of being mystic, ecstatic, minimalist, ritualist, aestheticist, devoted to others.' Organising weird parties became her speciality, creating unusual atmospheres, to make people 'feel different': for example, Edwardian splendour set in colonial Egypt, gardens illuminated by torches, fountains with bathing belles, wines with exotic fruit and colouring, Arabian food. Like a magician, she called herself Ali + Cia.

Though never a hippy 'stricto sensu', nor a feminist 'though I shared the attitudes, but without the militancy', participating in a large number of university and political organisations allowed her to have 'very intimate and prolonged communication, and sex too, with many different compañeros'. Not till she was twenty-eight did she marry. That involved persuading the man of her choice – it took her five years – that she was right for him, that he ought to give up his comfortable bachelor existence, even though she 'did not correspond to his idea of a wife'. He was no virgin either, 'far from it', but she told him he was a stick-in-the-mud and she forced him out of it. Paco is the only man in the world, she says, whom she finds 'one hundred per cent acceptable'.

But that did not prevent her, after ten years of marriage, from leaving him. A wonderfully attractive bohemian, a man of the theatre, who had been baptised in fire during the events of May 1968 in Paris, took to eating at her restaurant. 'Come and have a siesta with me,' he said. She began visiting him; they went into the country and made love. 'It was Paradise', passion of a sort she had never experienced. 'The greatest conflict of my life' followed. Alicia told Paco, 'I need to clarify my situation.' To reject the experience her new lover was giving her would diminish her personality. So for ten months she tasted passion. Then she decided her lover was 'not fully acceptable'. Back to Paco she went. Paco the mature, polite, elegant, scholarly master of detachment, for all his coolness, had been hurt. However, they were reunited. 'Paco never once mentioned the subject again.' Her marriage emerged strengthened; her admiration for Paco knows no bounds; 'he never gets angry'.

Nevertheless, Paco 'knows only part of me. We don't try to be too intimate, so as to keep a sort of mystery between us. . . . If you know too much, you become a prisoner.' The way to keep a marriage going is to avoid digging into each other's soul. Take great care not to speak too directly, not to hurt the other. If you must pour out your heart, find someone else. Alicia has found someone in a 'misogynist, unmarried but not homosexual' expert on patristic theology who is a friend of her husband too: he is a most 'spiritual man'; and Alicia goes with him into the country on picnics. 'I can talk with him about anything. I inspire him and he inspires me. He understands the minutiae of my arguments.' There is no sex involved.

Sex is a separate matter, a distinct activity, 'not to be ruined by an excess of intimate feelings or confidences, because then you become a slave to it'. That does not mean Alicia wishes to avoid intimate feelings. 'I have never been afraid of my intimate feelings. I have always enjoyed psychotropic experiences without panicking at the idea of losing contact with my inner self or with my body. I know which melody, which rhythm, which smell or caress or stroke will provide me with my desired intimate feeling.' Engaging in sex is thus comparable with cooking: both create pleasant 'intimate feelings', both enable one to create such feelings in another.

She distinguishes first of all 'pure sex'. In her youth she had this with a 'Tantric man', with whom she carried on an 'ultra erotic correspondence, with a profusion of illustrations' and whom she visited two or three times a year for the 'actualisation of all our fantasies': she called it emotional luxury. Sex combined with friendship is different, a wonderful combination, but sex rarely leads to friendship, though she would not 'reject it as a stimulus'. But if sex is confused with love, only conflict can follow, or marriage.

With the passing of the years, the frequency and variety of Alicia's sexual activities have diminished. Friends are less demanding; there is less time, less space. Younger partners are not to her taste; those of her own age or older are 'quarrelsome, possessive, ambitious, neurotic, unable to rest or to play'. Having a varied sexual life is 'wonderful: it keeps one in love with life'. Former lovers remain with her for ever, 'integrated into my way of loving'; she loves them still, even if she does not miss them; remembering past loves, seeing them happen again in the mind's eye, is exciting and almost like being in love.

There is no reason, she thinks, why there should be any limits to sexual activity: 'I have not yet found limits to my predisposition.'

Group sex and lesbianism, it is true, do not appeal, but she remembers the visit of a foreigner to whom she was 'very attracted: I could have played and enjoyed with her, she was so wonderful to me, so much in need of affection, just recently a widow; we probably had telepathic sex.' There are other forms of sex which do not appeal: for example, 'bland men, demanding or dependent on me', and 'vampires', and 'husbands who come home expecting their wives to bring them their slippers and whisky on the rocks: what a horror'.

Paco is the ideal man because he is never possessive, always respectful, happy, funny and, above all, 'not too present'. His great virtue is his solid independence. 'We can share our free time, or not, go out together or separately.'

'From infancy, I was brought up and trained to enjoy sensual pleasures, spending four months each year in a paradisiac, wild place (today the house is so big and beautiful but the surroundings have been destroyed), with fig, almond and olive trees, vineyards and tomatoes, and the sea, and freedom; all the family, friends; long, long days of exploring sensual pleasures.' If she were marooned on a desert island, she would choose to take a knife with her, 'to carve words on the trees, to kill animals and drink the blood, eat the flesh, build a hut for an isolated love affair'.

All Alicia's ambitions are private ones. Changing the world does not interest her. Apart from making Paco happy, she has no specific targets, except 'a desire to reach perfection in an oriental way', meaning developing her potentialities. Fame is worth striving for because it gives you a wider choice of interesting people to meet, but for no other purpose. Money is useful to widen your opportunities, but it is dangerous, because rich people tend to meet only other rich people, worrying about their pure silk dresses, all doing the same thing. However, professional success is not enough: it is too often combined with a hopeless private life. Five years running a restaurant, 'like a sister of charity, like a captain of a ship', have given her a great sense of achievement, but that is not enough.

When Paco was recently ill – fortunately he made a complete recovery – it made her think: What would she do if he died? Not work in an office, for sure. She would take two paying guests and look after them. Poverty does not worry her: it would not stop her from indulging her pleasure in thinking, in having a private life. Being alone does not frighten her either. The other side of her flamboyant sociability is her shyness, or her absorption in her own mental world. Just as the man who laughs most is often miserable, so the one who delights in party-going is often most alone. Alicia insists that she is a solitary person. Looking out of her window towards the country around Madrid, she sees nature completely indifferent to humans, and takes that as a model: the need to be indifferent to one's own worries, to be detached from oneself. But to be solitary does not mean that you have to be isolated. She is not; she alternates between going out with friends, to the cinema (she takes a picnic there, too, and eats throughout the performance) and being alone. Sociability is to her a sort of language – the more you practise, the richer the kind of communication you can have with people. But at a young age, she taught herself to cut her own hair, and has never been to a hairdresser since. That is the sign of her independence. Her hairstyles are always exotic, like nobody else's.

Krafft-Ebing, the expert on sexual perversions and Freud's colleague in Vienna, said that hunger and love govern all world affairs. But they both forgot about hunger and concentrated on the torments of love, which was unfortunate, because sex, food and drink have always been fellow travellers in the search for pleasure. If sexology had not become a separate scientific subject, if the search for knowledge had been organised differently, if there were professors of happiness who studied the passion for pleasure as a whole, in all its forms, a different outlook might have emerged. Physical urges are not despots and have frequently been disobeyed; tastes are not fixed for ever. The way to look afresh at desire is to consider what people want at table and in bed as part of a whole.

Gastronomy is the art of using food to create happiness. There are three ways of eating, and three corresponding ways of searching for happiness. To eat until one is full up is the first and traditional way, putting faith in old recipes and well-tried methods. The aim is to become contented, to be comforted, to feel cosy, to purr. This is the cautious approach to pleasure, with the motto 'Protect yourself from foreign bodies'.

Foreign bodies are not only the fly in the soup, but also everything that is unusual, forbidden, unfashionable, threatening. It was in the process of learning to eat that humans made their fear of foreign bodies a virtue and called it taste. Mental habits developed which mimicked the patterns set by eating, and fear of foreign bodies spread to many other aspects of life; routine, however boring, appeared to be the safest insurance policy. Much of history has consisted of wars against foreign bodies, because the first kind of happiness humans sought out was that which gave security. Nothing would ever have changed if caution had triumphed, but there were always nervous and lonely people who did not feel safe, regarding themselves as foreign bodies too, strangers in their own surroundings; contentment seemed impossible to them.

So a second way of eating was invented, treating food as an amusement, a form of permissiveness, a caress of the senses. The purpose was to seduce and be seduced, with the help of romantic candlelight, to create conviviality around delicious odours. In such circumstances, one's attitude to the world at large is only temporarily modified: one flirts with foreign bodies during the meal, but they do not affect how one behaves in the office. This kind of eating has suited the person who has despaired of being fulfilled by a quiet life, who yearns for distractions and surprises, who seeks a different kind of happiness in frivolity, in being jokey, cynical, ironical, refusing to be made permanently miserable by the big problems, like starvation and stupidity. The cooks who prepare food for them are like jazz musicians, improvising playful flourishes, never reaching a conclusion.

But of course it is impossible to be happy in more than a very superficial way while others are unhappy. When peace and quiet, or wit and detachment, began to pall, a different yearning was born, to make a personal, original contribution to life. The search for a third kind of happiness – which moderns call creativity – demanded a way of eating which corresponded. All invention and progress come from finding a link between two ideas that have never met, bringing foreign bodies together. For individuals who aspired to be creative, eating became part of the process of looking at the world in a more adventurous spirit. Creative cooks found qualities in food that nobody suspected were there, uniting ingredients that never used to mix. Creative diners are constantly engaged in losing their fear of strange foods, and of foreign bodies.

However, this does not mean that there are three kinds of people, each of them stuck with their habits. Creativity is the preoccupation of the master chef, consciously trying to innovate, but those who believe they are doing the opposite, endlessly reproducing the same grandmother's recipes, are sometimes creative without realising it. It is true that there are people who eat roughly the same food as their ancestors thousands of years ago, but variety creeps in all the same, however limited the menu may seem. Thus in Ghana a poor community, completely unknown to the world of culinary connoisseurs, eats 114 species of fruits, 46 kinds of leguminous seeds and 47 greens. In the Andes, a peasant can distinguish without difficulty between 300 varieties of potato, and he will cook his stew with as many as twenty to forty varieties, carefully balanced. Every time a recipe is not strictly followed, every time a risk is taken with changed ingredients or proportions, the resulting food is a creative work, good or bad, into which humans have put a little of themselves. The invention of a new dish is an act of freedom, small but not insignificant. There is still enormous scope for such acts, since humanity today eats only about 600 out of the hundreds of thousands of edible plants.

Children have usually been brought up either to be loyal to their family's taste or, more recently, to develop individual identities by asserting their own taste. But now some are being encouraged to treat tastes as they treat people, as worthy of being respected, recognised, understood, and not to erect high walls between those whom they will or will not speak to, between foods they like or do not like. The French schoolchildren whose syllabus now includes systematic lessons in the art of tasting are the pioneers of an important revolution. An open mind about food, and about the tastes of foreigners, inevitably modifies one's attitudes to one's neighbours.

The world was for long divided into three major empires, of roughly equal size, based on the three main staple foods, wheat, rice and maize. But what separated people even more was the sauce or spice they added: olive oil in the Mediterranean, soya in China, chilli in Mexico, butter in northern Europe, a whole variety of aromas in India. The Russians rioted in the 1840s when the government tried to persuade them to grow potatoes; being used to living mainly on rye bread, they suspected a plot to turn them into slaves and force a new religion on them; but within fifty years they were in love with potatoes. The explanation is that they added the same sourness – *kislotu* – which had always given savour to their food, and which was what they were ultimately addicted to. Every people puts its own scent on its food, and it accepts change only if it can conceal the change from itself, by smothering each novelty in its scent. Optimism about change, whether in politics, economics or culture, is only possible if this premise is accepted.

The Americans have used sugar as the taste which makes all novelties acceptable. Sugar, which does not smell and which has the magical power to make almost anything superficially palatable, has indeed united the world's taste more than anything else. Once a rare and divine medicine – honey was called the perspiration of the sky, the saliva of the stars – its production has increased fortyfold in the last hundred years: it is the culinary expression of democracy. Only when Latin American chocolate, previously spiced with chilli, was married to sugar (by Conrad van Houten of Amsterdam in 1828) did it capture the world's palate. In 1825 Brillat-Savarin, author of *The Physiology of Taste*, predicted that sugar was destined to be the 'universal flavouring'. At that time Goethe was paying 2.70 gold marks for one kilogram; sugar was the elixir of pleasure only for the rich, who spent more on it than on bread. Now the prophecy has been fulfilled: almost every packaged food contains sugar.

All culinary progress has been dependent on the assimilation of foreign foods and condiments, which are transformed in the process. Chinese food reached its apogee in the twelfth century thanks to the importations of adventurous merchants. The food of Europe was orientalised by the massive use of spices – it was almost Indian in the Middle Ages. Then it was Americanised by the introduction of the potato, the tomato, the Christmas turkey and other Native American produce. Fast food is neither American nor European, but an inheritance from the street vendors of the Middle and Far East. The *nouvelle cuisine* is the result of a graft of Japanese ideas on to French tradition. These importations have always been by minorities, against opposition. All innovation encounters opposition.

However, hunger is still being satisfied without full awareness of what it is one is hungry for. Some delicious foods have no nutritional value, others are disagreeable until a taste for them is acquired, others still do not stop one feeling hungry but stimulate one to eat yet more, to prolong the pleasure of eating, like a lover seeking to prolong an embrace. Trying to make sense of such behaviour can clarify a lot more than one's tastes in food – for example, how far one is interested by new sorts of pleasure, or innovation and creativity in general, whether one is willing to risk disappointment or failure, whether one wants to be brave and free more than to be applauded, whether one likes to discuss one's pleasures, whether one enjoys giving pleasure to others. Gastronomy is a branch of knowledge in its infancy, focusing not on self-indulgence but on exploration, not just on self-exploration but on the exploration of the whole of nature. It can look forward to ever-widening horizons of pleasure and understanding, even though it has its dark side, for it has done little to deal with the obscenities of famine and cruelty, and it will perhaps only receive proper recognition when it does. Nevertheless, forks and spoons have probably done more to reconcile people who cannot agree than guns and bombs ever did.

The pleasures of sex, however, have narrowed rather than broadened with the centuries. Sex is the miracle which makes humans, who are normally frightened by strangers, feel attracted to some of them. But it has so far failed to produce even a fraction of the flowers – of affection or understanding – which it might have done.

How to obtain in sexual relations something of the warmth and security, the sense of knowing where one belongs, that mother's cooking gives, was what the pagan religions taught. For them the world was one great self-sustaining sexual machine: the sky impregnated the earth with its moisture, and every copulation was part of this permanent process of self-renewal, not a sordid act, but an affirmation of kinship with the whole of nature. The Hindu god Shiva set an example by the delight he took in spreading his seed among women, and his followers could regard their sexual instincts as proof that they had something of the divine in them too.

The gratification that one was part of a whole was reinforced by the feeling that one could contribute personally to keeping the world going, for nature had to be encouraged as well as thanked. The Masai in East Africa did so by periodic Feasts of Love: for several months all restrictions of friendship and marriage were lifted, and people came together from hundreds of miles to fecund-ate the earth and animals and each other, in front of their priests: everybody made love to everybody else, only mothers and sisters excepted. These occasions were not orgies but a way of giving life a boost. 'Sex is exhausting work,' a Kikuyu woman observed to an anthropologist, 'one does not have time to talk.' But a pleasure is not diminished when it has been hard-earned.

The Chinese made sexual activity a source of comfort by placing it in the centre of their system of medicine and emphasising its essential role in safeguarding health and curing diseases, improv-ing the circulation of the blood and relaxing the nervous system. Men could strengthen themselves through frequent intercourse, which produced energy by uniting the male and female principles, but they had to take great care to give equal pleasure to women, in the same way that they kept the earth's soil in good heart, for women produced vital juices which prolonged life. The bizarre exaggerations of these doctrines obscured their deeper message. In *The Art of the Bedchamber*, the Han minister Chang Tsang describes how he attempted to live to the age of 180 years by sucking the secretions from women's breasts. But what virtually all ancient sex manuals insisted on was the importance of paying attention to women's desires. In Europe, ancient folklore advised that no conception was possible unless the woman experienced pleasure. Infertility, said Culpepper's *Directory of Midwives*

(1656), was caused by 'want of love between man and wife'. Those who today regard sexual activity as an essential part of healthy living have their deep roots in these pagan traditions, dedicated, as the Taoists put it, to 'the simple and joyous art of living only with a view to living'.

However, just as some people tired of mother's cooking and started exotic restaurants to amuse themselves in unexpected ways, so some sought amusement for amusement's sake in exotic beds. But whereas knowledge of food expanded, as a result of trade and travel, and was constantly added to, the erotic imagination quickly became repetitive. By roughly AD 450 the techniques of sexual pleasure had been comprehensively described in the *Kamasutra*, a summary of numerous much longer works, compiled by the ascetic celibate Vatsyayana. Though accounts of personal adventures embroidered on this work of learning and literature – notably Kshemendra (990–1065) in *The Harlot's Breviary* and Koka (?1060–?1215) in *The Mysteries of Passion* – the range of available pleasures remained virtually unaltered; for over a thousand years after Ovid and Lucretius, Europe had nothing to add. Pornography had its holy books; devotees tended to become addicted to a particular obsession. The fantasies that took up residence in the minds of lovers focused above all on conquest, domination and submission, as though the most common relationship available out of bed had built fences around the bed. There was no genuine rebellion in fantasies of being forced to do forbidden acts, or being seduced by forbidden admirers. Every generation imagined it found freedom in these fantasies, but it was simply tying the same old noose – or one of a small selection of nooses – round its imagination.

Thus the most exciting sexual experience a Chinese man could have, from about the tenth century onwards, was to glimpse the feet of a woman who had reduced them to three or four inches in length by preventing their normal growth in childhood.

> *I don't know when this custom began:*
> *It must have been started by a despicable man*

wrote a Chinese woman poet. The practice of footbinding was

introduced by dancers of the imperial court, and copied by the aristocracy as a mark of distinction. The middle classes then adopted it, to show their respectability, paying lip-service to chastity, and for hundreds of years thereafter it remained an unquestioned object of sexual desire, because sex does not like to question itself about what it desires. A woman with bound feet could neither work nor walk very far; she proved that her husband could afford to keep her at home in idleness. The unnatural hobbling walk excited men; manipulating the feet became the indispensable prelude to intercourse; sex manuals were published recommending eighteen positions in which intercourse could be combined with manipulation of the feet, and advising on different ways of grasping them for increasing degrees of ecstasy, kissing, sucking, nibbling and chewing, inserting one wholly in the mouth or eating melon seeds and almonds placed between the toes. Parents knew they could sell their daughters as prostitutes at a higher price with deformed feet, and girls were praised for bearing the extreme pain binding caused, which meant having broken bones for life. Tiny-foot beauty contests used to be held at festivals in Buddhist temples – the 'assemblage of foot viewing', originally to exhibit prospective candidates for the imperial harem. Though the Manchus, who conquered China in the seventeenth century, decreed the abolition of the custom and insisted that they were proud of their big feet, that was not enough to make women give it up, for sensuality is always complacent in its habits; bound feet were cherished as a pleasure said to be not inferior to that of intercourse itself, and men savoured the mixture of pity and delight which they felt at the sacrifice; the feet were kept hidden, and so remained as mysterious as the sex organs themselves. It was only the movement for women's liberation two centuries later that provided reasons for wanting to walk freely on one's own feet. As late as 1895, a French doctor reported that Christian Chinese were admitting in confession that they had lascivious thoughts about women's feet; and in the northern province of Suiyuan women continued to be fanatically devoted to footbinding, and to making beautifully ornate, tiny shoes, right up to the 1930s. This happened to be the particular form of cruel eroticism that China developed at the height of its prosperity, when it was a world leader in technology and the arts, because prosperity allowed it. Europe obtained similar satisfactions with the wasp-waist corset, even though doctors warned, from Roman times and the Renaissance to the nineteenth century, that it was a serious danger to health.

Instead of broadening the notion of pleasure, cycles of repression and permissiveness succeeded one another. The rich reproached the poor for their witty bawdiness, but were then fascinated by the habits they condemned, and borrowed them. The poor fell in and out of love with respectability. Uneven mountains of erotic art are reminders of alternating decades, and sometimes centuries, of licentiousness. In China, for example, though pornographic objects survive from at least 1000 BC, the first peak in their production occurred in the seventh century AD, when the empire expanded as far as Iran, Korea and Vietnam: prosperity encouraged erotic luxury, even though, or perhaps because, people were so deadly serious about their success; this was the period when China invented the examination system, which the whole world later copied, and it made no difference that China was ruled by an empress, Wu Chao, who was a bigoted and superstitious nun. Only in the thirteenth century was puritanism re-established as the official creed. Then in the eighteenth century, a renaissance of erotic art revealed fashionable prostitutes in elegant brothels as the centres of culture and entertainment, and Nanking became famous for its enormous 'pleasure quarters' of unprecedented lavishness. In the nineteenth century the Emperor Tongzhi died of syphilis, and in the Taiping rebellion (1850) 6,000 boy captives were castrated to be used as male prostitutes, complete with bound feet and heavy cosmetics. Most countries could produce a comparable history of waves of obsession and revulsion: sometimes puritanism has come from the left, sometimes from the right, sometimes from those in power and sometimes as a reaction against them. There is no golden age to look back to in matters of sex. In the 1950s the Chinese Communists systematically destroyed enormous quantities of erotic antiques in an attempt to wipe out the memory of these ambiguous traditions.

When in the 1930s and 1940s Kinsey investigated how Americans obtained their sexual gratification, he found that the rich and the poor had quite separate ideas, as though inhabiting different planets. The poor devoted themselves at an early age 'with single-minded commitment' to genital intercourse, making love before marriage seven times more frequently than the rich, using prostitutes three times more often; but with advancing years they became more faithful to their spouses than the rich, who, by contrast, began life with more prudence: in youth they masturbated twice as much as the poor, confining themselves more to petting, but as they grew older they cultivated the 'art of love', becoming preoccupied with breasts and foreplay, unlike the poor, who had doubts about experimentation and even about kissing, and considered nudity obscene. In other words, sex had become less straightforward the more prosperous people became. The advice given in the *Kamasutra* had been inspired by the leisured merchants of India's luxurious medieval cities.

The third form of sexual pleasure produces love and enduring sympathies, so it is creative, but it has always been treated as a mystery, from the earliest folklore to the present. Thus the fairy tales which teach West African children about sex present it as a game of hide and seek, without ready-made solutions: the popular children's stories about the adventures of Mr Penis and Mrs Vulva are tragic farces, which instead of urging obedience to custom, merely offer humour as the way to cope with inevitable difficulties. Parents are presented as grotesque; Mr Penis, who comes out of the trees, is a generous man, but Mr Testicles is selfish; the moral of each story is contradicted by the next; adultery is both idealised and punished; sex is both fun and cruel; and the fantasies do not necessarily excite desire.

The rest of the world has continued to treat sex as a puzzle, perhaps because it has conducted most of its business on the assumption that selfishness is the best way to prosper, and the forces of sex, which are capable of stimulating selflessness, therefore appear disruptive, best excluded from the public arena and relegated to the privacy of the home. Even Christianity, the religion of love, has been afraid of sexual love, confining it strictly to marriage, which Luther compared to a hospital that cured lust.

The foreign body that was a constant threat to sex as a creator of love was jealousy. Diderot defined love in his great *Encyclopedia* as 'the possession and enjoyment of another being'. It was this desire to possess – inevitable perhaps while property dominated all relationships – that made lovers so insecure, the fear of loss, the refusal to accept that a love had to be felt afresh every day. The roots of this attitude are to be found in the ancient Indian sex manuals which saw love as a combat, involving conquest – that too was inevitable when war decided everyone's fate. The *Kamasutra* ingeniously suggested that love was perfect only when both parties were victorious. But all too often there was questioning as to who loved whom most.

All or nothing, total subjugation: these military ideals limited the influence of sex on relationships outside the bedroom, preventing it from being as creative as it could be. For there is a vast amount of sexual feeling which never finds genital expression – unrequited love, attractions and sensations of various degrees of mildness, most of which go to waste. It is forgotten that in infancy sexual arousal results from a wide variety of causes, many of which have nothing to do with sex – unusual or frightening experiences, being chased, being beaten, going for a ride in an aeroplane – and only gradually is the range of stimuli narrowed: what emotions count as sexy is decided partly under peer pressure. Concentrating attention on the orgasm, on the moment of triumph and surrender, has restricted the idea of sexual pleasure, as too has the belief that sexual energy needs to be discharged, like ammunition that will explode in one's own face if not shot into its target. It is forgotten that among the Chinese and the Indians, there were many who advocated intercourse without ejaculation, that in many so-called primitive tribes, like the Dani in Indonesia, four to six years of sexual abstinence are observed after each birth, that among the Yoruba most women do not normally sleep in the same bed as their husbands, that before the easy availability of contraceptives, couples would spend hour upon hour kissing: penetration is not the only form of caress.

Ways of putting vague sexual attractions to useful purpose were investigated in the seventeenth and eighteenth centuries, particularly among the French, who developed flirtation and coquetry as arts of sociability, avoiding exclusive entanglements. At that time, in old-fashioned usage, a lover (*amant*) could still mean an admirer, and not necessarily a sexual partner; to 'make love' originally meant to court, rather than to engage in intercourse. Flirtation gave a new direction to courtly love, it was sex without sex, a prolongation of the preliminaries to sex which need never be consummated; but instead of idealising the loved one, it tried to understand him or her, and to probe how they could be mutually agreeable and stimulating. But most people were in a hurry to make conquests, and accused flirts of cheating, of pretending, of being incapable of love, of behaving as though permanently attending a masked ball. That hostility was understandable when marriage and procreation were regarded as women's principal function, but as soon as women and men are treated as independent personalities, whose opinions are worth discovering, then flirtation can be recognised as the first step in the creation of a relationship whose purpose is joint and mutual exploration. It may be that another word may replace flirtation, to indicate its expanded meaning as an adventure based on attraction but seeking to do much more than attract.

'One man does not have enough thoughts for one woman,' says a proverb of the Kung bushman tribe. Adultery and divorce have not been particularly imaginative ways of accommodating oneself to this fact. But the Congolese author Sony Labou Tansi who wrote that eroticism is the art of 'cooking love well' reminds one that there is a vast menu still to be discovered, and that much affection will continue to go to waste until it is. Cooking, of course, involves not only the senses, but also interest in all that is alive, and food tastes best if it fills those who eat together with benevolence, however temporary, towards one another.

History of food

The debates and publications of the Oxford Food Symposium, which I founded a dozen years ago jointly with Alan Davidson (former HM Ambassador to Laos, who resigned from the Diplomatic service rather than be posted back to London to work among the 'Whitehall warriors') and which brings together about 150 experts on gastronomy, provide a guide to the latest research; the periodical *Petits Propos Culinaires*, edited by Alan Davidson, contains critical reviews of the most intelligent writing on food. I give here only a sample of different sorts of books: Raymond Sokolov, *Why We Eat What We Eat*, Summit, NY, 1991; A. M. Barrett, *Neuropharmacology of Appetite Regulation*, Proc. Nutrition Soc., vol. 37, 1978; N. Tinbergen, *The Study of Instinct*, Oxford UP, 1974; F. T. Simoons, *Food Avoidance in the Old World*, Wisconsin UP, 1967; H. P. Kleyngeld, *Adoption of New Food Products*, Tilburg UP, 1974; Hiromitsu Kaneda, *Long-term Changes in Food Consumption Patterns in Japan 1878-1964*, Yale Univ. Economic Growth Center, no. 127, 1969; J. M. Weiffenbach, *Genesis of Sweet Preferences*, US Dept of Health, 1977; Trevor Silverstone, *Appetite and Food Intake*, Dahlem Konferenzen, 1976; John Burnett, *Plenty and Want: A Social History of Diet in England from 1815 to the Present Day*, Nelson, 1966; Chris Wardle, *Changing Food Habits in the UK*, Earth Resources Centre, 1977; D. W. Walcher, *Food, Man and Society*, Plenum, NY, 1976; D. S. Ironmonger, *New Commodities and Consumer Behaviour*, Cambridge UP, 1972; Michael Lipton, *Why Poor People Stay Poor*, Temple Smith, 1977; Trygg Engen, *The Perception of Odors*, Academic Press, 1982; C. M. Apt, *Flavor: Its Chemical, Behavioral and Commercial Aspects*, Proceedings of the Arthur D. Little Flavor Symposium, Boulder, Colorado, 1977; G. Blix, *Food Cultism and Nutrition Quackery*, Symposium of the Swedish Nutrition Foundation, Uppsala, 1970; Mark Nathan Cohen, *Health and the Rise of Civilisation*, Yale UP, 1989 (challenges the view that people have become healthier, with impressive prehistoric and anthropological evidence); John Yudkin and J. C. McKenzie, *Changing Food Habits*, McGibbon and Kee, 1974; A. I. Richards, *Hunger and Work in a Savage Tribe: A Functional Study of Nutrition among the Southern Bantu*, Routledge, 1932; Raymond F. Hopkins, *The Global Political Economy of Food*, Wisconsin UP, 1978; R. Robbins, *Famine in Russia 1891-2*, Columbia UP, 1975; Jean-Robert Pitte, *Gastronomie française: Histoire et géographie d'une passion*, Fayard, 1991; R. E. F. Smith and David Christian, *A Social and Economic History of Food and Drink in Russia*, Cambridge UP, 1984; Michel Maffesoli, *Aux Creux des apparences: Pour une éthique de l'esthétique*, Plon, 1990; William Kingston, *Innovation*, Calder, 1977; Sterling Ortman, *To Feed this World*, Johns Hopkins UP, 1976; R. N. Salaman, *The History and Social Influence of the Potato*, Cambridge UP, 1949; J. D. Haas and G. G. Harrison, 'Nutritional Anthropology', *American Review of Anthropology*, 6, 1977, pp. 69-101; J. H. van Stuyvenberg, *Margarine: A History 1868-1969*, Liverpool UP, 1969; B. A. Hennisch, *Fasts and Feasts*, Pennsylvania UP,

1976; K. C. Chang, *Food in Chinese Culture*, Yale UP, 1973; E. N. Anderson, *The Food of China*, Yale UP, 1988; T. C. Lai, *At the Chinese Table*, Oxford UP, 1964; Michel Jeanneret, *Des Mets et des mots: Banquets et propos de table à la Renaissance*, Corti, 1987; Piero Camporesi, *Bread of Dreams: Food and Fantasy in Early Modern Europe*, Polity, 1989; Barbara K. Wheaton, *Savouring the Past: The French Kitchen and Table from 1300 to 1789*, Chatto, 1983; Harold McGee, *On Food and Cooking: The Science and Lore of the Kitchen*, Scribners, NY, 1984; Georges Vigarello, *Le Propre et le sale*, Seuil, 1984; Om Prakash, *Food and Drink in Ancient India*, Munshi Ram, New Delhi, n.d.

History of sex

Again, I give only a sample: J. D'Emilio and D. B. Freedman, *Intimate Matters: A History of Sexuality in America*, Harper, NY, 1988; John Bancroft, *Human Sexuality and its Problems*, Churchill Livingstone, 1983; Herant Katchadourian, *Fundamentals of Human Sexuality*, 5th edn., Holt, NY, 1985; Oskar Koenig, *The Masai Story*, Michael Joseph, 1956; Pat Caplan, *The Cultural Construction of Sexuality*, Tavistock, 1987; David M. Halperin, *Before Sexuality: The Construction of Erotic Experience in the Ancient Greek World*, Princeton UP, 1990; M. Feher, *Fragments for a History of the Human Body*, 4 vols., Zone, NY, 1989; Ronald and Juliette Goldman, *Children's Sexual Thinking: A Comparative Study of Children Aged 5 to 15 Years in Australia, North America, Britain and Sweden*, Routledge, 1982; R. H. Van Gulik, *Sexual Life in Ancient China*, Brill, Leiden, 1961; Lina M. Fruzetti, *The Gift of a Virgin: Women, Marriage and Ritual in Bengali Society*, Rutgers UP, 1982; Akhileshwar Jha, *Sexual Designs in Indian Culture*, Vikas, New Delhi, 1979; Michael Allen and S. N. Mukherjee, *Women in India*, Canberra ANU, 1982; B. Malinowski, *The Sexual Life of Savages in North Western Melanesia*, Routledge, 1929; B. Malinowski, *Sex and Repression in Savage Society*, 1927; Roy Ellen, *Malinowski between Two Worlds: The Polish Roots of an Anthropological Tradition*, Cambridge UP, 1978; Derek Freeman, *Margaret Mead and Samoa*, Harvard UP, 1983; Lowell D. Holmes, *The Quest for the Real Samoa: The Mead/Freeman Controversy and Beyond*, Bergin and Garvey, Mass., 1987; R. Firth, *Man and Culture: An Evaluation of the Work of B. Malinowski*, Kegan, 1957; Hans Licht, *Sexual Life in Ancient Greece*, Routledge, 1932; Lawrence Birken, *Consuming Desire: Sexual Science and the Emergence of a Culture of Abundance 1871–1914*, Cornell UP, 1988; Francis L. K. Hsu, *Iemoto: The Heart of Japan*, Wiley, NY, 1975; Iwao Hoshii, *The World of Sex*, vol. 4, Woodchurch, Kent, 1987; Peter Brown, *The Body and Society: Men, Women and Sexual Renunciation in Early*

Christianity, Faber, 1988; I. B. Horner, *Woman under Primitive Buddhism*, Routledge, 1930; Charles Fowkes, *The Pillow Book*, Hamilton, 1988 (a history of erotic literature); John Byron, *Portrait of a Chinese Paradise: Erotica and Sexual Customs of the Late Qing Period*, Quartet, 1987; Howard Levy, *Oriental Sex Manners*, NEL, 1971; Howard Levy, *Chinese Footbinding*, Neville Spearman, 1972; Iris and Steven Finz, *Erotic Fantasies*, Futura, 1991; Patrick J. Kearney, *History of Erotic Literature*, Macmillan, 1992; Wayne Paul Robinson, *The Modernisation of Sex*, Elek, 1976; V. and B. Bullough, *Sin, Sickness and Sanity: A History of Sexual Attitudes*, New American Library, 1977; Peter Wagner, *Eros Revived: Erotica of the Enlightenment in England and America*, Secker & Warburg, 1988; René Nelli, *Erotique et civilisation*, Weber, 1972 (on Arab–European relations); Travaux de l'université de Toulouse le Mirail, série A, tome 24, *Les Visages de l'amour du dix-septième siècle*, Toulouse, 1984; Roger Bougard, *Erotisme et amour physique dans la littérature française du dix-septième siècle*, G. Lachurie, 1986; Gérard Clavreuil, *Erotisme et littératures: Afrique noire, Caraïbes, Océan indien*, Acropole, 1981; Suzanne Lallemand, *L'Apprentissage de la sexualité dans les contes d'Afrique de l'ouest*, Harmattan, 1985; Pierre Hanny, *L'Erotisme africain: Le Comportement sexuel des adolescents guinéens*, Payot, 1970; R. B. Symington, 'Sexual Behaviour of Rhodesian Africans', *Journal of Biosocial Science*, vol. 4, no. 3, pp. 263–75; William H. Hopkins, *The Development of 'Pornographic' Literature in Eighteenth-Century and Early Nineteenth-Century Russia*, Indiana Univ. Ph.D thesis, 1977; N. N. Bhattachayya, *History of Indian Erotic Literature*, Munshshiram Manoharlat, New Delhi, 1975; H. C. Chakladar, *Social Life in Ancient India: Studies in Vatsyayana's Kamasutra*, Greater India Society, Calcutta, 1929; Jayadeva, *Gitagovinda: Love Song of the Dark Lord*, trans. Barbara Stoler Miller, Columbia UP, 1977; Edward C. Dimock, jnr., *The Place of the Hidden Moon: Erotic Mysticism in the Vaisnavasahajïya Cult*, Chicago UP, 1966; Bernard Golse and M. Bloch, *L'Amour Chaos*, Hachette, 1987; 'La Folie Amoureuse' in *Dialogue*, no. 96, 1987; Sue Griswold, *Beautiful Merchandise: Prostitution in China 1860–1936*, Haworth, NY, 1982; Edward J. Bristow, *Prostitution and Prejudice: The Jewish Fight against White Slavery 1870–1939*, Oxford UP, 1982; Jacques Rossiaud, *Medieval Prostitution*, Blackwell, 1988; Allegra Taylor, *Prostitution: What's Love Got to Do With It?*, Macdonald, 1991 (an experiment); Stephen Owen, *Mi-Lou: Poetry and the Labyrinth of Desire*, Harvard, 1989 ('All the bright colours of physical things call out to one another, and how amid all this can a human being find rest?').

J. R. Planche, *Cyclopedia of Costume*, Chatto, 1976; C. Wilson,

Encyclopedia of Scandal, Grafton, 1987; Donald Schon, 'The Fear of Innovation', in Barry Barnes (ed.), *Science in Context*, Open Univ., Milton Keynes, 1982; J. Puisais, *Le Goût et l'enfant*, Flammarion, 1987.

7

*How the desire that men feel for women,
and for other men, has altered
through the centuries*

Is it inevitable that as women become increasingly adventurous
and have ever higher expectations of life, they will find men less
and less adequate?

In 1968 Patricia was a twenty-year-old law student who enjoyed
herself so much on the barricades that she failed her examinations;
but she is without regrets: 'It was fun.' Taking a job as a clerk, she
continued her protest as a trade union activist for ten years: that
was fun too. 'I loved it. I was passionate about it. But one gets old
quickly.' So at the age of thirty-five she went back to university,
then trained at the Income Tax School in Clermont Ferrand, and
now she is an inspector of taxes. Having met another inspector in
a previous chapter, it is useful to be reminded that two members of
the same profession are rarely exactly alike.

Being a sort of policeman, after having spent her youth demonstrating against policemen, gives Patricia much satisfaction: the
barricades did not go up entirely in vain. 'I am changing the way
the blood is collected. I am not Sherlock Holmes, nor Zorro. My
point of view is that there must be equality for everyone, so that
some do not get away with paying less, to the disadvantage of the
rest.' Out of 180 tax investigators in her region, there are only
twenty women, who hesitate to apply because the essence of the

job is confrontation with men. However, being a woman has not been a problem in her own experience: 'People are so worried by having their accounts examined that they see only the inspector, not the woman.' She takes on only powerful men – each year her task is to investigate twelve companies in the million-pound class. 'I am totally cool and not vindictive. It is not my justice that I impose, but what is laid down in regulations. Only rarely do I report that there has been bad faith – as in the case of the managing director who ran a Porsche secretly paid for by his company. On the contrary, I get emotional, almost sick when I feel I might cause the collapse of a firm. My purpose is not to make heads roll.' One legacy of 1968 has been that government officials do not bark at citizens as aggressively as they used to.

Has she stopped hating the rich? She comes from a comfortable background, as many of the utopians of 1968 did. When the tax officials went on strike, she did not join them: her salary is low, but magnificent compared to what people earn in other parts of the world. 'I have seen a girl die at my feet in New Delhi, and the passers-by simply walked round her.' She could earn twice as much if she became a private tax consultant, but it would mean longer hours, and the attraction of her job is that it leaves her the 'time to live'. 'I love my work. I don't want to do anything else. I deal only with intelligent people.' The salary pays for air tickets. There is no continent, except South America, which she has not visited, but never on organised tours. She is just off to Kenya; last year it was Thailand and the Philippines. Ideally, she would spend six months a year abroad, living the life of ordinary people, 'integrated' into another civilisation. 'I am a citizen of the world. If something painful happens anywhere, I feel it personally.' Racism is her great, intolerable enemy.

Patricia looks upon these aspects of her life as successful. But on the subject of personal relations between one man and one woman her message is a sad one. Her own parents had no message for her either: whenever she asked her mother for anything, she was told, 'Ask your father.' Her mother, with whom she has good relations, has since told her that she loved her husband and therefore never took decisions without consulting him; Patricia, who knows her own mind, thinks her mother was incapable of making decisions.

But decisiveness has been a pest in her private life. Arguing is fun too, and when she knows her facts she will not stop until she has made the man she is with surrender. 'I know a lot about economics and politics, and when I tell him he doesn't understand, he doesn't like it. Perhaps I could use my knowledge with less vehemence: I don't seem able to do it gently. That is because I want the man to win. I want him to have his legitimacy.'

The consumer society not only allows women to buy new gadgets and clothes all the time; it can also mean enjoying men and then discarding them like an unfinished meal thrown into the waste disposal. 'I now realise I have been consuming men.' Her taste for them has not diminished, but her palate has become ever more refined. It takes time to understand what it is that one wants from men. She does not need them to mend a fuse, or help her buy a car, least of all to fill in her income tax form. The object of her search is a man who can do things she cannot do, a man whom she can admire. No man has yet satisfactorily met that challenge, from someone whose job is to make men quake in their shoes. 'For eight hours a day I am not a woman, I am a civil servant. Men are worried by a woman who can stand up to a managing director; they want softness in a woman. I can't be soft in my work, I have to hide my softness, and I am marked by that. It is the price I pay for having interesting work.' She continues to fall in love with men. 'I enjoy it while it lasts, but it never lasts long. I am incapable of having a durable relationship. When I am in love, I become stupid, dependent. Unconsciously, I think I don't like that, so I try to make it fail. It turns into a power struggle.'

She does not quite think that she has made the wrong choices. What she did she had to do. Women were suddenly liberated in the sixties, and she wanted to experience the new freedom. 'It was the Age of the Pill and I had to profit from that.' There was a chance of a new sort of life for women. Now she sees it went wrong, but she does not know how it can be put right next time round.

The feminists went too far, she says, making demands, nothing but demands. It was a mistake. But on the other hand, 'We wanted to show that we were the equals of men.' Now she says women must go back to femininity; she disapproves of her feminist friends who refuse to cook, even when they invite a dozen people to dinner. On the other hand, she resents women getting on by playing at being women; she rejects 'men who expect women to believe what they say just because they are men'; she could not bear to be a housewife.

'My life as a woman-man has failed. I have been taken for a ride.' The idea of the couple, however, seems to her to be workable, except that she cannot make it work herself. She hands the buck to the next generation. They impress her by the caution they show with their affections, by the more relaxed atmosphere they have introduced into relationships, by their romanticism, their desire to make relationships more beautiful. Looking back, there was little beauty in the way her own generation met and had sex: 'I used to behave like a man. If I wanted to go to bed, I did. But it was cold.' In Katmandu she was invited to try drugs, and was proud of it, and fell into a coma for forty-eight hours, which completely cured her of the desire to try drugs. Now she says, 'Sex is not important; it is of no interest to me, unless it is part of something much broader.'

Now it is up to men to make an effort, instead of just watching women and smiling at them. The trouble is that too many of the men she knows are feeble: 'They lack gumption.' So whom does she go to for advice? To an astrologer: 'We need something: astrology replaces religion.' And the astrologer told her that she would always have trouble with men.

Florence is about fifteen years younger, but she has not quite found a solution either. Every member of her family has 'a passion'. Her father's is mountain climbing, which he loves partly because he is fascinated by risk, and he is willing to take risks because his daily life bores him, he wishes to get beyond what he is in daily life and to escape from the routine. That sort of withdrawal he learnt very young, because his own parents quarrelled all the time; he never quarrels with his wife. Some people are learning to have a sense of themselves without beating up those around them. Florence's sister's passion is horse-riding. Every other member of the family is totally consumed by one passion.

However, Florence has never been told by her parents what the direction of her life could be; they have never given her advice, and she has never asked for it, because she felt their advice would not suit her. She never speaks to them about the intimate subjects that matter most to her. All they want is for her to be happy – which perhaps means they do not know what they want. She says the best way to make children happy is to be happy oneself, happiness is contagious. All her father can do is to warn her when she becomes too big-headed, and she values that. But maybe he does not feel qualified to give advice, for he is not satisfied with himself; or at least that is her guess, because she does not really know him. How many people have known their parents well? Since the time that private lives have become more important than public lives, it has become impossible for traditions about how to live to be handed down from generation to generation.

If only her teachers had been able to give her some deep ambition, if only she had met one who was captivated by what he taught and could have captivated her; but her school career was a total flop, even though she got high marks. No university professor influenced her either. She became a teacher herself, but that made her despair of education: four-fifths of her colleagues were depressed; there were only one or two who cared about their work, to whom education was almost a religion.

Who else could give her a sense of direction? 'I was brought up in a world where girls talked amongst themselves,' but she has no taste for gossip. 'We were never told we could talk with men.' Discussions with them are always overshadowed by the possibility of seduction: 'Few accept the idea that true conversation is possible with women. As soon as a man sees that you like him, he assumes you want to be seduced. Going to bed can be useful in starting a conversation, but it is not necessary.' There is much less of it among the young women she knows. Her best friend is a former lover, with whom relations are much more pleasant now that there is no question of seduction. The way men look at her sometimes pleases her, but sometimes makes her despair. 'To speak with a man, you first have to win the right to speak with him.'

Her trouble with men is the same as her trouble with having a single 'passion'. No activity absorbs her completely, for her passion is discovery, travel, meeting the unknown, which means being free of a single obsession. There are too many sides to her personality. She appears to be a career woman, but she also likes sewing; she refuses to conform to the image her job implies; men can do that, but not women. If she had to choose between her career and her private life, she would prefer the latter. If she had a child, she would take time off for it, without sacrificing herself, however; too many men are stunted by feeling they cannot pay back the debt they owe to a mother who sacrificed herself for them. Men say they find her disconcerting. She replies that she has met only men who are frightened. They do not understand that it is not security that she wants from them. 'The most interesting things I have done have been in states of insecurity. When I decide to do something, I burn my boats. Security puts me to sleep.' A man who charms a woman thinks he has acquired her, whereas a relationship of love should involve a new charm every day. 'My need is not to have complete security in my affections. I need to be put in a state of danger.' So one of her hobbies is hang-gliding, but she wants to make other discoveries also.

Florence married young, lived with her husband for a year and then apart from him for three years. She liked the idea of a common home, but also of having a separate home as well: few men can understand that, or that she would want to go away for a few days; they expect her to account to them for her time; they judge her friends, though she does not ask them to speak to her friends; she cannot bear it when she wants to read alone and is prevented. It does not worry her that she is too demanding: 'I'm lucky I know how to live alone.' But she will not rest until she finds someone to love and be loved by; that is necessary for the inner peace which she needs to be able to do other things. Perhaps men do not like the idea that love is only a basis for doing other things. If only the ones she had met had not been so unwilling to take risks; it is fear which makes them lead lives which do not quite satisfy them. 'I do not want my story to be an average, ordinary one.' Her ideal man used to be like everybody else's – intelligent, beautiful, charming, humorous – but she now wants more than that, to be able to admire him profoundly. 'It is pretentious to say you have met no men of your own level, but I have met none who likes my contradictions. I am changing fast, but the men I have loved could not follow; they are men who have slowed down. I need a man to give me confidence, because sometimes I have doubts.'

It does not follow that one needs to live with one's husband every single day. Her husband would not have made a good father; and 'I am not so egoistic as to want a child alone'. He allowed himself to become financially dependent on her, and that was disastrous – dependence perverts a relationship; when little things go wrong, it makes one mean, and less loving. So her relationship with money has changed. She used to like window shopping and buying things: 'I was told that when unhappy, I should go and spend money; I could have become a person who owned fifty pairs of shoes.' But she has lost her taste for acquiring possessions: she now spends most of her money on travel and books, not clothes; she finds her modest earnings are enough to allow her to do what she wants.

Taking risks has meant that she has abandoned a series of safe jobs. Everybody advised her not to, but she ignores advice, because a job is like a love affair: if you do not love it, you should leave it. It is very important for a job to be a relationship of affection. After teaching, she became a journalist on a major provincial daily newspaper; but during her three years there, nobody ever told her what her purpose might be. Everything proceeded by routine. 'Routine kills.' There were able journalists whose talents were being destroyed, who knew they were writing rubbish, but had stopped caring, while others were not allowed to make full use of their talents. Nobody wanted to rock the boat, because the paper was a secure job for life. When she resigned she was told she must be mad; but she insisted one must leave a job as soon as one has lost one's enthusiasm for it. The whole of the industrial world would collapse if people listened to her: it is worth imagining how one would replace it.

Florence now lives in Brussels, a city launching itself into the unknown. 'Something will happen from this mixture of people,' but no one knows what. Courage is a way of taking pleasure in the surprises that chance breeds. She is setting up a press agency where journalists from all over Europe, with compatible temperaments, can work together. The search for compatibility allied with independence continues. New adventures are constantly complicating her life. She used to have a prejudice against the USA, disliking its 'taste for money and its obsession with efficiency'. Chance has made her visit it, and she has changed her mind. She knows how to do that. Certainly she met some narrow-minded people, but others who were not. 'I lost my complexes.' Efficiency impressed her. In Canada she met a marvellous man. For four days they talked. He was not afraid to say what he felt; he seemed authentic: 'He satisfied my desire for harmony and gentleness.' But she does not know what will become of this friendship, which appeals only to one side of her. 'He does not put me in danger. I need not to have complete emotional security.' Time will tell.

Then she went to the Lebanon, where there is real physical danger. She has become friends with a Lebanese family and is acting as a sort of godmother to their daughter, who is being educated in Europe. That is a new kind of relationship for her, a new taste of the varieties of friendship, to be the mentor of a very young adult, aged eighteen (Florence is twenty-seven), with ideas very different from her own.

Florence's menu for the future is to develop her imagination. They never taught her how to do that at school.

At the same time as women became more demanding in their expectations of men, scientists were discovering that the world of animals was not, after all, dominated by males. The big male baboon seemed to behave as though he dominated his tribe, but when the tribe was studied more closely, it became clear that a female decided where it travelled and who could sit next to whom. The male, it used to be thought, searched for a wife and then left her to look after the babies; but it was discovered that more frequently the female takes the initiative and in about 40 per cent of primates the male cares for the young. Females were assumed to be passive victims of their hormones, but it has emerged that these are produced as much by their own behaviour and that of those around them as by automatic, unavoidable processes.

Above all, the fact that not all nature is clearly divided into male and female has been emphasised, notably the many creatures who reproduce without copulation, and those whose sex can change over their lifetime, almost as though they decide on a change of clothes. There are species which are all female for most of the time, reproducing themselves without intercourse, until a crisis occurs and food runs out; then they stop producing offspring identical to themselves and start giving birth to males, whose function is to introduce different solutions to the crisis. Males are not necessarily stronger; females are not necessarily much interested in the company of males except when they share parenting; and they can be very difficult to please in their choice of a partner: a bird called the western grebe rejects 97 per cent of male advances, and it is, significantly, outwardly indistinguishable from the male except by its voice. There are many solitary animals indifferent to sex, like the female deer which lives with males for only one day each year. But it doubtless took many thousands of years for animals to develop these eccentricities and these varied forms of independence.

If this has any relevance to humans, it is scarcely surprising that women have found it so difficult to meet males who measure up to what they would ideally like males to be. Women's efforts to change men's attitudes towards them have a long history, but it has varied in its objectives, as changes have inevitably produced unexpected, unwanted results. To take one example, women have attempted to transform courtship, modifying the rules of seduction again and again, as a way of altering the relationship of married couples.

Though tradition has many faces, one of the most common was that girls should know no man until they married their husband, and indeed that they should not extend their knowledge after marriage either. In such conditions, the courtship of the male was modelled partly on commercial methods – negotiating with her parents – and partly on military practice – the siege, the parade of strength, the sowing of confusion with gifts and promises, until she said, 'I surrender.' But some women rejected their suitors, for no reason that men could understand, and that was like the invention of a new kind of armour, which made bows and arrows obsolete. Not all women were young when they married, obedient to their parents. In many periods, about one-third of weddings have been second marriages, because people died so early (thirty was once the life expectancy). Widows (and later divorcees) may well have shown the way.

Women used to be told that they would learn to love their husbands after marriage, however disagreeable they might find them at first. But some women began demanding that the possibility be proved before the wedding, or at least that the man should first convince them that he loved them. Once this happened, men lost control of courtship, because there was no technology for falling in love, no institution which could arrange it. The results can best be traced in the history of courtship in the USA, the least secretive, the most varied and probably the most influential world-wide. Americans were already complaining of a crisis of masculinity in the 1860s. But every generation, conscious only of how it differs from its parents and its children, forgets how old are most of the subjects of argument between men and women, how humans persist in banging their heads against the same wall.

William Alcott, in a book entitled *The Young Wife* (1833), wrote that there was 'a very general opinion' that 'the love of husband and wife must after marriage necessarily begin to decline'. Courting couples of the time have left letters bemoaning 'the almost universal unhappiness of married persons'. Brides in particular were frightened by the 'great and unknown duties for which I feel incompetent', not only domestic duties, but the need to turn their husbands into 'virtuous and happy' men. 'It is awful thus to bind myself for life.' 'Grateful joy' did indeed balance the 'fearful responsibilities' and the feeling that intimacy was often quickly lost. So two centuries ago women had already started working on modifying their links with men.

One way was for women to tell men exactly what they felt and everything that they thought: they called it 'candour'. Tradition kept the sexes apart, in two separate mental and physical worlds. 'Society allows no sincere hearty friendship' between men and women, wrote a bride in 1860, 'but I will not be insincere.' Another said, 'I can love you better from the fact that to no one have I made myself known as to you, to no one have I given such confidences.' But a man replied, 'Men are afraid of showing themselves in their true colour and as they are.' They had to husband their reputation in the outside world most of all, because success depended on reputation. It was a great risk to embark on intimate conversations, and to say as one boldly did, 'Let us communicate our ideas, our notions to each other.' But, wrote another, 'It is a fiction that any two human beings are constituted peculiarly and in all respects fitted for each other, perfect counterparts, capable of a mystical, absolute union.'

The romantics said, 'We must be alike.' However, women began discovering things about themselves in the process of opening themselves out to their suitors. 'Openness became almost an obsession for couples after 1800.' It was not easy for women, whose reputations depended on discretion and prudery. It took courage to demand that men give priority to the search for meaning in life: 'I don't want you to work so hard,' wrote a young Boston woman to her fiancé, and many millions have since repeated the request, when they did not demand the very opposite. To rebel against the idea that income must come before intimacy is revolutionary indeed. Already in 1850 a woman longed to be her suitor's 'spiritual doctor'. The closer they looked at one another, the more unique each became, the less did the stereotype of the male and the female seem helpful.

Ideals are always just beyond one's grasp. Some women worried about that, and their worries took them backwards. First they worried that they were not worthy of their suitor's admiration. Then they worried that they did not adore their suitors with sufficient passion: 'I feel I do not love you with that intensity of love of which I am capable.' Real love should, thought another, be 'without effort'. But, said a third, she still was not sure that she had found love 'with all my heart, so that the horrid old worry is gone'. In other words, they lost their nerve. Lack of confidence became the great oppressor. They relapsed into admiring apparently confident, masterful and assertive men, who seemed to possess the most precious of jewels, certainty. Then they were disturbed if the men proved to be weak and revealed that they wanted a 'woman upon whose experienced judgement one can place confidence'. The great problem became not just to meet the right man, but to do so at the right time, to have desires which harmonised at that particular moment, let alone for ever. Another method tried by some women to modify their relations with men was to increase their experience of them. The middle class learned how to do this from the working class, which was not corseted by the fear of meeting social inferiors. 'Dating' was originally a slang word of the poor, first used in 1896. By the 1920s most of the youth of America was obsessed with dating, which surveys showed to take more time than any activity apart from schooling. Until about 1945, the aim in dating was to go out with as many people as possible: the more sought after one was the better. There was less talk of love now. The goal was to acquire self-confidence by more certain means, namely by the number of dates one had. A college student boasted he had dated fifty-six girls in nine months.

Many imagined that dating was a struggle for power, since power was what adults talked about. It did indeed free youth from the control of their parents, since it involved going out to places of public entertainment, escaping from direct supervision at home. It did replace the system of 'calling', by which a suitor had to ask permission to call on a woman he coveted, and it was up to her whether she allowed him to, or up to her mother, who often organised the call. Men imagined that dating restored supremacy to them, since they paid for the entertainment and the drinks and the meals, that they were buying women for their amusement, spending on them as they spent on their cars. And many girls undoubtedly became convinced that dating as many boys as possible was the only way to be considered popular, and so to catch a popular husband, who would only want a popular girl; and they spent lavishly on cosmetics and clothes, which was the counterpart of the boys' expenditure. The American economy certainly benefited from this devotion to spending, which not long before had been very un-American.

Dating seemed to be the democratic method of finding self-confidence; it was a kind of vote, a perpetual election. Instead of worrying about elusive love, the young proved to each other that they were popular by winning a date. At dances, the most successful woman was the one who danced all night, but never twice with the same man. 'Never brush off a date,' advised the *Woman's Home Companion* in 1940, 'because he might come in handy for an off night.' To such an extent was it a masculine privilege to take the initiative, that when a woman at college asked a man for a date to a Saturday dance, 'he cut her off in mid-sentence and walked away'. Women had the privilege of deciding whom they would accept and how over-booked they would pretend to be. At the University of Michigan, the female students rated the males according to their dating value: 'A – smooth, B – OK, C – pass in a crowd, D – semi-goon, E – spook.'

After the Second World War, however, they again lost their nerve. Sixteen million young soldiers were sent overseas, a quarter of a million were killed, 100,000 married English, French and other foreigners. Panic struck that there were no longer enough men to go round, and worse, that the veterans were demanding mature, sophisticated women, such as they had met in Europe, and had no use for silly girls. Dating was therefore transformed. Now 'huntresses', 'hijackers', 'man-stealers' were admired instead. Going to college became a way of snaring a husband, 'getting a Mrs degree', and if by October one had not found a partner, one was a failure. The most common age of marriage for women fell to its lowest level ever, nineteen. Dating began at an increasingly early age in the schools. In 1959, 57 per cent of American teenagers were 'going steady'. The original aim of dating – widening one's experience – was abandoned in favour of a sense of security, and the price was high.

Back to femininity was the advice given by women's magazines in the 1950s (as it has been at varying intervals since then). 'Everybody can become a type that's in the current physical fashion if they have the courage to make themselves into that type,' the *Ladies' Home Journal* assured its readers. The *Woman's Home Companion* had a questionnaire to enable one to assess one's femininity: the first question was: Do you wear nail varnish? The second: Do you have career ambitions? From the age of twelve, girls wore padded bras to make themselves imitations of the ultimately feminine, the actresses with enormous breasts. The purpose of this was to simplify life, to enable each sex to have clear roles. In a poll organised by the *Senior Scholastic*, the vast majority of women said they preferred men who knew the proper etiquette to those who had a pleasing personality. By being feminine and submissive, these women were supposed to make men become more masculine, helping them to 'feel like men'. That seemed necessary when women competed with men at work, and the code of the old-fashioned gentleman seemed to be a way of avoiding competition in private life. But of course men who needed to have their masculinity boosted were soon found to be inadequate, and women decided that they had taken the wrong turning.

So it can be seen that though women have been able to change the way men dance around them, the dancers themselves are still awkward and difficult. Long before the sexual revolution of the sixties, Americans worried about their sexual identity; the University of Southern California was already offering its students courses on the 'six basic differences between men and women' in the 1950s. Women who complain today that men have no oomph or gumption have read too many history books about valiant knights in shining armour, and not enough about pretence and lies, and have forgotten how fragile are the roots of confidence. Besides, changing people is like decorating a house: as soon as one room has been done, the ugliness of the others becomes more apparent.

Meanwhile, it has also been discovered that the man on whom Don Juan was modelled was a homosexual. He loved chasing women, but did not like them; it was the chase that excited him. The history of homosexuals has hitherto been considered marginal to the relationship between men and women, but recently a mass of new information has come to light through scholarly research and franker autobiographies, revealing it to be more important than was ever imagined. Though the research has aimed above all at constructing an identity for this minority, it has produced discoveries of significance to the majority too. How differently have men behaved when they have fallen in love with other men?

Sexual relations between men have passed through four phases, but each new phase did not put an end to its predecessor, so all four now co-exist. Originally, homosexuality was a conservative force, strengthening established institutions, as much a ritual as a source of pleasure. It was an integral part of pagan religion, whose gods enjoyed all forms of sex; and only when sex ceased to be a divine amusement was homosexuality seriously persecuted. Sex and magic used to be part of the same mystery, and shamans often tried to be simultaneously male and female, dressed as such, to increase their powers over both. When soldiers ruled and worshipped martial qualities, they sometimes used homosexuality as a method of strengthening the military caste: the Japanese samurai went to war accompanied by a young male sexual partner; ancient Celtic warriors 'offered themselves to other men without the least compunction' and were offended if their advances were rejected. Many societies made homosexual relations an obligatory temporary phase in the process of growing up, and in some places that applied to women also (but written historical evidence on lesbianism is less abundant). Among the ancient Greeks, the older partner in a homosexual relationship was supposed to be providing a course in citizenship and a preparation for married life to the younger one, who was expected to feel no pleasure or desire but gratitude, though in practice the ideal was often corrupted. When the relations between the two sexes were dominated by problems of property, homosexuality was an alternative way of increasing status and a form of conspicuous luxury: some Chinese emperors kept powdered and rouged boys as well as wives; the Norman king of England William Rufus surrounded himself with young male courtiers with long hair, wearing women's clothes; and under Queen Elizabeth I, fashionable rakes walked the streets with a mistress on one arm and a catamite on the other to show off their prowess. Modern boarding schools continued, without realising it, homosexual traditions common to numerous tribes all over the world. There have been times – particularly times of conflict – when men with feminine traits have been admired: thus Napoleon patronised the sculptor Canova, who loved to portray effeminate young men, and the castrato singer Crescentini, who gave a feminine voice to virile roles; that tradition continues in the success of modern pop stars of indeterminate sex. Mussolini, by contrast, had a stadium built surrounded by sixty statues of naked musclemen, all brawn. Homosexuals have alternated between these ideals as much as heterosexuals.

Homosexuality has been more or less accepted in about two-thirds of human societies at some time or other, and it has occasionally concerned large sections of the population. For long it was tolerated even by the Catholic Church: in 1102 Saint Anselm, Archbishop of Canterbury, demanded that the punishment for it should be moderate because 'this sin has been so public that hardly anyone has blushed for it, and many, therefore, have plunged into it without realising its gravity'. The crusading king Richard the Lionheart was no less pious because of his sexual preferences. The Church was originally more concerned with stopping its priests having sexual relations with women; when it waged a campaign against that, homosexuality became even more common, especially in monasteries, where Saint Aelred of Rivaulx exalted it as a way of discovering divine love.

It was only in the twelfth and thirteenth centuries that a mass repression of homosexuality began in Europe, as part of a campaign against heresies of all sorts, which developed into the terror of the Inquisition. This is the second phase. Penances were no longer judged to be enough: in 1260 France began the persecution by establishing the penalty of amputation of the testicles for the first offence, of the penis for the second and death by burning for the third. Hitler's attempt to exterminate all homosexuals along with the Jews was the culmination of that story. Persecution made homosexuality no longer conservative, but dangerous and clandestine. 'Away with hypocrisy. Discreet debauchery means little to me. I want to enjoy everything in broad daylight.' So wrote the poet Abu Nawas (787–814), praising masturbation. But openness had ceased to be possible. Oscar Wilde maintained that if everyone expressed their desires openly and freely, the world would be rejuvenated, but it was unlikely to happen, because even the bravest were afraid of admitting what they were really like; so, given that persecution of nonconformity was the rule, he advocated, as the second-best course, that people should discover the satisfactions of pretending and concealing, acting a part and making a fool of oppressors and critics. Genet argued that there was no point in trying to be oneself because it was impossible to know when one was truly authentic. Dominique Fernandez added that if there was no discrimination against homosexuals, he would be delighted as a lover of liberty, but he would lose much of the pleasure he has in being a homosexual, which, for him, means being a pariah, reinforcing his sense of being different from others by being interested in matters which cannot be discussed openly: 'Sex,' he says, 'is not what interests me most in homosexuality', and, as a novelist, he cannot envisage writing about happy homosexuals. Candour and secretiveness thus emerged as equally balanced competitors, and desire as the accomplice of the forbidden. The thrill of heterosexual adultery, as its latest Cambridge analysts have emphasised, also comes to a considerable extent from secrecy and risk. It is too simple to assume that personal relations would be trouble-free if everyone were candid, even if everyone could be candid; something more is required, namely a desire to understand what one is being candidly told.

The third revolution in Western homosexual history started in the nineteenth century, when homosexual activity was classified no longer just as a sin, but as either a disease, or a sign of a defective upbringing, or the result of a genetic disposition. Until then, the men who had visited London's transvestite 'molly-houses', which flourished in the eighteenth century, were not considered to be homosexuals, any more than are Tahitian men who use the male prostitute to be found in every village. The word homosexual was invented only in 1869 (when Freud was thirteen years old) by the Viennese writer Benkert, in the hope of avoiding persecution, by showing that homosexuals constituted a 'third sex' independently of their will, and that they could not therefore be accused of vice or crime; hitherto, the names by which they were known had been jocular, not medical classifications, and the word 'gay' revives that tradition.

This segregation created enormous anxiety, as homosexuals probed into their deepest feelings, sometimes with remarkable artistic results. But gradually individual isolation was mitigated by the discovery that they were not as alone as they imagined; it emerged that many of humanity's greatest achievements were due to people who had passionate attachments to members of their own sex, including eminent founders of modern science (Newton, Halley, Robert Boyle), of computer technology (Turing, who was condemned to be chemically castrated), of economics (Keynes), of art (Leonardo, Michelangelo, Botticelli and Caravaggio down to Francis Bacon), of philosophy (Wittgenstein), of music (Beethoven, Schubert, Tchaikovsky), of literature (Proust heads a long list), of children's writing (Hans Christian Andersen), and even the creator of the archetypal American hero, Horatio Alger, even world conquerors like Alexander the Great and Julius Caesar ('a husband to every wife and a wife to every husband'). Renaissance Florence, known as 'the new Sodom', was a precursor of San Francisco. Other minorities have also had famous ancestors, but what is relevant about these attachments is that they reveal infinitely varied combinations of sexual desire and other profound longings.

Michelangelo wrote, 'I am, of all who were ever born, the most inclined to love persons. Whenever I behold someone who possesses any talent or displays any dexterity of mind, who can do or say something more appropriately than the rest of the world, I am compelled to fall in love with him, and then I give myself away to him so entirely that I am no longer my own property but wholly his.' For others, the fear of old age seems to have been dominant, so that youth appeared as the saviour: for example, Robert Louis Stevenson's solution was not only to seek inspiration from his young stepson Lloyd Osborne, but also to write adventure stories for boys perpetuating a fantasy which eliminated all thought of decay; and a whole genre of popular 'romances', from H. Rider Haggard onwards, had the same purpose. Freud himself, who saw some homosexuality in everybody, wrote, 'In my life, woman has never replaced the comrade.' But the search for the comrade, in infinite variety, male or female, was seldom the whole story.

Coming out of the closet, proclaiming the independence of the Queer Nation, was the culmination of this phase. That made it seem, superficially, that homosexuals were necessarily interested only in their own kind. But those who could fathom their own deeper motives often wanted more than just to be themselves, having as their ultimate goal to escape from themselves, seeking partners from other cultures, from another generation, from a different social class. Awareness of this marks the beginning of the fourth phase. Edward Said, writing on Orientalism, has said that virtually every European writer who travelled to the Orient after 1800 searched for a different, freer, less guilt-ridden sexuality than that available in Europe. Arab love as well as war enabled T. E. Lawrence to 'quit me of my English self'. Desire was the magic which made possible transitions into unknown territory. It allowed the normally contemptuous prosperous classes to see their so-called inferiors, and themselves, in a different light.

When Aids transformed attitudes to promiscuity, more urgently among homosexuals than among heterosexuals, some began investigating new kinds of relationships, 'based on esteem rather than passion or convention', new forms of love-making, 'non-penetrative sex', drawing nearer to women's conception of the erotic. Edmund White, one of the most lucid homosexual writers, says, 'Our childlessness, our minimal responsibilities, the fact that our unions are not consecrated, even our very retreat into gay ghettoes for protection and freedom ... have fostered a style in which we may be exploring, even in spite of our conscious intentions, things that will some day be for the heterosexual majority.' Friendships between homosexuals and heterosexuals, where desire ceases to be aggressive, have become a significant new relationship.

The heterosexual majority discovered the meaning of this phase in 1993, when *Les Nuits Fauves* won France's equivalent of the Oscar, the César, for both the best film and the best first film of the year. It is full of 'explicit' scenes of homosexual love, and yet the press was unanimous in praising it: the Minister of Culture almost gave it official blessing, declaring that 'a whole generation has recognised itself in it'. Its author said, and the public understood, that it was not a film about a homosexual man who falls in love with a girl, but about 'a man in pieces, digging deep into all his contradictions'. Cyril Collard, who wrote it, directed it and played the lead role of Jean, became a cult hero, not just a new version of James Dean, not just a restless rebel without a cause, but one who went a stage further, able to understand his problem and to talk about it in a way that made sense even to those who were not in the same tragic impasse as himself. Collard was desperate to escape from aimlessness: 'I felt I was passing through life like those American tourists who rush through countries "doing" as many cities as possible [and] I was absolutely alone.' He made love to as many men as possible, but each encounter was only 'a tragedy endlessly repeated'; his desires were 'like islands, events never linked to each other'. Though the erotic excitement gave him a brief sensation of 'omnipotence', 'my descent into hell was nothing but a game of shadows ... the bottoms, breasts, genitals and stomachs I fondled belonged to nobody.' He hardly ever had a conversation with these lovers. The immediate satisfaction of desire was what mattered, but all he could achieve was orgasm, 'a brief moment of happiness'. Soon he found it impossible to sleep without having an orgasm first, and saw himself as 'hooked on sex like a cocaine addict on his drug'; he took cocaine to increase his desire and to postpone the orgasm; suffering reminded him that he was alive, but that was not enough; he was horrified to observe how easily sex lapsed into violence, how violence became exciting, which made him disgusted with himself.

However, Collard was more intelligent than James Dean: beginning as a mathematician, he had studied engineering, given it up to become leader of a rock band, had two novels published by France's top editor, Françoise Verny, made several films, still continued to read books, and then became an actor when several stars refused to tarnish their image by participating in *Les Nuits Fauves*, judging it to be too frank. Collard's significance is that he was determined not to be ashamed of being so many different persons, even while constantly doubting whether there was talent in any of them, and he made no secret of the fact that his film was autobiographical. When he half falls in love with a girl who complains that he does not love her enough, he admits that she was only 'a sandwich between me and me'. When she eventually settles for another man whom she does not love but with whom she can at least share in building some kind of future, he is bold enough to challenge the universally preached creed that love is the solution to all problems.

Love is revealed to be like freedom, only a starting point: then many more choices have to be made. Collard refused to reject anything without first looking at it carefully, and at the same time he refused to be categorised in any way, even sexually. He goes a stage further beyond saying that he is a homosexual or a bisexual or anything else, that he is usually a man but sometimes feels female, or that no single image of femininity pleases him, just as his girlfriend, however besotted with him, sometimes needs a change from him. To be fully alive, for him, means to participate in the 'planetary battle', to be 'a part of history' through the intermediary of the love of another person, but in the hope that others would join in the battle, and that together they would find something worth fighting for; it is not simply being Me, nor is it simply You and Me. This declaration of purpose, that what matters most is a noble purpose, is still vague; it will need many films to spell it out. Cyril Collard died of Aids, aged thirty-five, two weeks before his award could be presented to him.

Comparing how women have tried to influence men in courtship with how men have courted each other makes sexual desire emerge in a new light: it is not just an irresistible hurricane, nor a snake responding to the piping of only certain kinds of snake-charmers. It is too simple to say that each individual has to discover, helplessly, by what he or she is 'turned on'. Desire is no more inexplicable than taste. Over the centuries, it has been extraordinarily flexible and versatile, serving opposing causes, playing many different roles in history, like an actor, both comic and tragic, sometimes simple roles, reproducing hackneyed stereotypes, and sometimes experimental, complex ones, deliberately mysterious. This suggests that other alliances, other excitements, are also possible.

For that to become clearer, it is necessary to get to know better the ghost that has haunted so many relationships in the past, the desire to have power over others.

Courtship

Evelyn Shaw and Joan Darling, *Strategies of Being Female: Animal Patterns, Human Choices*, Harvester, 1984; Sarah Blaffer Hrdy, *The Woman That Never Evolved*, Harvard, 1981; D. Crews, 'Courtship in Unisexual Lizards', *Scientific American*, 1987, pp. 72–7; Ellen K. Rothman, *Hands and Hearts: A History of Courtship in America*, Harvard UP, 1987; Beth L. Bailey, *Courtship in Twentieth-Century America*, Johns Hopkins UP, 1988; *Cupid's Guide to the Wedding Ring*, reprinted Black Pennell Press, Greenock, 1988; Anon., *Conversational Openings and Endings*, Bentley, 1891; C. Nyrop, *The Kiss and its History*, translated from the Danish by L. Sands, 1901 (learned and instructive); Nicholas J. Perella, *The Kiss: Sacred and Profane*, California UP, 1969 (the 'French kiss' was known in sixteenth-century France as 'baiser à l'Italienne'); James T. Y. Liu, *The Chinese Knight Errant*, Routledge, 1967 (showing loyalty only to 'the one who appreciates you'); Jacques Gernet, *Daily Life in China on the Eve of the Mongol Invasions 1250–1276*, Allen and Unwin, 1959; Jacqueline Huppert-Laufer, *La Féminité neutralisée?: Les femmes cadres dans l'entreprise*, Flammarion, 1982; Helga Dierichs and Margarete Mitscherlich, *Des Hommes: Dix histoires exemplaires*, Des Femmes, 1983; Lawrence Stone, *The Road to Divorce: England 1530–1987*, Oxford UP, 1990; P. Ariès and G. Duby, *Histoire de la vie privée*, 5 vols., Seuil, 1987 (a mine of information); Jane Lazare, *On Loving Men*, Abacus, 1978; Yan de Kerorguen, *Le Plaisir chaste*, Autrement, 1984 (on changes towards sex in France in the 1980s); Pascal Lainé, *La Femme et ses images*, Stock, 1974; Germaine Greer, *The Female Eunuch*, MacGibbon and Kee, 1970; Germaine Greer, *The Obstacle Race*, Secker & Warburg, 1979; Germaine Greer, *Sex and Destiny*, Secker & Warburg, 1984; Betty Friedan, *The Second Stage*, Summit, NY, and Michael Joseph, 1982; Ezra Storland, *The Psychology of Hope*, 1969; Robert J. Stoller, *Observing the Erotic Imagination*, Yale UP, 1985; K. S. Srinivasin, *The Ethos of Indian Literature: A Study of its Romantic Tradition*, Chanakya, New Delhi, 1985; Jo Ann McNamara, *A New Song: Celibate Women in the First Three Christian Centuries*, Harrington Park Press, NY, 1985; Dana Vannoy-Hiller and W. Philliber, *Equal Partners: Successful Women in Marriage*, Sage, 1989; Gordon Thomas, *Desire and Denial*, Grafton, 1986 (on priests); Julia Kristeva, *Les Samurais*, Fayard, 1990 (on intellectuals); Michel Crépu, *La Force de l'admiration*, Autrement, 1988; M. Foucault, *The History of Sexuality*, Viking, 1986ff.; Didier Eribon, *Michel Foucault*, Flammarion, 1989; Michael Young, *The Ethnography of Malinowski*, Routledge, 1979; Frederic Errington and D. Gewertz, *Cultural Alternatives and a Feminist Anthropology: Culturally Constructed Gender Interests in Papua New Guinea*, Cambridge UP, 1987; Unni Wikan, *Behind the Veil in Arabia: Women in Oman*, Johns Hopkins UP, 1982; Andrea Dworkin, *Right Wing Women*, Women's Press, 1983.

Homosexuality

David F. Greenberg, *The Construction of Homosexuality*, Chicago UP, 1988; Dominique Fernandez, *Le Rapt de Ganymède*, Grasset, 1989; Wayne R. Dynes, *Encyclopedia of Homosexuality*, St James Press, 1990; Wayne R. Dynes, *Homosexuality: A Research Guide*, Garlan, NY, 1987; Martin Duberman, *Hidden from History: Reclaiming the Gay and Lesbian Past*, Penguin, 1989; Jonathan Dollimore, *Sexual Dissidence: Augustine to Wilde, Freud to Foucault*, Oxford UP, 1991; Peter Binton, *Talking to . . .*, Third House, 1991 (conversation with Edmund White);

How the desire that men feel for other men has altered

Edmund White, *Genet*, Gallimard and Chatto, 1993; Wayne Koesten-baum, *Double Talk: The Erotics of Male Literary Collaboration*, Rout-ledge, 1989; Bret Hinsch, *Passions of the Cut Sleeve: The Male Homosexual Tradition in China*, California UP, 1990; T. Watanabe, *Love of the Samurai: 1000 Years of Japanese Homosexuality*, GMP, 1989; John J. Winkler, *The Constraints of Desire: The Anthropology of Sex and Gender in Ancient Greece*, Routledge, 1990; Eva Cantarella, *Bisexuality in the Ancient World*, Yale UP, 1992; Eve Elvin, *Sex and Society in the World of the Orthodox Slavs 900–1700*, Cornell UP, 1989; James W. Saslow, *Ganymede in the Renaissance*, Yale UP, 1986; Bruce R. Smith, *Homosexual Desire in Shakespeare's England*, Chicago UP, 1991; Alan Bray, *Homosexuality in Renaissance England*, GMP, 1982; R. P. Mac-cubbin, *'Tis Nature's Fault: Unauthorized Sexuality during the Enlight-enment*, Cambridge UP, 1985; Laura Engelsten, *Sex and the Search for Modernity in fin-de-siècle Russia*, Cornell UP, 1992; Antony Copley, *Sexual Moralities in France 1780–1980: New Ideas on the Family, Div-orce and Homosexuality*, Routledge, 1981; David M. Halperin, *One Hundred Years of Homosexuality*, Routledge, 1990; Jeffrey Weeks, *Coming Out: Homosexual Politics in Britain*, Quartet, 1977; K. Weston, *Families We Choose: Gays and Kinship*, Columbia UP, 1991; Ken Plum-mer, *Modern Homosexualities*, Routledge, 1992; Gilbert Herdt, *Gay Culture in America*, Beacon, Boston, 1992; Warren J. Blumenfield, *Homophobia*, Beacon, 1992; Mary Searle Chatterjee, *Reversible Sex-Roles: The Special Case of Benares Sweepers*, Pergamon, 1981.

8

How respect has become
more desirable than power

Dreaming, sleeping and forgetting: has any statesman ever claimed to be a specialist in these arts? Only the mayor of Strasbourg, Catherine Trautmann. Still in her thirties, elected to preside over the parliamentary capital of Europe, is this woman suggesting that the music of politics will never sound the same again?

Her adventure began with a thesis on dreaming, sleeping and forgetting with special reference to the Gnostics. This was a sect which flourished around the same time as Jesus Christ, and whose essential belief was that every individual is a stranger in the world; even God is a stranger, because His creation is imperfect and He does not feel comfortable in it either. It was a sect of optimists, however, who were convinced that everybody would find salvation, or at least all those who unravelled the symbolism in which the world was wrapped up, and who discovered the rituals necessary to triumph over evil. Christianity competed with Gnosticism and then borrowed ideas from it; later William Blake, Goethe and Jung were among the great minds who drew inspiration from it. Catherine Trautmann thinks the Gnostics have a lot to say to modern people who, like them, are ill at ease in an unjust world. They were marginals, and she is a marginal too (that is to say, she does not feel herself to be part of an established order, which claims that things are the way they have to be). The Gnostics had 'a certain detachment', which she also seeks to cultivate. They tried to see beyond appearances, to find a hidden meaning in what seemed meaningless, to carry out 'an exegesis of the soul'; and she, too, says what interests her most is not the obvious, but what is forgotten. They believed that apparent opposites were not necessarily different, and tried to transcend the differences between male and female: that strikes a chord in her, whose first political instinct was to be a feminist intent on changing the way people treated each other.

However, to achieve her aims, she made a deliberate choice. Instead of enrolling in the feminist movement, she joined the socialists. To improve the world, she said to herself, one must not hold aloof, but participate in the mainstream. She decided to become 'a marginal integrated into society', changing it from within. A marginal she remains, but that now means that she keeps her freedom within society, not allowing marginalism to become selfishness or pride. Still, marginals do not forget their dreams.

As a child, she was always saying to herself that she must not forget what she learned; but as an adolescent she read Freud, and realised that forgetting is not always accidental. The irrational side of human nature intrigued her. On the one hand she was determined to develop into the sort of person she wanted to be, and carefully drew up a list of her objectives. On the other hand, she could not convince herself that she would ever unravel the mysterious processes which lead a person to one course of action rather than another. Her thesis on the Gnostics does not explain her present politics; it was an exercise in an academic framework, but it was also an attempt to discover what she was searching for, and she has not become a conventional politician, because she is still in the process of 'unravelling', trying to make sense of herself and of others.

When she was elected mayor, at the age of thirty-eight, her daughter said, 'You wanted to be mayor for a long time and you never told me.' She replied, 'I didn't know that was what I wanted.' But a friend said, 'You can't pretend that it is an accident that you are mayor. Can't you see you have been aiming for it all along?' No, said Catherine Trautmann, 'I didn't realise I was.' It is not easy to know what one is striving for. She asks herself, What is my aim now that I am mayor? There is no simple answer.

Her family comes to mind straight away, before she can enunciate some grand political principle. One of the first goals in her life is to have a successful partnership with her husband. Politicians do not normally begin by talking about their private lives, though that is the one interest they share with all their electors, except those carried away by more lonely ambitions. The agreement she made with her husband, when she married at nineteen, was that neither would ever limit the freedom of the other. She 'loves politics', she says. It is a passion, like an affair. 'My two daughters have accepted this very well, because I tell them that politics is very important in my life.' This means she sees them less than she might otherwise wish. 'I am not a superwoman.' Her husband, her parents and a circle of friends have stepped in to create a network of affection around the children. That is not something that happens naturally. She knows how difficult it is for a working mother to find a crèche: her own failure to find one was the spur which led her into politics in the first place.

However, even with all this goodwill and patience, a marriage can easily fall apart. A woman, she says, can be very demanding in wanting to be listened to; her insistence can be 'brutal'. One day, 'I said to myself, Stop it. You are asking too much. Married relations have a tendency to become theatre, plays acted over and over again. . . . You reach scene 3, act 5. . . . You realise you are putting on a performance. You are a victim of habit and you let yourself be carried away.' The key that lets you off the stage is the decision that you must never allow contempt to enter into your life. 'Contempt is the worst of all things, it is a symbolic way of killing a person. It revolts me.'

Her unconventional conclusion is that there can be no question of politics giving her one key, one dogma, one solution to all problems. Having studied theological disputation in ancient times, she is struck by how similar modern politicians are to the divines of the past in the way they think. 'That makes it impossible for me to go on naively repeating an ideological speech.' Politicians may form separate parties with others whose opinions they roughly share, but within each party there is always conflict. She likes grappling with these conflicts, finding stratagems to cope with them, provided there are rules to the game, as in a sport. The search for power cannot be the goal, because 'people in power lose a part of their identity: there is a constant tension between yourself and the public office you hold', between the individual and the traditional way of exercising authority. She wants politicians to remain individual human beings. The ones she likes most are 'untypical' politicians. What matters most in the kind of politics she favours is the continued pursuit by politicians of their own self-understanding, their 'spiritual development'.

They must not expect success, because every victory is provisional, merely one step, never the final one. Politics is a never-ending apprenticeship, requiring one to adapt oneself to the fact that other people are different. That is its reward, the discovery of the diversity of humanity: 'Being in politics is a wonderful way of observing the variety of life.' And of course life is full of failures: 'It is important to recognise one's failures: the test for politicians is how well they can accept their failures.' Women have been frightened by politics, she says, because they perceive it as a 'hard' world; but in fact they have an advantage over men; women are 'double-sided'; they view the world both as public and as private, which prevents them from getting lost in abstractions. 'Women have more freedom as politicians; men accept a lot of things from them that they won't tolerate from each other; and with women there is an expectation of new ideas, of change.'

In her youth she too was frightened, and not just of politics: 'timid, anxious in the company of others'. As a young mother she was worried because she was not sure how to deal with her children, or how to answer their questions. So her ambition became 'to get beyond my timidity'. She has always felt herself to be a solitary person, which may seem to contradict her image as happy in her work and at home; but she has discovered that there is value in the sense of being alone. 'Solitude is my internal pillar, my secret garden. No one enters except those who are closest to me.'

Catherine Trautmann takes care to remain a double person. Granny Marmalade was her nickname as a student. It is still her favourite hobby to concoct jams and preserves out of quince, pumpkin and tomato, to her own recipes. She enjoys making clothes and 'unusual objects', works of art out of bits and pieces. Artists with a bitter or sarcastic humour please her most, the surrealists and the great caricaturists. At home, she does not make speeches. When she is with her husband, she is not the mayor.

Queen Elizabeth I of England said, 'I know I have the body of a weak and feeble woman but I have the heart and stomach of a king.' To have the stomach of a king is no longer an adequate ambition. The model of the Strong Man who can inspire obedience is out of date. Catherine Trautmann's weaving together of private and public life suggests that politics can have a different texture. To her opponents she is, of course, just another rival to be ousted, and one side of her is indeed fighting the traditional war of politics, but her less obvious side gives hints of new possibilities in human relations.

To be king: that was once the universal dream, not only of politicians, but of fathers who ruled over their children, husbands who treated their wives as servants, bosses who could almost say Off with his head, officials who forgot their haemorrhoids by imagining that their shabby chairs were thrones. In real life, for the last 5,000 years, the vast majority of humans have been submissive, cringing before authority and, apart from short-lived outbursts of protest, sacrificing themselves so that a small minority could live in luxury. They would have grown tails, but for the fact that most of them found someone else over whom they could themselves play the tyrant, someone weaker, someone younger. Inequality was accepted for so long because the bullied found victims to bully in their turn. The powerful leader was admired because he incarnated dreams of authority which humble people secretly cherished and tried to enact in their private lives. But now the obsession with domination and subordination is beginning to be challenged by a wider imagination, hungry for encouragement, for someone who will listen, for loyalty and trust, and above all for respect. The power to give orders is no longer enough.

In the past, outward signs of respect – the raised hat, the low bow – proved that people accepted and acknowledged their submission to the powerful. Now, however, the quality of the personal relationship between two individuals has come to count for more than rank or status. Though politicians have installed themselves in the palaces of kings, they are the least admired of the professions, far below doctors, scientists, actors, even poorly paid nurses and teachers. It is not surprising that women have on the whole not wanted to be politicians of the traditional kind. Every time a politician makes a promise and does not keep it, all would-be kings become a little less credible.

Two worlds exist side by side. In one the struggle for power continues almost as it always has done. In the other it is not power that counts, but respect. Power no longer ensures respect. Even the most powerful person in the world, the President of the United States, is not powerful enough to command everybody's respect; he probably has less than Mother Teresa, whom nobody is obliged to obey. Traditionally, respect was converted into power, but it has now become desirable for its own sake, preferred raw rather than cooked. Most people feel they do not get as much respect as they deserve, and obtaining it has become for many more attractive than winning power. Attention is focused on family life, where the aim is no longer to have as many children as possible, which was once the way to become rich, but to create bonds of affection and mutual respect, and extend them to a circle of chosen friends. It is no longer the clan or nation which decides whom one should hate and whom one should court. The powerful are mocked more than they have ever been, even if they are still feared. Modern government, which tries to control more aspects of life than kings ever did, is constantly being humiliated because its laws seldom achieve what was intended, are evaded and twisted, seldom succeeding in altering mentalities, which decide what really happens, seldom able to resist speculators or global economic trends.

Imaginations are beginning to work in a new way. It has ceased to be admirable to treat people like animals, whose domestication was once humanity's proudest achievement. Cows were taught to work day and night to produce 15,000 litres of milk a year, when once their daily yield was little more than a single pint. Sheep learnt to grow 44 lb of wool a year, when once a mere 2 lb sufficed to keep them warm, and in the process they started to bleat continuously, to behave like sheep, which they did not use to do. Pigs have been transformed from free-ranging, pugnacious foragers of the forest into docile wallowers in their own urine, forced to have such unaccustomed close contact with others, to gobble up their meals in a few minutes – whereas once the search for food was a never-ending preoccupation – that they have no choice but to alternate between sleep and aggression, biting each other's tails. Even sexual behaviour has been transformed: some animals have become much more randy, others have almost lost interest; some, reared in all-male groups, have established stable homosexual relationships; bulls, fed on high-protein diets, relieve their tension by masturbation. Some animals have been bred to retain their juvenile characteristics for life. Since the eighteenth century, when inbreeding came into fashion, many have become more uniform, more stereotyped than they once were. It was usually only when animals became commercially useless that pleasure was taken in their companionship: but it has been only recently that humans have begun to question whether the way to show affection to dogs is to breed them deliberately into grotesque and painful shapes.

This is how people discovered what power meant: the ability to make others behave as one wanted them to. That used to inspire enormous respect. The experience of domestication showed that living beings were capable, under pressure, of a vast range of behaviours and temperaments and that they could be made to contribute to their own enslavement, becoming attached even to owners who maltreat them. Few noticed how the slave master was often enslaved by his victim. For soon humans began trying to domesticate one another, breeding for subordination and domination. When they learnt to domesticate plants also, they became the first casualty of their invention. Once they got involved in ploughing and hoarding their crops and weaving and cooking in pots, once they specialised in different crafts, they found themselves compelled to work for a minority intent on monopolising the good things of life, landowners who organised the irrigation, priests who arranged for the rain to fall and warriors who protected against marauding neighbours. The first theology of which there are records, that of Sumeria, stated that humans had been expressly created to relieve the gods of the need to work for their livings, and if they did not, they would be punished with floods and droughts and starvation. Soon kings claimed to be gods, and priests demanded an ever higher price for their consolations, assuming ownership of more and more land. Noblemen and gangs of warriors intimidated those who tilled the soil, sparing their lives only in return for a part of their produce, imposing a truce on violence in return for help in pillaging foreign countries. Thus an élite accumulated power, enabling it to live in high luxury, and to stimulate the flowering of the arts, but civilisation was for many little more than a protection racket. Under this system, respect went mainly to those who lived at the expense of others. There has never been enough respect to go round, because so far only small quantities of it have been cultivated.

The Romans, who ran one of the most successful of all protection rackets, made it possible for a few hundred thousand of them to give up work and to receive free food from the government, paid for by tribute exacted from the foreign 'protected' territories which made up their empire. However, the cost of rackets has always gone up with time, as more people obtain a share of the profits, as the administration becomes more cumbersome and as armies become more expensive, because citizens have usually ended up preferring to pay mercenaries to do their fighting for them. The more prosperous a civilisation, the more people it attracts from across its borders, eager for pickings, and the more it has to spend defending itself, or buying them off; it invents ever more complex arrangements in order to survive, and eventually these become too complex and the civilisation ceases to function. The Soviet Union became apoplectic when it ended up spending most of its budget on defence.

It was only in 1802 that domination and subordination among all living creatures began to be studied scientifically. At the very moment when Napoleon was creating dukes and barons, and re-establishing hierarchies, the blind Swiss naturalist François Huber described how bumble bees also lived in a strictly hierarchical order. In 1922, the year in which Mussolini became prime minister, Schjedelrup-Ebbe showed how even starving hens always allowed their leader (the 'alpha' hen) to eat first and did not dare interfere until it had finished; how if it was removed, the hens still did not eat, but waited until 'beta' had had its fill, and so down the line. The pecking order of hens was revealed to be as rigid as in an army, to such an extent that when taken away for a few weeks and then returned to their original flock, each immediately resumed its old rank. The reward was that the flock lived in peace, did not fight over food and produced more eggs. The price was injustice. Those at the bottom of the hierarchy not only got less to eat, but had fewer offspring, suffered from stress, deteriorated physically and in moments of danger – when food ran out, when the population became too dense – were made scapegoats and mercilessly attacked. The same principles were observed in other creatures: the children of dominant rabbits, wolves, rats, tended to become dominant too; baboons had aristocratic dynasties. Nature seemed to be saying that equality was impossible, and that only the strong can hope to be respected.

In the 1980s, however, it was discovered that aggression, which used to be regarded as the essential characteristic of animals, was not what it appeared to be. Making peace after a fight was a skill cultivated with quite as much attention. When dominant and subordinate chimpanzees were, for the first time, observed as individuals and not just as species, they were seen to be constantly engaging in angry or violent confrontations, but within forty minutes, no less than half of them were kissing and stroking their former enemies. Sometimes a crowd of them gathered round to watch the reconciliation and to applaud the kiss. This did not mean that they were not aggressive, for without the aggression there would have been no reconciliation, nor that they all made peace in the same way. The males, after fighting amongst themselves, make peace twice as often as females who have fought females, as though power, for males, depends on forming alliances, which are never permanent; the friend of today may be an enemy tomorrow; and exchanges of help on a tit-for-tat basis involve no promises for the future. President Tancredo Neves of Brazil unwittingly put into words what male chimpanzees do all the time when he said, 'I have never made a friend from whom I could not separate, and I have never made an enemy whom I could not approach.'

Female chimpanzees, by contrast, are much less concerned with status, do not make obeisance to each other. They do not behave like soldiers saluting officers, as the males do; their coalitions are of a small circle of family and friends, whom they choose for emotional reasons, and not on the basis of importance in the hierarchy. They distinguish between friend and foe more sharply than males, often having one or two absolute enemies with whom reconciliation is out of the question.

The link between love and aggression has been observed also in the chimpanzee custom of hardly ever punishing their children, and in its result, that they do not maintain close ties with them either, unlike rhesus monkeys, who are much more aggressive, and who treat their daughters harshly but develop lifelong bonds with them. What female chimpanzees are good at is establishing peace between males: for example, one of them might bring two male rivals together after a fight, sitting between them so that they do not have to face each other, allowing both to groom her, and then slip away to let them groom each other; sometimes she looks over her shoulder to make sure they are at peace, and if not, returns to put the arm of one round the other. While the females stimulate affection, the males call a truce to hostilities by developing common interests, or pretending to. For example, they find an object and call in everybody to come and look; everybody comes and then goes away, except the old adversary who pretends to be enthralled, until eventually they touch, groom each other and are friends again, or rather temporary allies until the next fight.

These discoveries are about chimpanzees, not about humans. Even if the latest discovery is that chimpanzees eat leaves containing antibiotics when they are ill, and other kinds of leaves, with contraceptive, oestrogen-like properties, when they wish to reduce their families, they remain chimpanzees. But this new knowledge makes it clear that humans have misinterpreted what they call their animal inheritance. They are no longer faced with the simple choice which has dominated all history, that they should either be 'realistic' and behave as if life is a struggle of brute force, or else withdraw into utopian dreams and imagine that all will be harmony if only aggression were outlawed. Many, perhaps most of them, still believe the 'realistic' view, as expressed by Heinrich von Treitschke (1834–96): 'Your neighbour, even though he may look upon you as his natural ally against another power which is feared by you both, is always ready, at the first opportunity, as soon as it can be done with safety, to better himself at your expense.... Whoever fails to increase his power, must decrease it, if others increase theirs.' But Treitschke is now revealed to have been a little boy who longed to be a soldier and who, being almost totally deaf, had to content himself with becoming a professor, dreaming of powerful leaders leading powerful nations, waging war to show their contempt for other nations. Contempt can now be seen to be a perverted way of begging for respect. It is not a method which works. War is no longer regarded as the most noble activity. And yet politicians have not given up using the metaphor, 'fighting' for their principles, 'defeating' their rivals. A new language has not yet been found for 'winning' respect.

The business world has been quicker to sense the need for it. Its hero used to be the aggressive manager who terrifies his employees and at the same time convinces them that they are enjoying doing what they are told. Aggression has remained a virtue in the business vocabulary, though it has undergone cosmetic surgery, so that power has emerged with a younger face, proclaiming that it is really a domestic game, in which everybody who tries hard enough can win. However, managers have increasingly stopped seeing themselves as giving orders, or even making decisions, and have instead come to believe that their function is to encourage their staff to find solutions for themselves. Faced with the entry of women into business, they have become more transparent: the human weaknesses lurking behind the façade of power have become more visible. When the veil between public and private life is removed, the powerful are naked. So increasingly respect counts for as much as power in the balance sheets.

Meanwhile, psychiatry has been portraying the power-hungry as sick, suffering from an allergy to disagreement. Hitler (an admirer of Treitschke) was a classic case, who tried to eliminate not just disagreement, nor just enemies, but doubt itself, persuading his followers to obey him without question, explaining that conscience was 'a Jewish invention, a blemish, like circumcision'. In politics, there is no room for trust, added Stalin, basing his authority firmly on terror. However, neither of these two dictators was satisfied with the mountains of power he accumulated. Hitler constantly complained that he was disobeyed, that even the army 'has consistently tried to impede every action I have thought necessary'. And Stalin, though he bribed, cheated and forced his enemies, by all forms of torture, to acknowledge that he was right and that they were wrong, could never obtain enough reassurance that he was the genius he would have liked to be. Both remained desperately hungry for respect.

One of the most important promises of democracy is that it will provide respect for everybody. Athens ensured this not only by giving the vote to all citizens, but by rotating offices by lot, so that anybody, however uneducated or humble, might become president for a day. No Athenian demeaned himself by being an employee of a fellow citizen. But this mutual respect only worked because Athens was a protection racket also, depending on an empire and slaves and women to keep their wonderful philosophical discussions going. Democracies have still not found a way to eliminate the gradations of disrespect created by money, education and appearance.

So it was to religion that individuals most frequently turned in search of the respect they yearned for. All the world's great churches agreed that every human being, however humble, had a spiritual dignity. The exactions of rulers, the insults of employers and the humiliations of daily life seemed less intolerable when they touched the outer self, leaving intact the consolations of inner convictions. And when religion did not suffice, other creeds, like stoicism, socialism, liberalism and feminism, reinforced the defences of human dignity. The major changes in history have resulted less from revolutions displacing kings, than from individuals ignoring kings and giving their allegiance to spiritual values instead. That is still happening. The prophecy that the twenty-first century will be a religious one is not a prophecy, but an acknowledgement of what has happened quite regularly in the past. It does not mean that politicians are replaced by priests, but that people switch off from the vast mundane pressures which they cannot control. Instead they turn their energy to their private lives: sometimes that leads them to be selfish, but sometimes they react to the animosities of the big world by seeking more nurture, more generosity, more mutual respect.

The Roman Empire is a good example of a gradual collapse of security, efficiency and values. Emperors continued to reign, but individuals gave their allegiance privately to religions which kept them from total despair. Christianity did not automatically provide an inner conviction of one's worth with baptism, or there would have been a far greater rush into it: it probably converted no more than half a million individuals in each generation in the first few centuries of its existence, and many Christians were not fully convinced that either God or their fellows did respect them, for conversions often took place because Christian miracle cures of disease seemed more amazing than pagan ones. It was only after a period of membership in a community of converts, exchanging charity and encouragement, that Christians began to feel valued for themselves. But then they quarrelled, becoming allergic to disagreement, using force to obtain respect, becoming persecutors, allies and imitators of the powerful. Every time churches have become too greedy for power, behaving like governments, the faithful eventually switch off and turn to a new consolation or a new idealism.

There is nothing unprecedented in the current uncertainty about where humanity's daily meal of respect is going to come from. It is not the first time that official resources of respect have crumbled, causing people to scurry back to old beliefs, and to nibble at new ideologies. The great religions grew out of a search for a meaning in life very similar to that which is taking place today; and they won after competing with hundreds of sects and crazes now forgotten. The present-day movement for human rights, women's equality and the sanctity of the environment springs from the same sort of yearnings which the great religions tried to satisfy between twenty-five and thirteen centuries ago. It does not offer total freedom from doubt, nor the certainty that it is possible to win respect simply by being a member of the winning side, following obediently behind leaders, nor the hope that there will be a society one day in which everybody agrees on what is respectable, for disagreement is beginning to be accepted as inescapable, and even as a virtue; but that is not very different from the attitude of the most modern kind of religion, which has little in common with the old dogmatic clericalism. It is significant that the mayor of Strasbourg, who says that her greatest disappointment is that democracy has failed to find a really new way of dealing with power, combines a background in a modernised religion and in the search for more respect between the sexes.

Respect cannot be achieved by the same methods as power. It requires not chiefs, but mediators, arbitrators, encouragers and counsellors, or what the Icelandic sagas call peaceweavers, who do not claim to have a cure for all ills, and whose ambition is limited to helping individuals to appreciate each other and to work together even when they are not in complete agreement, ensuring that disputes do not become suicidal. The difficulty in the past has been that such people have often demanded too high a price, and have ended up demanding obedience.

The majority of women who have gone into traditional politics have been more or less disappointed, and have felt that however high they rose, real power eluded them, that they were forced to play a game controlled by men who sacrificed normal living in order to enjoy power, and that they had to deal with bureaucrats who, however devoted they were to the public service, were running a factory producing frustration, a nineteenth-century factory whose ideal was impersonality. It has been proved time and again that skilled manipulators of power can be dislodged only by skilled manipulators of power, who perpetuate the same system. The current weariness with old-fashioned politics represents not lack of interest in the common good, but near despair at the difficulty of contributing to it, and at the regularity with which idealistic leaders have made compromises with hypocrites, despite themselves, or with the dogmatic, despite their principles, because the struggle for power is merciless, and cannot be fought without allies. Once they are involved in that struggle, even pious Christian bishops, and Buddhist monks, and Confucian scholars, have ceased to be holy or selfless.

So fighting to get new laws, or to obstruct opponents, has never been a wholly successful strategy, or an adequate programme for those who value generosity. What ordinary people can do for themselves to increase their mutual respect, without repeating the mistakes of the past, I shall discuss in the next chapter.

Domestication of animals

E. S. E. Hafez, *The Behaviour of Domesticated Animals*, 3rd edn., Bailliere Tindall, 1975; Peter J. Ucko and G. Dimbleby, *The Domestication and Exploitation of Plants and Animals*, Duckworth, 1968; Yi-Fu Tuan, *Dominance and Affection: The Making of Pets*, Yale UP, 1984; S. Bokony, *History of Domestic Mammals in Central and Eastern Europe*, Akademie Kiado, Budapest, 1974; Hans Kruuk, *The Spotted Hyena: A Study of Predation and Social Behaviour*, Chicago, 1972; Maurice Caullery, *Parasitism and Symbiosis*, Sidgwick and Jackson, 1952; W. P. Rogers, *The Nature of Parasitism*, Academic Press, NY, 1962; Robert Delort, *Les Animaux ont une histoire*, Seuil, 1984; Keith Thomas, *Man and the Natural World*, Allen Lane, 1983.

Power

Ann Ruth Willner, *Charismatic Political Leadership*, Princeton, 1968; Richard Sennett, *Authority*, Secker & Warburg, 1980; Douglas Yates, *Bureaucratic Democracy*, Harvard, 1982; P. F. Piven, *Why Americans Don't Vote*, Pantheon, NY, 1988; Burdett Loomis, *The New American Politician*, Basic, NY, 1988; Lisanne Radice, *The Job of a Backbencher*, Macmillan, 1987; Michael Rush, *Profession of the British M.P.*, Exeter Univ., 1989; *Fifth Survey of British Social Attitudes for 1988*, on 'Trust in the British establishment' (or lack of it); G. R. Searle, *Corruption in British Politics 1895–1930*, Oxford UP, 1987; Gunnar Boalt, *The Political Process*, Stockholm, 1984 (sportsmen were the most highly admired (47 per cent) in this Swedish enquiry, followed by 'champions of peace' 32 per cent, scientists 28 per cent, actors 26 per cent, music groups 26 per cent, authors 25 per cent, and only then politicians, equally with financiers, 24 per cent; José Tharakan, *Political Attitude of the Non-Voters in Switzerland*, Zurich thesis, 1983; André Bercoff, *La France des seigneurs*, Laffont, 1989; David A. Baldwin, *Paradoxes of Power*, Blackwell, 1989; Jan Winiecki, *Resistance to Change in the Soviet Economic System*, Routledge, 1991; Sharon Kelly Heyob, *The Cult of Isis among Women in the Graeco-Roman World*, E. J. Brill, Leiden, 1975; Philip Brook, *The Origin of Citizenship in Ancient Athens*, Princeton,

1990 (on mixing); Anne Phillips, *Engendering Democracy*, Polity, Cambridge, 1991; Yolande Cohen, *Femmes et contrepouvoirs*, Boreal, Montreal, 1987; Mike Savage and Anne Witz, *Gender and Bureaucracy*, Blackwell/The Sociological Review, 1992; Marilyn French, *Beyond Power: On Women, Men and Morals*, Cape, 1985; Judi Marshall, *Women Managers: Travellers in a Male World*, Wiley, Colchester, 1984; Michael A. Goldberg and John Mercer, *The Myth of the North American City*, UBC Press, Vancouver, 1986 (on Canadian attitudes to power); Jean Claude Lugan, *La Petite ville au présent et au futur*, CNRS, 1983; Sabine Chalvon-Demersay, *Le Triangle du 14e: Des nouveaux habitants dans un vieux quartier de Paris*, MSH, 1984; William M. Kurtines and J. L. Gerwirtz, *Moral Development through Social Interaction*, Wiley, NY, 1987 (on the Manville School of the Judge Baker Guidance Centre in NY); François Jullien, *La Propension des choses: Pour une histoire de l'efficacité en Chine*, Seuil, 1990; Bryce Taylor, *Assertiveness and the Management of Conflict*, Oasis, Leeds, 1989; Doris A. Graber, *Verbal Behaviour and Politics*, Illinois UP, Urbana, 1976; Alain-Gérard Slama, *Les Chasseurs d'absolu: Genèse de la gauche et de la droite*, Grasset, 1980; Alain-Gérard Slama, *L'Angélisme exterminateur*, Grasset, 1993; Patricia Hewitt, *The Abuse of Power*, Robertson, 1982; Peter J. Wilson, *The Domestication of the Human Species*, Yale UP, 1988; Paul Wheatley, *The Origins and Character of the Ancient Chinese City*, Aldine, Chicago, 1971; Ruth Glass, *Clichés of Urban Doom*, Blackwell, 1989; Norma J. Chalmers, *Industrial Relations in Japan: The Peripheral Workforce*, 1989; Michael Blaker, *Japanese International Negotiating Style*, Columbia UP, 1977; W. Dean Kingley, *Industrial Harmony in Modern Japan*, Routledge, 1991; F. C. Jaher, *The Rich, The Well Born and the Powerful*, Illinois UP, 1973; David Mercer, *IBM*, Kogan Page, 1988; Peter Hennessy, *Whitehall*, Fontana, 1989; John Dickie, *Inside the Foreign Office*, Chapmans, 1992 (showing 62 per cent of diplomats saying they are frustrated and half wanting to leave); Jean-François Kahn, *Esquisse d'une philosophie du mensonge*, Flammarion, 1989.

9

How those who want neither
to give orders nor to receive them
can become intermediaries

What is Thérèse thinking, as she shows you to your table in La
Vieille Alsace in Strasbourg, and helps you order your meal? She
gives the impression that she knows exactly what to say, what to
do, all energy, a brisk walk, a direct look, friendly solicitude. If
you have been there before, she can remember more about you
than you ever told her. How are you to know that she is not a
waitress waiting for better times, that she is not calculating the size
of the tip you can afford to give, that she has a master's degree in
the history of art? Thérèse has been doing this job for fifteen years
because she has a purpose in life.

Her self-assurance seems to be the essence of her character: in
fact it is one of her main preoccupations. Once upon a time she
was a girl completely crushed by timidity, a spoilt only child. Then
one day she escaped, having fallen in love with a man eleven years
her senior, who plunged her into a completely different sort of life.
Together, they opened a restaurant, L'Arsenal: soon the two small
dark rooms, in a street no tourist ever heard of, became a favourite
resort of the local intellectuals, artists and journalists. She knew
nothing about food, not even how to make coffee. He cooked; she
served and learned. Above all, she welcomed the customers, and
gradually she became the ever-so-modest star of the place, who

gave it its peculiar atmosphere. 'I was drawn out of my innermost self. Now there are two sides to me.' Her shyness hid behind a new conviviality. Running a restaurant encouraged her to try to be agreeable at all times, to be tolerant, to reject first impressions of people, to show friendliness without being obsequious. In return it made her feel liked. Her trouble, however, is that she never feels she is liked or loved enough: behind her poise, her insecurity gnaws at her. After a day of socialising, she likes to withdraw, be by herself, read. In private life she is much more reserved and, she claims, 'difficult'.

Gradually the meaning of her work dawned on her: she was dedicating herself to curing people of the incapacitating shyness she knew so much about. 'I discovered people want to meet, but they always need some other person to arrange it.' She began cultivating the art of discovering who would mix with whom. Bringing together two great painters, for example, was unwise: 'It is war.' Artists never feel sufficiently appreciated; she understands exactly the taste of that unquenchable thirst. One of the regulars at the Arsenal was Tomi Ungerer, who had had exhibitions all over the world but whose greatest ambition was to be loved as a favourite son in his native Alsace, an ambition now fulfilled.

Thérèse has published two books on the city's historical monuments and she is also a painter; she feels a very strong need to do something creative, but, having no illusions about her talent, she has concluded that her need for creativity is not as profound, not as 'vital', as that of great artists. So 'I prefer to recognise art in others, to give them a sense of appreciation'. Her speciality is to 'recognise beauty'. There are intermediaries in business, there used to be intermediaries in marriage, but the social intermediary is a profession still in its infancy.

As she grew up, and began making discoveries, her lover ceased to be a god: 'He lost his authority.' She moved out, but continued to work at L'Arsenal, to the puzzlement of her friends, who could not understand why she did not make a clean break. But loyalty, gratitude and a detached friendship pleased her, friendship in all its varieties. When suddenly he died, her past did not vanish; he lives in her still; but for him she might now be a quiet mouse in a museum. She has married his opposite, a man two years younger than herself: her friends are all marrying younger men the second time round. A person needs many kinds of relationship.

There are about fifteen people with whom she has a 'real friendship', some from her university days, others are customers who have become friends, dining at the restaurant once a month to renew the bond: 'It is part of the protocol.' With three or four she goes on excursions to discover towns she does not know: but some she could not travel with, though she adores them, because they have neither the same instincts nor the same income.

She works on her private life. It is conscious, creative labour, not easy to do in a provincial city where people gossip and do not understand why someone should choose to move out of the narrow path of conformity. Though widely admired for the way she went on to preside over a grand restaurant in the centre of town, but one which she does not own, Thérèse is very conscious of being an 'outsider', a 'marginal'. But who then are the insiders?

Dreams of long journeys abroad haunt her, with a group of artists, creating beauty, dreams of a balanced life (no longer on her feet thirteen hours a day, smiling), dreams of having her own restaurant, or of setting up a department in the city hall, to bring people together who have never met. . . . Her starting point is that she has an inner wound, but simply to accept its existence is not enough: she has to transform it by doing something positive.

Her way out of timidity is to be an intermediary helping others out of theirs. But it is not only people who suffer from timidity who are finding their vocation in becoming intermediaries.

Surrounded by high waves, in the middle of the North Sea, a team of French engineers are constructing the first offshore oil-rig platforms. Nobody is sure whether it will work, but everything has to be done fast. The boss tells a woman in her early twenties that there is a problem at the top of the tower, sixty metres high. She climbs it without blanching. She knows he is only testing her, that it is a sort of initiation, and she accepts the challenge, because her ambition is to become a member of what she calls 'the industrial elite'. Her grandfather and father were in top jobs; she was told only boys could become engineers: 'I wanted to prove them wrong.' Catherine Delcroix has never suffered from timidity.

Her husband is a member of the elite, too, the manager of a large electronics factory. Her daughter aged six is already showing signs of joining the élite, competitive and passionate about getting first prize at her smart school, to which she was sent 'for elitist reasons, I do not conceal it, to give her a better chance in life'. Having become deputy managing director of a major engineering firm, she confessed: 'I work to earn my living, but above all to be socially recognised. A housewife is not recognised: however useful the work she does for society, she is regarded as unemployed. If one is not paid, one is nothing.' She is doing more than avoiding 'the intolerable position of being supported by my husband', and ensuring that nobody can make decisions on her behalf; she wants a kind of freedom which gives her the right to do the work she likes doing, the right to be listened to when she speaks. Her career seems to say, 'Only the people at the top are truly free.'

'I have always refused to think that there are any obstacles in my way just because I am a woman.' After graduating in maritime engineering, she spent four years pioneering new technology on Mobil's Helmdale oil field: 'It was like living in the Far West, among cowboys, and I loved it. I was unusual in being a woman, but I simply had to use that peculiarity, in a good way. That meant especially making sure that no one should have a hold on my private life, and obtaining respect from the men.'

Becoming a member of the elite involves knowing what is wrong with it, but taking the view that it has the power to find solutions, and not being discouraged when the solutions produce unexpected disasters. The joy of being an engineer is building something out of nothing. 'You start with an idea from a client, who is not sure what he wants, and two or three years later you make it come true. This is what has kept me going as an engineer.' She has confidence in technology: it is true engineers have ruined, for example, the Aral Sea: 'One asks oneself how could they have made such terrible mistakes?' Her answer is, 'One must try and avoid repeating the mistakes. The growth of industry is inexorable; its purpose has been to make profits, but also to raise the standard of living, to improve communications, to yield fascinating new knowledge.' Maybe people are no better off from the psychological point of view, no worse either, but it is impossible to go backwards, to return to the land. People must simply apply their intelligence to ensure that well-meaning sophisticated plans do not produce disasters. Technical solutions are possible. Governments are not just obstacles in the way of initiative; they have a good side, because they do not put profit first. Thanks to technology there is no need to queue in the morning to buy a theatre ticket for the evening; you can book by minitel. 'People are too nostalgic.'

However, now, 'because I am a good engineer, I do something else'. Being at the top no longer means giving orders, but rather being an intermediary. As a manager she spends much of her time negotiating with trade unions. 'I used to believe that truth is a technical truth. Now I realise there is a human truth.' She was never trained to think about human behaviour; she is learning management by doing it, not by reading books on management, which could usually say what they have to say in a couple of pages. Being a member of the management élite puts one at loggerheads with others in it too. The people with most power, she thinks, are those who have been trained to look at the world as a series of systems, not of individuals, and who can therefore produce systematic solutions: that is why graduates of the National School of Administration and not engineers run companies. She prefers opportunism, listening and being adaptable is her method. 'I want relations to be cordial, even if my reputation is that I am sometimes a little distant. That is false; when I am walking in a corridor I am thinking about other things.'

Three years later I find her no longer organising and negotiating with 300 engineers, but promoted to be technical director of her whole group, with the task of making it more efficient in the face of economic crisis. Now the problem is that everybody in the firm has different ideas. 'I am astonished by how freely the engineers speak, which we never did when we were young. They have opinions, which we did not. The young ones want to become managers straight away, and to make others do the practical work for them.' Increasingly engineering is concerned as much with people as with technique. Catherine Delcroix has ended up as an intermediary between people and the facts of life, ensuring that her engineers maintain their morale, and that the young ones do not lose heart because recession offers fewer opportunities for grand adventures.

Being a woman is sometimes an advantage, she says, because it allows her 'to make jokes', but sometimes a disability, because 'a woman has less authority'. A Texan client, after shaking everybody's hand except hers, assuming she was a secretary, then asked a stupid technical question, and 'I didn't restrain myself from showing him how technically ignorant he was. He took it well, realising how insulting he had been; but even if I had been a secretary, that would not have been a good reason for not saying good morning.'

The price of being at the top is working very long hours and organising everything around one's profession: when offered an attractive job which involved her being in Paris while her husband worked in Nantes, she accepted, and they lived separately for four years. 'I spend very little time with my daughter.' There is no point in complaining. If she worked part-time, she would never get such an interesting job. Women who can afford help in the home should be grateful that they have only psychological problems. Those who cannot afford help have no solution: she is very sorry for them, but that is the way life is. Fortunately, there are women who prefer not to take employment and who like looking after children. 'It is good we are not all the same.'

Neither too confident nor too sceptical, Catherine Delcroix regrets that she has little religion: 'I would like to be certain about more things.' She feels some unease about having put her daughter in a Catholic private school: 'I would not like it to give her too many certainties.' However, Catholics teach 'spiritual values, like self-sacrifice', as other religions do too, and she believes in spiritual values. A woman said to her, 'Thank goodness we have faith, otherwise we wouldn't do anything.'

Catherine Delcroix's faith in success is as perfectly balanced as an oil rig, a genuine feat of careful engineering. It includes nothing absurd or impossible, but just enough of the personal to give it elegance. What would she like to do that she has not done? Write books. It is only here that her confidence reaches its limit: 'I don't feel I have enough talent; I don't have enough strong ideas; I wouldn't like to write like Jules Romains; I couldn't even write like Colette, who had no ideas and could only produce delicious descriptions of life. My books would have no value.' But she adds, 'Writing is the most interesting creative art of all.' She finds time to read. Even in her first year at engineering school she got through the whole of Proust in three months; and ever since she has read the classics, one author at a time. 'It helps my understanding of the world.' On holiday, skiing, gathering mushrooms, 'there is always something new to discover', but 'I don't know whether I could live free all the year round'.

The élite, today, is an intermediary between nations. Brought up in Toulouse, people assume she is a Toulousaine. No. 'I come from no particular region of France; I belong to no region, neither did my grandfather or father, who moved around where their work took them.' Her husband comes from a family very rooted in the north of France, and she felt a foreigner when she married into it. Her culture and language, she says, make her French, not European; she has never met anybody who calls himself, above anything else, a European; but she has no difficulty with people from foreign countries: 'Indeed, one becomes fossilised unless one has such relations. I would not worry about taking a job in New York or Timbuktu. I would not mind living abroad. In fact I prefer working with foreigners. It is more amusing, it offers discoveries.' For seven years she was employed by a German company's subsidiary in France. 'I like German culture and I can speak the language.'

Sometimes the élite has also been the intermediary that gives a helping hand to those who are less fortunate. Fifty years ago, her grandmother, married to a factory manager, devoted most of her spare time to good works among the poor; but today that is discredited as 'paternalism'. Today, in the suburbs just beyond Catherine Delcroix's offices, there are slums, illiteracy and street violence, but a director of an industrial firm, fully occupied rushing from appointment to appointment, has no time to think about, let alone solve, the problems of society. 'In this firm, we deal with the firm's immediate problems. I have not thought about what we could do about external social problems. I have never had this kind of conversation with my colleagues.' But now the head of the firm has organised a conference on the relations between industry and government; it is not only parliament which is an intermediary between the citizen and the state. The well-to-do have become timid in their relations with poor people, who are strangers and to whom they dare not speak.

There used to be a clear division between what the timid and the brave could achieve. All the most prized rewards went to those who gave orders, while those who did what they were told were more or less despised. But there is a third kind of activity in which the timid and the brave can be on equal terms. Intermediaries can achieve more than their own personal talents permit. Mice can sometimes move mountains. That is why being an intermediary offers more hope than trying to dominate others, or fighting to have one's merits recognised. Dropping out is not the only alternative to the rat race. However, to appreciate why so few people have thought of themselves as intermediaries, even when that is what they were, it is necessary to dig deeply around the roots of ambition.

Until recently, most ordinary people were expected to work in one of two ways, as peasants or as artisans. That, said Luther, was how they best 'pleased the Creator'. However, priests tried something different themselves. They were the first to become intermediaries, and they won enormous prestige negotiating between human frailty and divine strength, even when they were not themselves particularly brave. Then merchants set themselves up as intermediaries too, but they did less well: for a long time they were held in suspicion, because they lacked magic powers and did not know how to inspire ordinary people's imagination. When famine was an ever-present danger, they were blamed for shortages and for selling at exorbitant prices. Their god, Hermes, was a trickster and thief. Plato laid it down that it was impossible to engage in trade and be virtuous at the same time, though his Academy was founded by a merchant. St Thomas Aquinas said merchants were bound to have trouble attaining salvation, because the temptation to sin was inherent in their occupation. In China, merchants were officially classified at the very bottom of the social scale, inferior to both peasants and artisans; in India, only members of a segregated caste were allowed to engage in the dirty business of moneylending. Everywhere, retailers were despised because they had to be obsequious to customers, no matter who they were. When Napoleon called the English a nation of shopkeepers it was like calling them a nation of pimps. It has taken intermediaries about twenty-five centuries to be appreciated.

That happened rather suddenly. A new vision of the universe was needed before it was possible. Intermediaries are another example of how seeing a problem in a different context alters one's attitude to it. Nobody knew until the nineteenth century how two substances could combine to form a third one. The guess was that they must have something in common, an affinity, a sympathy – objects were spoken of as though they were alive. Newton called this affinity 'sociability'. It was as though objects could have love affairs; Goethe borrowed from the chemistry of his day to call one of his books *Elective Affinities*, meaning that couples were made for each other. Fontenelle marvelled at the way a substance, having united with another, then quit it to unite with a third: the adultery of objects was as mysterious as that of humans. It was only in 1835 that Baron Berzelius of Stockholm introduced the word catalyst into chemistry, having observed that these combinations often required the presence of a third party. He did not know how catalysts worked. But the third party became suddenly vital.

The idea of catalysis gives intermediaries a new status. Previously, they were mere links or hyphens, supplying needs felt by others. As catalysts, by contrast, they have an independent existence and purpose: they can create new situations and transform people's lives by bringing them together, without having any arrogant pretensions themselves. To be a catalyst is the ambition most appropriate for those who see the world as being in constant change, and who, without thinking that they can control it, wish to influence its direction.

Until the businessman could be imagined in such a creative role, he remained modest, tagging behind those who had more prestige, forgetting his ambitions as soon as he was rich enough, retiring to the countryside to become a landowner, dreaming of marrying his daughters to aristocrats, masquerading as an art collector, admiring other occupations more than his own. His central ideals were his own prosperity and that of his family. When he played a public role it was to support causes which boosted his self-esteem but still did not give him a central, independent importance. His choices remained erratic: sometimes he backed Protestantism, which said usury was all right, and sometimes royal absolutism against the nobility, which was the enemy of all upstarts like himself; currying the favour of governments, he did not disdain to collect taxes which he hated, if he could keep a cut of the takings. In Japan, even when he combined with others in large firms and exerted a powerful influence on the state, he did so indirectly. In all cases, the businessman remained discreet.

The first world religion to show any enthusiasm for commerce was Islam. The Prophet Muhammad had been engaged in trade, and his first wife was an important businesswoman in the very mercantile city of Mecca. The Koran said, 'Merchants are the messengers of this world and God's trustees on earth.' 'The markets are God's tables,' added Al-Ghazali. The Muslims were the first to produce a book in praise of commerce, *The Beauties of Trade*, by Ja'far b. Ali ad-Dimishqi, in the twelfth century, arguing that trade is 'the best of all gainful employments and the most conducive to happiness'. Islam's extraordinarily rapid expansion over half the globe was a commercial as well as a religious victory. Baghdad and Cairo became the most luxurious of all cities during the European Dark Ages, and a visit to them was the equivalent of tasting the delights of Paris five centuries later.

So Muslim merchants had no need to defend mercantile values against religion. On the contrary, they intermarried with the clergy-scholars (*ulama*), who engaged in trade also. An example is Abdallah al Sarqawi (Rector of the Azhar in Cairo from 1793 to 1812) who, after starting life in poverty, grew exceedingly rich in this most senior of religious posts, because he was also a businessman, while his wife excelled as an investor in property, shops and baths. Trade in the Muslim world was a branch of sociability, to be savoured as much for the pleasures of human contact, conversation and haggling as for the monetary profits.

In the eighteenth century, Europeans also began to change the merchant's image; intellectuals now saw him as an ally against aristocracy. Voltaire portrayed English businessmen as models of adventurousness and honesty. Playwrights presented them as the silk threads uniting nations by trade, bringing peace and prosperity, truly 'universal men'. The French Revolution rewarded them with freedom from government interference, but even after it, merchants, except the super-rich, still preferred to push their sons into state jobs or the professions. Only in recent years has the expansion of the service sector (another incarnation of intermediaries) suddenly made true in the West what the Sincere Friends of Basra, who wrote an encyclopedia in the tenth century, observed in their own world, when they said, with some exaggeration, 'Everybody is either a craftsman or a merchant.' Today the majority of jobs in Western countries are in services, and more than half of these are held by women. The idea of being an intermediary has become less remote.

However, its past aberrations still haunt it. Even in the USA, which liked to repeat that America's business was business, rejecting aristocratic values and finding it acceptable to compliment a priest by telling him that his sermon was businesslike, the intermediary had trouble in becoming the national hero. The reason was that the money-makers, having uneven imaginations, continued to model themselves on kings, building empires, forgetting that America was intended to be, and originally was, a land of independent farmers and small entrepreneurs. Big business changed that, transforming most people into salaried employees. Americans were originally thrifty; big business persuaded them to become consumers instead, insatiable, richer but more vulnerable. The business moguls and tycoons (titles of the emperors of Hindustan and Japan) preached that 'every American has the right to do anything short of direct robbery or murder', which was how the *New York Graphic* defended the unscrupulous millionaire Cornelius Vanderbilt. When his son William H. Vanderbilt was asked in 1882 whether he ran his railroads in the public interest, he replied, 'The public be damned. We like to do everything possible for the benefit of humanity in general, but we do first see that we are benefiting ourselves.'

So millionaires were not treated as gods in America, but only as proof that anybody could be rich, justifying faith in a democracy which unfortunately was a long way from making everybody rich.

The mere creation of wealth was never enough: ideal Americans also had to be philanthropists; and yet when John D. Rockefeller died in 1937, there was a public debate about whether he would go to heaven or not, despite his vast charitable benefactions; something else was needed in addition. Though businessmen influenced governments, no businessman as such has ever been elected president, the nearest being Herbert Hoover, who was a mining engineer. The great depression of the 1930s shattered the reputation of businessmen as miracle workers. After the Second World War, they felt obliged to launch one of the largest advertising campaigns in history, spending a hundred million dollars to persuade the nation that they incarnated the ideology of Free Enterprise. The campaign was a flop, because Americans wanted security too. In 1958, 700 corporations combined to offer the masses the reassurance they seemed to be asking for: *Management Creeds and Philosophies*, published by the American Management Association, put forward the new image of Business Responsibility, caring about every social issue, from the handicapped to symphony orchestras. Gradually businessmen felt their way towards a new role as mediators between conflicting interests. But suspicion of them was slow to decline.

American management science has been a constant search for a business ideal that is more than that. A whole variety of formulae were invented to guarantee higher profits, but there has always been an elusive goal beyond them. Multinationals, having become more powerful than many nations, had to find a purpose other than their own aggrandisement. The ruthless robber barons were replaced by managers offering efficiency, by leaders offering inspiration, then by manipulators of networks who could achieve a hitherto impossible consensus. 'Management by objectives', the creed of the 1950s, 'organisation and development' (in the 1960s), 'corporate culture' (in the 1980s) were successive slogans designed to create a sense of unity among employees. But in the 1990s the idea of directing a firm's energies in royal style, by issuing orders, is discredited: 'Every individual must feel unique'; autonomy for each little group has been found to be more productive; the manager's job is to thrive in a chaotic world he cannot control. He is at last reconciled to being, openly, an intermediary.

Of course, there are skeletons in every historical cupboard. Intermediaries have often been corrupt, criminal, greedy, abusing almost every profession, exploiting almost every class, and it is foolish to idealise them. There is no occupation, no relationship, which has an unblemished record; there have even been saints who used their influence to manipulate others. No institution has ever remained honest without personal virtue, which each individual has always had to cultivate carefully like a rare plant liable to become extinct at any moment. But being an intermediary does provide an opportunity for people who feel that they have limited talents or resources to add something positive to the world, particularly a world which seeks equal respect for everybody.

Intermediaries have often adopted that vocation because persecutions or exclusions prevented them from following other careers, so for a long time their lot did not seem enviable. The Armenians and the Lebanese, for example, have produced supreme intermediaries, at the cost of immense suffering. The Greeks became the world's leading shipowners because the Ottoman occupation shut off their opportunities on land. Jews were the major importers and exporters of ideas and goods across the curtain that separated Islam from Christianity in the Middle Ages: Westerners have largely chosen to forget what an immense force Islam was within a century of the Prophet's death (AD 632), dominating not only Spain and North Africa, but Asia right up to the Punjab; and the skills of the Jews benefited both civilisations. Moses Maimonides (1135–1204), born in Cordoba and long resident in Old Cairo, where he was court physician to the Sultan Saladin and his seventeen sons, was not only the most revered exponent of Judaic belief, but also a supremely rational and courteous reconciler of faith with science, of divine creation with an everlasting world; Islamic judges sought out his opinions and Christian universities used his works as textbooks. His *Guide for the Perplexed* offered a way out of 'irresoluteness': 'Character consists in keeping out of the way of fools,' he wrote, 'not in conquering them. . . . I seek no victory for the honour of my soul.' That modesty, that rejection of the military dream of annihilating enemies, is the essential attitude of the intermediary. Maimonides made it his principle to avoid angry disputes: 'Even when men insult me, I do not mind, but answer politely with friendly words, or remain silent. . . . I do not maintain that I never make mistakes. On the contrary, when I discover one, or if I am convinced of my error by others, I am ready to change anything in my writings, in my ways, or even in my nature.' He encouraged women to study, seeing books – which are of course intermediaries too – as humanity's essential nourishment.

The westernisation of the world greatly expanded the numbers of intermediaries living between two cultures or two economies. The Portuguese empire in the East relied on native middlemen called *compradors* (known in the Middle East as dragomans and in China as *mai-pan*). When in 1842 the British won the Opium War against China, they likewise sought out native 'experts in barbarian affairs', who spoke pidgin English, to manage their business locally. What distinguished these men was that they were merchants who were content to remain merchants (unlike the traditional Chinese merchants who used their profits to buy themselves the status of 'gentry-merchants'). They sent their sons to Western-style schools, refused to have their daughters' feet bound, wore Western clothes, neglected to master the Confucian classics, became cosmopolitans and opened the door to new ideas, though they were usually more imitative than creative in using those ideas. They subsidised Sun Yat Sen and the overthrow of the Chinese empire in 1911. But they did not become European, counterbalancing their services to Western capitalism by a strong nationalism. They have played a major role in history, but unobtrusively. Of course, governments have been suspicious of intermediaries whose interests cross national boundaries, and at various times it has been a criminal offence even to speak to foreigners. However, as travel becomes a universal pastime, a new era is opening for intermediaries.

Most advances in science have been the result of intermediaries venturing beyond the boundaries or the paradigms of their disciplines, uniting insights which come from different kingdoms of knowledge. Musicians have probably been the most important intermediaries of the emotions, bringing together people whom mere words divide.

Chemical catalysts are still something of a mystery: exactly how they get two separate substances to interact is not fully known. It used to be thought that they remained unchanged during reactions, but it is now believed they absorb a small portion of the substances they transform, and in so doing lessen the amount of energy needed to allow a reaction to begin. It was only in 1926 that the catalysts in living cells – enzymes, which are the indispensable regulators of chemical reactions in the body, controlling the digestion of food and the release of energy – were proved to be actual substances, not just properties of cells. In the gradual process of discovering how their activity is controlled, it emerged that sometimes an enzyme is inactive, until converted into activity by another enzyme. That is how blood clots: two enzymes have to combine.

Intermediaries may need other intermediaries to set them off. That is a new way of looking at the world, as a series of minute interreactions in the presence of others. It means that force is no longer in total command. It means also that the humble or the timid can contribute to great adventures without being too concerned as to who is superior to whom: a minute ingredient can have as much effect as a large one. Intermediaries inject an element of the unexpected into human affairs, which can have negative as well as stimulating results; and they are always tempted to demand too high a price for their efforts. But they flourish when they please all parties equally, when they oppress nobody.

The historical model of the intermediary is Maecenas, who died in 8 BC and whose name became synonymous with generosity. He was a rich Roman businessman who greatly increased his wealth by using his connections with the government, arranging the emperor's marriage, reconciling him with his rivals, negotiating peace with his enemies, using great charm, simplicity, cordiality, treating everyone he respected as an equal: 'Sleeplessly vigilant in emergencies, but in his relaxation from business more luxurious and effeminate than a woman,' said a contemporary. His pleasure was to encourage the poets of his day, even though everyone laughed at his own literary efforts. Aware that he owed his fortune to his friendship with the emperor, he bequeathed it all to him when he died. His method was essentially personal: he enjoyed the company of those he helped; it was a mutual relationship. That is what distinguishes intermediaries: they work at the individual level. It is impossible to have an army of intermediaries to destroy disagreement. Not everybody can be a leader, but everybody can be an intermediary.

Maecenas, however, is an incomplete model. The name of the mathematician and engineer Archimedes of Syracuse (287–212 BC) needs to be coupled with his. He is anecdotally remembered for jumping out of his bath and running naked through the streets of Syracuse in delight, shouting 'Eureka: I have found it', when he suddenly understood why his body felt lighter in water; but he deserves to be remembered outside mathematics too, because he used rational thinking to make difficult tasks easy, and small implements to move great weights, as in the famous screw he invented, in the lever, in the catapult, in the cog wheel. 'Give me a place to stand on,' he said, 'and I could move the earth.' Intermediaries follow that principle: the way for the weak to move the strong is not by force but by modifying their relationship, changing the angle of approach. When the Romans invaded Sicily, and a soldier entered Archimedes' house to arrest him, the mathematician asked him to wait while he finished solving a problem: the soldier was impatient and ran a sword through him. The trouble with the method of intermediaries is that it requires a great deal of patience and, above all, an ability to cope with fear.

P. D. Curtin, *Cross-Cultural Trade in World History*, Cambridge UP, 1984; Alice Teichova, *Historical Studies in International Corporate Business*, Cambridge UP, 1989; Johannes Hirschmeier and T. Yui, *The Development of Japanese Business 1600–1973*, Allen and Unwin, 1975; David Nye, *Image Worlds: Corporate Identities at General Electric 1880–1930*, MIT Press, 1985; P. L. Payne, *British Entrepreneurship in the Nineteenth Century*, 2nd edn., Macmillan, 1988; Katrina Honeyman, *Origins of Enterprise: Business Leadership in the Industrial Revolution*, Manchester UP, 1982; Hélène Verin, *Les Entrepreneurs: Histoire d'une idée*, Vrin, 1980; Janet L. Abu-Lughod, *Before European Hegemony: The World System A.D. 1250–1350*, Oxford UP, NY, 1988 (on medieval trade and cities); Yen Ping Hao, *The Comprador in Nineteenth-Century China*, Harvard, 1970; Susan Mann, *Local Merchants and the Chinese*

Bureaucracy, 1750–1950, Stanford UP, 1987; Denys Lombard, *March-ands et hommes d'affaires dans l'océan indien et la mer de Chine 13–20e siècles*, EHES, 1988; David G. Lo Romer, *Merchants and Reform in Livorno, 1814–68*, California UP, 1987; Oladipo Yemitan, *Madame Tinubu: Merchant and Kingmaker*, Ibadan UP, 1987; J. M. Thomasseau, *Commerce et commercants dans la littérature*, Bordeaux UP, 1988; André Raymond, *Artisans et commercants au Caire au 18e siècle*, Damas Institut Français, 1974; Maxime Rodinson, *Islam et capitalisme*, Seuil, 1966; Leila Tarazi Fawaz, *Merchants and Migrants in Nineteenth-Century Beirut*, Harvard UP, 1983; J. D. Tracy, *The Rise of Merchant Empires: Long-Distance Trade in the Early Modern Period, 1350–1750*, Cambridge UP, 1990; C. D. Sheldon, *The Rise of the Merchant Class in Tokugawa Japan*, Augustin, NY, 1958; Jennifer Alexander, *Trade, Trad-ers and Trading in Rural Japan*, Oxford UP, Singapore, 1987; Alfred D. Chandler, jnr., *The Visible Hand: The Managerial Revolution in Ameri-can Business*, Harvard, 1977; Richard Scade, *Reluctant Managers: Their Work and Life Style*, Unwin Hyman, 1989; S. M. Lipset, *The Confidence Gap: Business Labour and Government in the Public Mind*, Free Press, NY, 1983; Leon Hollerman, *Japan Disincorporated*, Hoover Institution Press, Stanford, 1988; Hiroshi Tanaka, *Personality in Industry: The Human Side of a Japanese Enterprise*, Pinter, 1988; Julia Davies and Mark Easterby Smith, *The Challenge to Western Management Develop-ment*, Routledge, 1989; Ronnie Lessen, *Managing Corporate Culture*, Gower, 1990; William Byrt, *Management Education, International Survey*, Routledge, 1989; Philippe d'Iribarne, *La Logique de l'honneur: Gestion des entreprises et traditions nationales*, Seuil, 1989; Earl F. Cheit, *The Business Establishment*, Wiley, 1964 (for earlier history); Sigmund Diamond, *The Reputation of the American Businessman*, Harper, 1955; Calvin A. Kent, *Environment for Entrepreneurship*, Lexington, 1984; John P. Kotter, *The General Managers*, Free Press, 1982; George A. Steiner, *The New CEO*, Macmillan, 1983 (all this for changing fashions); David Osborne, *Laboratories of Democracy*, Harvard Business School, 1988; Bennett Harrison and Barry Blackstone, *The Great U Turn: Cor-porate Restructuring and the Polarising of America*, Basic, 1988; David Vogel, *Fluctuating Fortunes: The Political Power of Business in America*, Basic, 1989; Peter Halbherr, *IBM: Mythe ou réalité*, Favre, Lausanne, 1987 (disillusioned); Renaud Sainsaulieu, *Sociologie de l'organisation et de l'entreprise*, FNSP/Dalloz, 1988; Franck Gauthey, Indrei Ratiu, Irene Rodgers, Dominique Xardel, *Leaders sans frontières: Le défi des différ-ences*, McGraw Hill, 1988; Gérard Bordenave, *Ford en Europe*, 1988; Alain Ehrenberg, *Le Culte de la performance*, Calmann Levy, 1991; Ros-abeth Moss Kanter, *The Change Masters: Corporate Entrepreneurs at*

Work, Unwin, 1988; Serge Moscovici and Willem Doise, *Dissensions and consensus*, PUF, 1992; R. Francès, *Motivation et satisfaction du travail*, PUF; Renaud Sainsaulieu, *L'Identité au travail: les effets culturels de l'organisation*, 3rd edn., FNSP, 1988; David Le Breton, *Passions du risque*, Metaillé, 1991; R. G. Streets, *The Impact of Service Industries on Underemployment in Metropolitan Economics*, Lexington, 1987; P. D. Anthony, *The Idea of Work*, Tavistock, 1977; Dawliffe Hall Educational Foundation, *Work versus Family*, 1991 (on experiments in part-time work); S. N. Eisenstadt and E. Ben Ari, *Japanese Models of Conflict Resolution*, Kegan Paul, 1980; K. Kressel, *Mediation Research: The Process and Effectiveness of Third Party Intervention*, Jossey Bass, S. F., 1989; Deborah M. Kolb, *The Mediators*, MIT Press, 1983; Susan J. Pharr, *Losing Face: Status Politics in Japan*, California UP, 1990 (on decision-making, 'privatising conflict'); Roland Calori and Peter Lawrence, *The Business of Europe: Managing Change*, Sage, 1981; James J. Hunt, *Leadership: A New Synthesis*, Sage, 1991; Craig R. Hickman, *Mind of a Manager, Soul of a Leader*, Wiley, 1990; Elizabeth Chell, *The Entrepreneurial Personality*, Routledge, 1991; M. Masayuki Hamabata, *Crested Kimono: Power and Love in the Japanese Business Family*, Cornell UP, 1990; Nick Oliver and Barry Wilkinson, *The Japanisation of British Industry*, Blackwell, 1988; Alan Rosenthal, *The Third House: Lobbyists and Lobbying in the States*, CQ Press, 1993 ('lobbyists aim at compromise so that they can fight another day', but grassroots movements 'less likely to compromise'); Lawrence S. Rothenberg, *Linking Citizens to Government: Interest Group Politics at Common Cause*, Cambridge UP, 1992 (on the problems).

Jack C. Ross, *An Assembly of Good Fellows: Voluntary Associations in History*, Greenwood, Westport, 1976; '*Espaces et temps associatifs*', issue of *Revue de l'économie sociale*, April 1988; Sofres, *Enquête Associations*, Jan. 1988 (showing more optimism about what associations can achieve in personal relations than in the organisation of work); Raoul de la Grasserie, *Des Intermédiaires sociaux*, 1908; Gilbert Moinier, *Le Rôle des intermédiaires dans la société moderne*, Paris thesis, 1924 (complaining about the increase in their numbers); Aix Colloque, Université de Provence, *Les Intermédiaires culturelles*, cyclostyled 1978 (on pitfalls, and on healers, singers, spokespersons); I. Epstein, *Moses Maimonides*, Soncinollen, 1959; M. R. Hayoun, *Maïmonide ou l'autre Moïse*, Lattès, 1994; A. J. Herschel, *Maimonides: A Biography*, Farrar Strauss, NY, 1982; M. Maimonides, *Guide for the Perplexed* (Eng. translation) 1919; E. J. Dijksterhuis, *Archimedes*, Munksgaard, Copenhagen, 1956; A. M. Laulan, *La Résistance aux systèmes d'informations*, Retz, 1986; Howard Good, *Outcasts: The Image of Journalists in Con-*

temporary Film, Scarecrow, Metuchen, NY, 1989; S. R. Lichter, *The Media Elite*, Adler, NY, 1986; T. R. Hewitt, *Advertising in Britain*, Heinemann, 1982 (for its origins); Oliver Schwartz, *Le Monde privé des ouvriers*, PUF, 1990; *Dialogue* has many relevant articles, *Cahiers du Grif* also, from the feminist perspective; Joel A. Tarr, *Technology and the Rise of the Networked City in Europe and America*, Temple UP, 1988 ('cities evolve to facilitate communication'); Michel Leiris, *African Art*, Thames and Hudson, 1967 (art as operational rather than for contemplation, as intermediary between the divine and the mundane; weaving as symbolic of creativity); Howard Becker, *Art Worlds*, California UP, 1982; Leo Spitzer, *Lives in Between: Marginality in Austria, Brazil and West Africa, 1780–1945*, Cambridge UP, 1989; Alfred W. Crosby, jnr., *The Columbian Exchange: Biological and Cultural Consequences of 1492*, Greenwood, Westport, 1972; Christopher Butler, *After the Wake: An Essay on the Contemporary Avant-Garde*, Clarendon Press, 1980.

10

How people have freed themselves
from fear by finding new fears

Women who are caressed by the smooth knitwear of Givenchy do not feel the hand of Nina touching them, though it is she who has sewn their garment and put on the buttons. There is no chance that they might ever catch a glimpse of her if they visit the great designer's showrooms, because she works in a factory over 200 miles away. Nina indeed has never been to Paris, let alone been invited to watch the models parade her clothes. 'If I were given a chance to attend a fashion parade, I'd be uncomfortable, because I'm a nobody; I'm only a worker, *une petite*, I count for nothing, I earn the minimum wage. I'd feel inferior in front of people with money, because they're superior to me. Money's everything. I'm uncultured. I don't know how to make conversation. It's only high society that goes to fashion shows.' But Nina is an elegant woman, warm and lively beneath her reserve.

She longs to escape from her grim factory, which she entered at seventeen, and she is already twenty-nine. 'I'm bored stiff with it.' Does she get no pleasure at all from her work? Not much, apart from the pride of being distantly associated with a great name. Sociologists may say that workers enjoy the social life of factories, but Nina complains precisely that there is no human contact in hers. She has borne it because she does not know where else to go.

'I have no diploma. I can do nothing else. I never liked studying. I think I could do other things. I was talented at school – not a genius, but I could have gone on to higher education.' She regrets that she left early in order to earn money. But then she draws back. 'It must be very hard at university: I wouldn't be capable of it.' Psychology interests her; she reads a lot of books, or at least 'easy books because I'm incapable of reading difficult books'.

I say many psychology books are incomprehensible to anybody.

'I didn't know that,' she says.

'It is not your fault that they are impossible to read,' I say.

'That never occurred to me.' Besides, she has trouble writing. 'To think and write at the same time, that's something I can't do. I couldn't do it at school. They told me I lacked imagination. [No teacher's insult is ever forgotten.] It takes me a very long time to write, because I'm afraid of making mistakes.'

Nina denies that she has any ambitions – and hates herself for saying so. 'I find myself ordinary.' Her mother, who also works in a garment factory, tells her she ought to be content with what she has, because many people try to better themselves but fail. 'I'm afraid of failing,' says Nina. 'Failure haunts me.' If the impossible dream did become reality, if she did go and study psychology, what would happen if she failed? If she gave up her job, and found herself with no work, what would she do? In Roanne, where old factories are closing down, and few new ones are opening, a job is almost life itself. 'I'm contented, by force of circumstances.'

And yet how can she be content walking past shops, unable to afford the clothes she would like, even though she has a job? 'I am tempted, and I cannot afford to buy.' It is not a fortune that she dreams of: her wage is the lowest an employer is legally allowed to pay. She would be content with just double that, which would bring her roughly to the national average. And yet, would she be content? She left school to earn money, thinking that 'with money one can do anything', but now she says, 'It isn't true.'

She is like a bird in a cage, fluttering her wings, banging against bars, collapsing and endlessly repeating her attempt to fly away. After talking of her ambitions, she remembers that ambition is ruled out for her, because she is always anxious, she has always been so: 'I brood, I ruminate a lot, about little nothings. I keep thinking about the same things, and about fear, and about the fear of death, which frightens me a lot. I was even worse when I was young. I keep asking myself why we should come on earth and then have to leave it. Why? It isn't fair. I can't accept it. But it's life. Losing a person one loves is horrible. I'm a Catholic; I shouldn't think this way; but I'm sceptical about life after death. I lost my grandfather two years ago – the first loss I've suffered. I find it awful not being able to see him again.'

Fortunately, she has her parents, but her psychology books have confirmed her in the belief that she is an anxious person because they are anxious too. 'All this comes from my childhood.' Her father is a tiler, now retired, who has had a very hard life, and whose suffering so depressed him that he would sit in a chair for months without speaking: 'It was terrible.' He has been in hospital, and that has affected her too. But she insists, 'I am not unhappy, because there are people worse off than me.' To the question why some people have happier lives than others, she has no answer.

She would like to have a family, but 'I'm reserved, suspicious, jealous'. The man of her dreams would be completely reliable. There is no shortage of other sorts who are attracted by her good looks. When she goes to public balls, men ask her to dance, buy her drinks, talk about themselves and then suggest they go to bed. 'I'm not a girl of the streets. These men aren't interested in me, but in sex.' The trouble with men is that they are liars. Sometimes they are already married and do not reveal it. Sometimes they make promises, and then she finds them going out with other girls. 'I want a man who belongs to me. And I don't forgive. I hold grudges. I haven't found the man of my life. And I won't marry the first man who comes along, because marriage is for life.' It would not matter if he was poor: someone on the fringes of show business told her that the rich are corrupted by money. It is 'sincerity, kindness and loyalty' that she wants.

Friends she does have, 'very few', whom she recognises by their support when things go badly, and who are not just interested in her company for the fun. These friends confide in her, tell her 'their most intimate secrets' and obtain comfort from her; she is proud of that. Honest, heart-to-heart talk warms the soul: 'My parents don't talk much.' She was afraid to talk to me, but she also wanted to; and she waited a long time for me until I had finished talking to others.

However, her conclusion was 'My life is a failure'. I interpret that to mean 'Only if I escape from my fears will I begin to live as I want to.'

The history of fear over the centuries shows that liberation from fear has from time to time been achieved, by two methods. The first has been with the help of fear itself, by escaping from one fear to another, which contains more hope. The second has been through curiosity about something quite different, which has temporarily blotted out the awareness of danger.

The supposedly fearless Vikings provide an early illustration of how this happened. They were the terrorists whom Europe feared most between the eighth and twelfth centuries. They managed to brave the seas to pillage, ransom and create havoc from Constantinople to Lisbon and Dublin, even though they carried inside them all the usual fears of poor peasants, as well as the loneliness of Scandinavia's long nights. They set out on these voyages of dangerous adventure because they felt an even more unbearable fear than did their neighbours who stayed behind, for they were tortured by the thought that their name and reputation, and not just their body and soul, might vanish into nothingness. To live just so that they could retire to paradise, which they believed offered only a ceaseless alternation between battles and feasting, did not seem worthwhile. Everyone has to die, it was written in the Sayings of the Most High (the god Odin):

> *Wealth dies*
> *Kinsmen die*
> *And you will die too:*
> *But I know one thing*
> *Which never dies:*
> *The verdict on every dead man.*

So an immortal reputation became the Vikings' goal: nothing

could be worse than to be forgotten, the respect of others was for them the sweetest form of riches. Death in battle ceased to be frightening, when it was seen as an opportunity to show self-control in the face of danger, and accepted with equanimity, regarded as trivial compared to the glory that could be won by dying with dignity. The Vikings were brave from fear of being despised, and that fear made them forget all other fears.

But this induced another fear, not of sinning, for the Vikings did not try to earn respect by being saints or sages, but of saying the wrong thing. They did not put their faith in gods, whom they imagined to be preoccupied by difficulties of their own: self-reliance was their ideal, obstinacy and stoicism in the face of all challenges, imperturbability. Their first commandment was therefore to hold their tongues. When King Harald the Ruthless (1015–66) wanted to pay a supreme compliment to someone, he described him as unmoved by sudden events: 'Whether it was danger or relief or whatever peril loomed, he was never in higher or lower spirits, never slept less or more, and never ate and drank save according to his custom.' The Viking theory was that there could be no useful purpose in showing fear, which would mean that they had lost their independence.

When they arrived in Normandy, they were asked what they wanted and they replied, 'We come from Denmark and we want to conquer France.'

'Who is your chief?'

'We have no chief; we all have equal authority.'

Would they agree to swear allegiance to Charles, King of France?

'We shall never submit to anyone, whoever he may be. We shall never accept any servitude.'

In fact they did have a chief, Rollo, but they considered him as only the first among equals, having chosen him themselves, because they thought he could best lead them to victory. When they found a haven for themselves in Iceland, they set up one of the most astonishing republics ever known, a sort of democracy reconciling the fear of losing their self-respect – which obeying a king would imply – with respect for others. If one of them lost a quarter or more of his possessions, they agreed that the rest would make good half his loss, but each person was limited to paying out only one per cent of his own wealth, and nobody could be compensated more than three times. They took their decisions in a general assembly, at which wives (who kept their own surname) and children could be present. They had left Scandinavia because they disliked being ordered around by kings, and so they created a society based on the fear of being ordered around and on the deeply egalitarian conviction that every person could win everlasting fame.

Iceland's extraordinarily rich literature shows how some of them achieved that goal, to remain individuals, each one different. If it is true that 'Viking' comes from the word meaning 'to withdraw', which some believe and others dispute, they were the first people to be proud to be 'marginals'. They paid a high price, in violence, for their freedom, the men with their quick passions, their unhappy loves, their jealousy of their fame, constantly protecting themselves from criticism with sword or satire, the women sometimes acting as 'peaceweavers' in the endless disputes about honour, but also proud and vindictive, urging the men to bloodshed. Nevertheless the Vikings did establish that there was no need to be beautiful or unconquerable to be famous. The god Odin was neither: one-eyed, fragile, impenetrable, devious, using ruse and magic as much as force to survive, with even his masculinity sometimes in doubt, dependent on women to keep him informed of what went on in the world, Odin was responsible for the unpredictable. The Vikings could face all other fears ultimately because, as rebels, they discovered how to turn the natural fear of the unpredictable into a source of inspiration. That is the clue they bequeathed, which nobody noticed.

Today, the Vikings' fear has become an epidemic: an ever-increasing proportion of humanity is worrying about what others think of them, no longer about how their ancestors might react looking down from heaven, nor what will be said about them in the history books, but how every act, every day, will be criticised and judged, both by those who know them and by those who do not. Creating a false impression is the modern nightmare. Reputation is the modern purgatory. The more a society thinks of itself as democratic, the more reputation matters in it and the more fear of other people's criticism, however petty, becomes obsessive: an American survey claims it is the fear which is most troubling of all. It is no accident that advertising and public relations have become the basis of business, politics, entertainment and even religion.

Harold Macmillan recalled how he ignored all dangers in the First World War, until one day he found himself separated from his troops and suddenly understood how they had made him appear brave. 'When one is in action, especially when one is responsible for men under one's command, proper behaviour, even acts of gallantry, are part of the show. One moves and behaves almost automatically, as a member of a team or an actor on the stage. But now it was all over; I was alone and nobody could see me. There was no need to keep up appearances, and I was frightened.'

On top of the fear of being found out, modern society is of course also tormented by the fear of crime, of being burgled or raped, or of going out alone at night, the fear of unemployment, illness, drugs, immigrants, war. So it is worth looking into the deeper roots of these fears, which meet up with other roots one does not expect to find.

For most of history, a person whose life was made miserable by fear was not thought of as having much chance of becoming courageous. Courage was seen as an exceptional gift, found in knights and martyrs, while ordinary people were assumed to be too weakened by poverty to conquer fear, and heroism in fighting the ordinary disasters of daily life went unacknowledged. So far, being a member of a civilisation has never freed humans from more than a few fears, for civilisations have always had a sense of being surrounded by hostile forces, and they have specialised, each in a different way, in concentrating attention on a certain number of dangers, promising protection from them, but seldom eliminating them, offering no more than an alleviation of unease by providing more or less plausible explanations of these dangers. New fears have perpetually sprung up to replace the ones that have gone out of fashion, just as cancer and Aids have replaced tuberculosis and syphilis.

It was civilisation which turned the gently murmuring sea into a dangerous abode of demons and monsters, and which prophesied that the waves and black storm clouds would soon sweep humanity away: that was Dürer's terrifying picture of how life on earth would end. Of course, people had some cause to be afraid of the sea, for as late as 1854, in a single year, the British fleet alone suffered 832 shipwrecks. But civilisations have trained the imagination to transform occasional disaster into a constant nightmare. Knowledge has not extinguished unreasonable fears, because it has also supplied new ideas for possible future catastrophes. The dead were raised up in imagination to haunt the living and to wreak revenge. That explained why things were going wrong, but the price was constant fear that the spirits were not being propitiated in the right way, or that tradition was being disobeyed. Learned experts never stopped adding frightening explanations, but did not apply to themselves their idea that the 'cold' moon, 'pale with fury', turned humans into lunatics.

Natural disasters were made much more dreadful by being attributed to supernatural forces. Fear of the devil was deliberately, almost lovingly, encouraged by those who claimed to understand how the world functioned. In Europe there was a great epidemic of terror at his machinations in the eleventh century, another in the fourteenth, and panic in the sixteenth, when the Germans were particularly affected: 'In no country in the world has the devil exercised a more tyrannical power than in Germany,' one commentator wrote in 1561. The advantage of attributing misfortunes to the devil was that he gave one the feeling that one understood one's misadventures; but his presence also produced emotional crises when one saw him too often, wherever one turned. Many harmless people were persecuted for being his agents. Dangers were multiplied by the mere thought of the tricks he played. Today the world may appear to be more densely populated than it was five centuries ago, but that is to forget the millions of devils, gnomes, goblins, monsters and wicked fairies who used to haunt it. They set up a habit of mind that there must always be someone to blame, some evil force to fear and to attack. Satan still exists for 37 per cent of the British, for 57 per cent of practising French Christians and for 20 per cent of non-practising ones; those who saw him incarnated in Marxism have still to decide who the successor will be.

Purgatory illustrates the way cures for fear created new fears. In order to diminish the terror of damnation in hell, the Catholic Church began suggesting, from the twelfth century, that sinners could purge their offences in a less awful place; but that only transferred worries to the pains of purgatory. Then the Church diminished the fear of purgatory by granting indulgences, which shortened one's stay there, but which gave an awesome power to the clergy. Charlatans took to selling indulgences, for the demand exceeded the supply; people then had to worry whether the indulgences would be effective, which made them think of the torments of purgatory with even greater frequency. To quench these fears, the Church encouraged processions, fraternities, benedictions, exorcism. More and more saints were set up as specialists in warding off disease (no less than ten of them, for example, could eventually be approached to cure syphilis). From the fourteenth century onwards, each individual was led to believe that a personal guardian angel watched over him or her, however sinful. But all this protection only increased consciousness of the dangers from which one needed to protect oneself.

When the strain became intolerable, there was an explosion. At a stroke, the Reformation abolished all these safeguards against fear, in the hope of abolishing fear. All sincere believers who repented were assured that there was a place for them in heaven. This was one of the world's most important revolutions, a revolution against fear, which lasted several centuries. The Christian God's image was completely changed, from a terrifying and angry tyrant demanding total obedience, to a merciful, infinitely kind father. The threat of eternal punishment was abandoned. Most Christians threw both hell and purgatory into the dustbin.

However, when religion stopped frightening people, they invented new fears to frighten themselves, as though they valued fear as a necessary part of the sensation of being alive. It was the only way of being alive that they could imagine. Since the eighteenth century, security has become, almost universally, the official goal for this life, but an unattainable goal, a paradise, ever harder to locate, invisible in a cloud of doubts. The American Constitution proclaimed the right to security, which meant the right to have no fears, but in vain. Psychoanalysts declared security to be necessary for the attainment of normality, but few people believe they are entirely normal. Insecurity has become the commonest complaint of our time. Odin the Unpredictable is no longer admired.

It is true that after 1762 when the first insurance company, the Equitable Life, was founded in London, it became possible to eliminate fears about the financial consequences of natural disasters. Living in towns, illuminated at night and guarded by police, diminished (for a time) the fear of violence, while prosperity and the welfare state reduced the number of those who feared famine, homelessness, illness, unemployment, old age. Nevertheless, the present generation spends far more money insuring itself against these fears than ever its ancestors paid churches or magicians for protection.

Children have shown in their nightmares that the old fears are not forgotten, that the ogres of the night still live, even though adults have become preoccupied with new ogres. The civilised have developed, meanwhile, with the help of literature and medicine, ever more subtle kinds of unease, greater consciousness of their own inadequacies, fear of both failure and success, refinements in the art of self-torture. Every time a religion or a political creed collapses – as most recently in Russia – and people are left spiritually naked, they are more than ever vulnerable to fear.

And every time a new disease is discovered, hypochondriacs add it to their list of fears. Medical students gave the lead, and have remained the worst victims; then popular health manuals spread the gospel of diseases hiding in every crevice, just like sin. Cotton Mather (1663–1728), who studied medicine at Harvard, before being involved in the witchhunts of Salem, confessed, 'I was myself troubled with almost every disorder that I read of in my studies, which caused me to use medicines upon myself that I might cure my imaginary maladies.' In the nineteenth century invisible germs replaced invisible spirits; doctors, dependent on rich patients, were said to be deriving five-sixths of their income from the treatment of imaginary complaints. Even the most gifted men found hypochondria irresistibly consoling: Tolstoy's wife wrote, 'Hour after hour, from morning to night, he thinks of nothing but his body and how to look after it.' Many marriages were doubtless strengthened in the process, like Darwin's, whose wife was able to be endlessly maternal. 'We are so fond of one another because our ailments are the same,' said the Goncourt brothers, valuing sickness as a sign of sensitivity. Hypochondria diluted fear with hope, and provided an alternative way to Know Thyself. The American woman who got into the textbooks for having made 548 visits to 226 physicians and 42 medical students, obtaining 164 different diagnoses, never despaired.

Trying to eliminate fear, rather than simply exchanging one fear for another, has produced strange results. In the course of the last hundred years, at the same time as the human race moved into cities, the Norwegian rat became a permanent resident of scientific laboratories. It has been transformed, by careful breeding over three hundred generations, into a more docile, less fearful creature, and this has been achieved even though almost its only function now is to suffer stress, run through mazes, press buttons to avoid electric shocks, and have bits of its body amputated. It has thus been proved that fear can be diminished even when dangers have grown more terrible.

However, a remarkable physical change has taken place in this rat as a result of its being less frightened. It now mates at any time of the year, which it never used to when it was wild and aggressive. Coping with danger has been replaced by copulation. Its adrenal glands have shrivelled to half or a quarter of their former size, and their functions have been taken over by the sex glands. Sexual activity as an alternative to hostility is not peculiar to rats: it has been observed in some monkeys too, who, when they use sex as a way of making peace, do not ejaculate – almost an equivalent of human sex with contraception. But more sex has apparently not made the rat happier; the more offspring it has and the more densely populated its abode, the more difficult its life becomes. And the female rat has been more fully emancipated from fear than the male, because the male persists in having a highly competitive hierarchy. This does not involve the female, which seldom fights either males or other females. The male rat remains a victim of the rat race, determined to dominate, but nearly always having to accept submission to, and fear of, the one who wins.

By contrast, cats, which lead solitary, independent lives without hierarchies, do not show any difference in fearfulness between males and females. It is only when hierarchies involve both sexes, as happens with chimpanzees and humans, that females become as fearful as males or more so. Though the Norwegian rat lives in laboratories as in a hotel, receiving regular meals, and though it has seemingly discovered the principle 'Make Love not War' of its own accord, it could not bring itself to abandon its social pretensions, and so it has not found the formula for a life without stress.

It used to be believed that the stomach was the seat of the emotions, but what actually happens in the stomach when fear is felt – and one vomits, or has butterflies, or has any of the many disabling sensations that the stomach can produce – only became clear in the 1950s. The hero of this discovery – patients too deserve to be remembered for their bravery – is known only by his first name, Tom, an Irish American who in 1895, at the age of nine, drank a bowl of clam chowder so scalding hot that his oesophagus was destroyed; unable to swallow, he had a permanent hole made in his stomach into which his meals could be poured. He chewed food for pleasure, but then spat it out into the funnel inserted into this 'fistula'. Apart from being unable to eat in public, he led a normal life as a stage manager, plumber, mechanic and sewer worker, until he was given a hospital job by Professor Stewart Wolf of Oklahoma, who spent many years peering into the hole. There had been similar investigations, particularly of wounded soldiers, for over a century, but this was the first time that the relationship between the stomach and the emotions was studied. Because Wolf followed every detail of Tom's private life and discussed his every thought, the stomach revealed itself as far more interesting than the heart, the supposed seat of emotion, whose monotonous pumping has nothing particularly human about it.

Every trace of fear could immediately be spotted in Tom's stomach. When, for example, his daughter installed a telephone, worry at the expense this would entail made his stomach bleed. It was not just frightening events which sent the stomach into a rage of activity, or made it go on strike, refusing to work, turning as pale as a terrified face, but also mere conversation about an unpleasant event, even a distant memory. The stomach reveals that humans conceal an enormous amount of emotion, which churns in their insides. It not infrequently attempts to commit suicide, letting its acids burn away the stomach wall; emotions have been observed to cancel the effect of medicines, which the stomach refused to absorb when it was angry. Other patients with fistulas of the same kind have since confirmed the sovereignty and sensitivity of the stomach: an Italian tax collector, to give but one example, showed a perfect concordance between the ups and downs of his private life, alternating between rage, excitement and terror, and the waterfalls of acid in his stomach. The stomach makes sure that people know when they are frightened.

However, sometimes, it can be and has been made to keep quiet. Humanity's triumph over its stomach first seemed possible in the eighth or seventh century BC, in *The Teaching of the Great Forest* (*Brhadaranyaka Upanishad*), which said, 'All fear is groundless.' The Indian war against fear took almost 2,000 years to develop its tactics fully, for it was only in the tenth to twelfth centuries that physical hatha yoga was invented, in the northern region later famed for its austere Gurkha troops. Yoga, as a system of preventive medicine based on a programme of physical and mental exercises, made it possible to control voluntarily functions which are normally automatic; as part of a spiritual discipline, Hindu or Buddhist, it prepares the way for the extinction of individual self-consciousness. 'Fear,' said the book, 'arises when there is another.' So if there is no other person in the world, there can be no fear. Exercises, under the guidance of a guru, taught the disciple that his individuality was an illusion and his soul part of the universal soul. The price of banishing fear, in other words, is to cease to be a person in the normal sense. The true follower gives up every possession, often surviving in a state of semi-starvation, triumphing over the stomach, increasing the pause between each breath, between exhaling and inhaling, to as much as five minutes, concentrating on a single object of meditation, until the mind 'dissolves'. Yoga is not for the weak, but many millions have calmed their nerves and their stomachs with milder, simplified versions of it. Moderating the symptoms of fear makes it easier to decide what it is one will fear.

The importation of these ideas into the West was limited by their association with mysticism and with doctrines like reincarnation; their over-simplification in the 1960s discredited them. But now that they are being used in the rehabilitation of the paralysed, who can be taught to modify their heart rate and blood pressure at will, so that they can learn to move again, the relations of the mind and the body, of fear and the stomach, may no longer appear as unchangeable as was once believed.

The view that fear is a natural animal reaction, about which nothing much can be done, is unfair to animals, many of whom are at times surprisingly lacking in fear. It may well be true that some individuals are born more fearful than others, or that the temperaments of babies are established at a very early age. The latest research, for example, claims that blue-eyed children have a particular tendency to being 'inhibited, which may be due to pre-natal presence of higher levels of norepinephrine' and that they remain inhibited beyond childhood, 'with unprovoked bouts of anxiety'. But that is no more a cause for despair than the discovery, a century ago, that some children were better at passing intelligence tests than others. It took some time to discover that intelligence quotients were not enough by themselves, that success depended on how the intelligence was used. Similarly, the division of humanity into introverts and extroverts was followed by the realisation that creativity was not guaranteed to either, that both qualities were needed for it and that the combination did not necessarily have to be in the same person: the meeting of two individuals has often triggered more originality than genius in one. Again and again, chance encounters and minute details have given very different fates to people who appear to belong to the same category.

'In the past few years,' writes a psychologist, 'important advances have been made in our ability to reduce human fears. Under controlled conditions, it is now possible to produce substantial and lasting reduction of established fears, even life-long fears, within thirty minutes.' But modern medicine cannot agree on the best remedy for fear, hesitating between drugs, the retraining of habit and the resolving of unconscious conflicts, and it will doubtless never be able to deal with fear as it dealt with smallpox: it can often attenuate fears, but each individual needs a different answer, and there is no knowing whether that answer will work, or for how long an alleviation will last. Moreover, medicine cannot stop people inventing new fears. It is not surprising that the enormous amount of scientific information accumulated about fear – each decade produces over 10,000 new studies, in the English language alone – has not lessened the demand for magic and unorthodox healers. Many people prefer to remain ignorant, to stick with the fears they know, to shut their eyes to statistics which tell them what they ought to fear. The psychologist Martin Seligman of Philadelphia, who specialises in converting pessimists into optimists, arguing that fear is essentially a sense of helplessness or a loss of hope, and that optimism can be taught, has observed that even in the USA not everybody wants to be an optimist: those who hold power, particularly just below the top echelon, tend to be pessimists, because it suggests wisdom; there are many advantages in cultivating fear. Nobody can cure those who want to perpetuate their fears, for fear that something even more terrible might happen to them. And some fears, in mild form, have been put to good use. Being claustrophobic is inconvenient, but being intellectually claustrophobic can be the beginning of originality.

The most important discovery of science about fear, however, is that its physical symptoms, in terms of the chemicals produced to defend the body, differ only in degree from those of curiosity. That makes it easier to understand how people racked by fears have been able to escape them or forget them under the influence of curiosity, of a preoccupation with some goal which so absorbs them that they behave as though they were profoundly courageous. Because the deep roots of fear are linked to those of curiosity, the fact that virtually nobody is born uniformly brave in every sphere of life does not have to mean that people are forever stuck with their own brand of cowardice. It is possible to surprise oneself. In former times, such an opinion would have sounded absurd, just as it used to be thought absurd to believe that a poor person could be the equal of an aristocrat. Only in an age of universal education and travel is it possible to appreciate the crucial importance of curiosity, which I shall discuss in my next chapter.

In 1991, a study comparing depression in men and women claimed that men predominantly sought relief in distractions, and so appeared to suffer less, while women favoured rumination, turning their troubles over endlessly in their minds, making themselves more miserable as a result. The supposedly emotional sex was found to give more time to thinking than the supposedly rational one. The contribution of women to the limitation of fear has only begun, for rumination does not necessarily lead to gloom, nor to panic, where thoughts go round in circles, stirring up dangers into imminent catastrophes and terror at the prospect of being afraid. Rumination can also hold fear at a distance, and choose what it ruminates about. Nobody is free who does not have that choice.

Vikings

Peter Foote and David M. Wilson, *The Viking Achievement*, Sidgwick, 1970; Judith Jesch, *Women in the Viking Age*, Boydell Press, Woodbridge, 1991; Regis Boyer, *Le Christ des barbares*, Le Monde nordique, Cerf, 1987; Lucien Musset, *Le Peuple scandinave au moyen âge*, PUF, 1951; Jon Johannesson, *A History of the Old Icelandic Commonwealth*, Manitoba UP, 1974; Aurélien Sauvageot, *Les Anciens finnois*, Klinckseick, 1961; Kirsten Hastrup, *Culture and History in Medieval Iceland*, Oxford UP, 1985; Kirsten Hastrup, *Nature and Policy in Iceland, 1400–1800*, Oxford UP, 1990 (an anthropological analysis of history and mentality); Eric Graf Oxenstierna, *The World of the Norsemen*, Weidenfeld, 1967; Maurice Gravier, *Les Scandinaves . . . des origines à la Réforme*, Lidis-Brepols, 1984; Jean L. Briggs, *Never in Anger: Portrait of an Eskimo Family*, Harvard, 1970.

Medicine of fear

Isaac M. Marks, *Fears, Phobias and Rituals*, Oxford UP, NY, 1987; S. J. Rachman, *Fear and Courage*, W. H. Freeman, NY, 1990; J. A. Gray, *The Psychology of Fear and Stress*, 2nd edn., Oxford UP, 1987; James C. Ballenger, *Clinical Aspects of Panic Disorder*, Wiley-Liss, NY, 1991;

Rebecca C. Curtis, *Self-Defeating Behaviour*, Plenum, NY, 1989; Martin Seligman, *Learned Optimism*, Knopf, 1991; Susan Nolen-Holksema, *Sex Differences in Depression*, Stanford UP, 1990; Margaret T. Gordon, *The Female Fear*, Free Press, NY, 1989 (61 per cent of female city dwellers in the USA fear to go out alone at night in their own neighbourhoods); Susan Baur, *Hypochondria*, California UP, 1988; C. S. Gelernter, 'Cognitive Behavioural and Pharmacological Treatments of Social Phobia', *Archives of General Psychiatry*, 48, Oct. 1991, pp. 938–45 (showing 69 per cent improvement by phenelzine drugs, 24 per cent by cognitive therapy, and 20 per cent with placebos, but the cognitive therapy is more long-lasting); Gerald L. Klerman, 'Birth Cohort Trends in Rates of Major Depressive Disorder', *ibid.*, July 1985, pp. 689–719 (showing increase through the twentieth century); Dwight R. Kirkpatrick, 'Age, Gender and Patterns of Common Intense Fears among Adults', *Behavioural Research and Therapy*, vol. 22, no. 2, 1984, pp. 141–50 (in Indiana, at that time, 30 per cent of women did not fear spiders or snakes, 28 per cent of female adolescents feared they may be homosexual, while men feared God, taking written tests and the effects of masturbation); John G. Carlson and A. R. Seifert, *Perspectives on Self-Regulation and Health*, Plenum, NY, 1991; John V. Basmajian, *Biofeedback: Principles and Practice for Clinicians*, Williams and Wilkins, Baltimore, 1989. This is a small sample: the number of facts about fear is terrifying.

History of fear

Jean Delumeau, *La Peur en Occident*, Fayard, 1978; Jean Delumeau, *Le Péché et la peur*, Fayard, 1983; Jean Delumeau, *Rassurer et protéger*, Fayard, 1989; J. J. Barloy, *La Peur et les animaux*, Balland, 1982; Elie Griguer, *Guide des dangers et des risques quotidiens*, Lefeuvre, 1980; George Pickering, *Creative Malady*, Allen and Unwin, 1974; Denis Duclos, *La Peur et le savoir*, La Decouverte, 1989; James M. Burns, *Roosevelt: The Lion and the Fox*, Harvard UP, 1963; Vivian Worthington, *A History of Yoga*, Penguin, Arkana, 1989; Jean Varenne, *Le Yoga et la tradition hindoue*, Culture, Arts, Loisirs, 1974 (Eng. translation, *Yoga and the Hindu Tradition*, Vehic, 1976; Georg Feuerstein, *Encyclopedic Dictionary of Yoga*, Unwin, 1990.

How curiosity has become
the key to freedom

'What a privilege to be able to talk about oneself,' says Mauricette, aged sixty-seven, a retired garage owner, but her talk is about how interesting she finds other people. When she recalls her Parisian grandfather, who was a horse-cab driver, she is sure he must have had a life of wonderful adventure: what could be more exciting than to meet people one does not know, to take them to a destination one could not guess? When she talks of her father, a commercial traveller in electric light bulbs who in 1919 set up one of the first shops to sell spare parts for motor cars, she acclaims him as a pioneer. He died when she was in her early twenties; she abandoned the philosophy thesis she was writing to take over his firm; doubling its size, adding a garage, she may have been one of the first women car mechanics. Curiosity has always been her prime inspiration.

After forty years in business, however, the challenge of having to satisfy customers every day began to pall. When she noticed her employees suggesting changes to her – whereas previously it was always she who used to urge them to change, and who repeatedly transformed the shop window – she realised she was no longer a pioneer. When she found her once faithful customers deserting her for the sake of the few francs they could save by buying a spare part in a supermarket, when she began waking up in the morning thinking, 'Damn, I've got to go to work', she knew a phase in her life was over.

So she sold the firm and at the age of sixty-three started a second life. 'Retirement – I hate the word – leaves one all alone in one's own bubble. Mine was too small for me, because I am curious about everything, I am like a collector of butterflies. I was made for commerce, because I am convivial; though the days were sometimes long, they were never empty; I was always learning something. I enjoyed contact with my customers. If there had been vocational guidance in my youth, I should never have embarked on philosophy, though I do not regret it, for it has widened my outlook. I wanted to try my hand at every job, to know everything. The shop had been my life only during the day. I never learned to cook, never shopped, always had help at home, always went out in the evening. I love celebrating, eating, using my hands, but also my mind, and I am curious about both religion and psychology.'

Now she has replaced work with several distinct sets of friends. Her oldest friend is a hairdresser, who began very modestly and gradually built up a business which was the largest and most fashionable salon in the city: 'She was a real boss.' Mauricette admired her for being everything she herself was not, always perfectly dressed and made up: 'When I wear a scarf, it just covers my neck, but with her it is like a cloud, wonderfully chic: to me she represented femininity, which attracts me, while I represented seriousness and equilibrium to her.' This is the friend with whom Mauricette has travelled all over the world, including China, Japan, Sri Lanka and the USA. Her own country does not interest her, because what she seeks in travel is to get away from the familiar, to experience amazement, which is no easy ambition when 'television wears away the capacity to be astonished'. But recently the hairdresser also retired, and became a different person, no longer a smart, sociable woman of the world, but a recluse, unwilling to go out. Mauricette cannot understand why, for she herself has become more outgoing than ever.

One set of friends are for having fun with, going out to restaurants and trips to Belgium, and a different set are for serious talk about serious subjects, with regular meetings to discuss

books: Teilhard de Chardin at the moment. 'I am involved in a
spiritual quest, though untroubled by anxiety.' The Catholicism in
which she was brought up was superficial, an excuse for socialising; but now that she has both more and less time, she is beginning
to think about eternity. 'I am not ready for death – though I am not
afraid of it – because I have not finished growing, nor finished my
search for God, nor my discovery of other people. I am no sister of
charity, but I like people, within limits; and I am curious about
them.' Does her pleasure in meeting new faces when she does
philanthropic work mean that she is really doing it for selfish
reasons? She worries occasionally about that, and occasionally
about being comfortably off in a world full of poverty; but she
tries to make her curiosity useful to other people. Returning to
the university to study 'morphopsychology' – which teaches what
people are like – enables her to meet younger students overwhelmed by problems of family and unemployment. She helps out
at the library for the blind, and enjoys talking with those who use
it. On other days, she is a volunteer at the railway station, assisting
the down and out. Her most fulfilling work, however, is at a home
for prostitutes – 'I knew nothing about them, paying no more
attention to them than to stray dogs in the street, but when I
discovered this home by chance, I became very interested by how
one becomes a prostitute, a double person. I look after two of
them, and learned how parents kick their children out when there
are too many mouths to feed, knowing they will end up in brothels. I treat these prostitutes as people, I do not judge them. One
of them said to me, "You have laughing eyes, and that does me
good." That is because I am conscious of being happy. Many
people have reason to be happy, but do not know it.'

The voluntary work, says Mauricette, has transformed her
appearance. 'I have an austere face, but now I smile in the street. I
used to smile when I had my garage, but it was a mask, the mask
of the shopkeeper being affable to customers. But when I smile
now, it is not a mask; rather, it is when I do not smile that I am
wearing a mask, when I present myself with my selfish mask on, to
people who have always known me that way – it would be too
difficult to change: the world of commerce is selfish. Now what I
regret in my past is the sins of omission: I passed by things which

should have made me explode, and I did not notice. I am trying to improve.'

All Mauricette's friends are widows, spinsters or divorcees. She never married. 'I sometimes think about the soul-mate I never met.' But no man ever seemed capable of helping her to blossom. 'I like going out with men, but I don't like being stuck with them, and have never loved a man passionately. Needless to say, I am neither a virgin nor a martyr.' Some of her friends would be shocked if they heard her saying that she would have liked to be a man, though she is happy to be a woman. Her reason is that she would like to know what men think, what they feel, for example, when they make love. Their behaviour she finds unbelievable, their emotions bizarre: how could Nazi men have executed people while playing Mozart? Men are often cowards too: 'They do not have the courage to leave their wives when they would prefer to be with another woman.' She sometimes speculates about how she might have behaved if she had married a man who became nasty to her: 'I do not mean who was unfaithful to me – unpleasant though that would be – but who deceived me morally, who used me, who obscured my vision. I ask myself whether I could have been a murderess, whether I could have killed him, whether I would destroy a person who tried to destroy me, who prevented me from living.' Leading a full life is much harder for women, she says, since society is organised for men, and women need greater will power.

Loneliness does not trouble her, nor regret at having had no children, because the shop was her child. In her seventeenth-century house, she enjoys the company of all those who once lived there, who in some way still live there. Her memories are not so much her own, as those of other people; she does not look at her family photos in the evenings, because they seem cold to her; nor does she think about her own past. Instead, every night, before going to sleep, she reads for about two hours, travel, biography, history, psychology, which forge new links with the world she does not know. 'We all of us have many facets, but only a few of these see the light. When I die, I shall say: Where are the other facets of myself?'

Life is becoming more and more like a shop, into which one can go 'just looking', to try on the clothes it sells, even if one has not got the money to pay for them. Actors spend most of their time doing just that, for themselves, and on behalf of their audience, finding out what it feels like to be someone else, and discovering that there is a piece of that someone else inside one. Everybody is to some extent an actor, but few have the chance to play many roles. Professional actors are most admired where freedom is most highly valued, because acting is an instrument of freedom, which enables people to realise that they are not imprisoned in themselves, but can understand others and be understood by them.

Charlotte Kady is still only a starlet, beginning to be known for her roles in Bertrand Tavernier's films. Previously she won more local fame as the presenter of a children's show on French television. She wants to be more famous still, because that is the only way to get a wider choice of roles. But what role is she capable of playing?

So far the world knows her mainly as a bubbling, happy girl. 'I created that image myself. I pretended to be fresh-faced, healthy in all respects, with no problems. But inside me there was anxiety.' Her childhood was indeed largely happy, except that she was enclosed by what felt like a barbed-wire fence. Her father was Algerian: a doctor, trained in France, who chose to settle in Lyon and marry a French medical student, and who during the violence of the Algerian War of Independence was forced into hiding for nine months at the very time of Charlotte's birth. At school, racism was the fence the children built around those who were different. 'I lied, pretended I was Turkish, not Algerian, because I had a grandmother who was Russian-Turkish. I was afraid I would be found out. I suffered enormously from the anti-Arab talk of people who thought I was on their side.' Even her mother's family refused to accept the marriage with an Arab. But of course it is useful for an actress to have two origins.

She had to fight for the freedom to be an actress. As a child, disguise and imitation were her favourite games, but in the respectable world of the provincial middle class to say that one wanted to earn one's living on the stage was still almost to declare an ambition to be a prostitute. Her teachers thought she should be a mathematician, but she spent her leisure reading every actor's biography she could find. So already she was two persons in one body. To find out what else she could be, or what else there was inside her, she announced that she wanted to study in Paris.

Every great city is a giant theatre, where provincials come wearing masks and try out new roles. Paris has been particularly successful because it has always encouraged its inhabitants to invent new roles. As the *Grand Dictionnaire Larousse* wrote in 1874, its 'high culture' was the product of 'continuous interbreeding and mixture'; today, still, only one in four of its inhabitants were born there. Since Charlotte's sister was an official in the Senate, she could live with her, which was respectable and her parents agreed. But in Paris, she studied not only at the university (becoming fluent in Russian, German and English), but also at a drama school, and so she graduated into advertising, the theatre and television. 'I still don't know how gifted I am, whether I will be an average actress or an excellent one. I won't find out till I have had more experience. I feel I have everything inside me, but it does not depend just on me whether I can draw it out. I need to be recognised by others. And that is partly a matter of chance.' Indeed she got her first job by chance: she was accompanying a friend to an audition; the director chose her, though she was sitting at the back and not an applicant. 'It has always been chance with me.'

Not only has she had a double life as a student, but with men she has played the role of both husband and wife. The first man she lived with was an actor who could not bear watching her succeed while his own career flopped. He said they should share their earnings – which meant her earnings. She agreed, on condition that he tried harder to get work, but increasingly he became accustomed to depending on her, as though it was natural that only she should be working. He spent her money, freely, buying luxuries of all sorts, using her credit cards, running her bank account into debit. 'Then I ceased to be blind and stopped loving him.' When she announced, 'I'm leaving, I don't love you', he threatened suicide, and stood on the parapet of the sixth floor till she promised not to abandon him. Only when her work took her away for three months did she finally gather the strength to go. 'He said he couldn't live without me, and grabbed me by the throat. I screamed. Fortunately people heard and came to my rescue.' Now, four years later, he is a changed man. 'I always told him that so long as we stayed together, he would never do anything but stagnate, and that our separation would do him good. He has had to fight for himself, and it has worked.'

Her second lover was more carefully chosen, a medical student who lived in another city and who belonged to a different world. 'I wanted to avoid being swallowed up in the typical life of a couple.' But he was determined to show her off, to make her behave like a wife who had a job only occasionally: 'He was unable to understand that my work is my passion. We were on different wavelengths.'

Being on television every day, being greeted by people in the streets, having articles about her in the papers, gave her, she says, not a swollen head, but maturity. Her partner in her programme, Emmanuelle Bataille (who later also had a part in Tavernier's film *These Foolish Things*), became her closest, inseparable friend. Their relationship is 'the equivalent of a love, as strong as that of a couple, without the sex, which makes me think that a relationship with a man has to be much more than sexual'. Emmanuelle is everything Charlotte is not: abandoned by her parents, shattered by a difficult childhood and by a marriage that ended in divorce, pessimistic, convinced she was ugly and uninteresting, she saw through Charlotte's outward jolliness, spotting the timidity and anxiety behind the rosy girl image: 'I am going to strip you of your mask,' she said. In return Charlotte helps Emmanuelle fight her moods. 'We are like two sisters, very intimate.'

Charlotte is a practising Catholic, but under Emmanuelle's influence her religion has become double too. Emmanuelle is a Buddhist as well as a Catholic. Charlotte has been convinced since infancy that she had a previous life; reading history books was like a reminder of a past she had lived through; in particular she believes she was a courtesan in the court of Louis XIV. 'If Mozart was able to compose at the age of five, he must have had the soul of someone else inside him. . . . As a girl, I thought I might have been a fly in a previous life, because it must be so terrible to be a fly.' It is not on the experience of a single life that she draws for her inspiration as an actress. Does that mean her life is decided in advance? 'We are in this world for a purpose, we have a mission to accomplish, we are endowed with qualities we can develop.'

Superstitious: she readily agrees she is so, very much, though she prefers to call herself 'intuitive', adding, 'I believe completely in clairvoyance.' Many years ago a clairvoyant made predictions about her which came true. More recently, another one astonished her by speaking of details of her private life which no one knew about: 'I am convinced they see things we cannot understand.' The prediction was that the man in her life would be someone much older than herself. Bertrand Tavernier is twenty years older, and it is with him that she now lives.

Tavernier has almost never made a film in which he does not scowl or growl at the Catholic Church, but she talks to him about her faith and he listens, saying nothing. He used to wait outside when she went into a church to pray; but after his father's death, on one occasion, he lit a candle and prayed beside her. He was very close to his father, another subject they do not discuss. But this relationship is working because Charlotte considers herself Tavernier's equal, even though he is so much more famous than she. Some people say she must be looking for a father, or that he is bound to crush her: but this is not so; she does not feel at all crushed 'simply because I am an obscure actress; I am often stronger than he; I feel strong where he has weaknesses'. She gives him what he lacks; his reserve is balanced by her openness; he says he owes an enormous amount to her. 'Because there is equality, we have an equilibrium.' None of the women he knew after his divorce from Colo O'Hagan could accept the fact that he was famous and they were not. He was worried that their relationship would suffer if Charlotte accepted a three months' engagement in a travelling repertory company, but she was determined to make her career on her own. They telephoned each other every day. Absence made them appreciate what they were missing. 'Leading two lives means you have twice as many things to say to each other.'

Actors cannot complain that they are too complicated; they almost have a duty to be complicated, to sustain their imaginations. 'Being an actress,' says Charlotte Kady, 'means drawing nourishment from the variety of life.' She contrasts herself with her older sister, who has an important official post and who is brilliant when she talks about problems of government, but whose confidence vanishes when the conversation turns to private life, about which she has little to say. The world of administration tries to separate itself by high walls from that of intimate feelings. To pass laws to announce what freedoms there should be is one thing. To discover what it feels like to be free is something quite different.

The most successful remedy for fear has been curiosity, but not just any sort of curiosity. Being interested in one's work, in a few hobbies, in a few people, leaves too many black holes in the universe.

A new lighthouse beaming curiosity appears every time a baby is born, and the world seems interesting again. The earth welcomed 78 million new lighthouses last year, but how many of them will have lost their fire two, three or four decades from now? In the past, very few humans have fulfilled their early promise or been able to function at more than a fraction of their capacity. They have always put on blinkers, and built walls around themselves, to keep curiosity in check. Individuals wanting to assert their independence have always had to fight to tear down the obstacles in curiosity's way. If being able to take an interest in anything is one of the conditions of freedom, then clearly freedom is not a fruit that is simply plucked from a tree.

The first obstacle to being curious has been the tradition that it is dangerous. Mythology is full of divine punishment striking down those who wanted to know too much; and in the Bible it is said, 'He that increaseth knowledge increaseth sorrow' (Ecclesiastes 1:18). Even the jovial, illegitimate cosmopolitan whom contemporary Europe has chosen as its hero, Erasmus of Rotterdam (1466–1536), the enemy of all dogmatism and war, insisted that curiosity must be restricted to the élite and that it must not be allowed to contaminate 'female chatterboxes'.

The first declaration of the rights of curiosity was made by Descartes (1596–1650), a profoundly restless man, who constantly moved house in search of peace of mind, from town to town, from country to country, but also a man deeply attached to the pleasures of lying in bed, having been granted the unique privilege by his schoolteachers, because of his delicate health, of getting up late: he kept this habit all his life, valuing bed as the best place to think, and it encouraged him to make thinking the very essence of being human. He made the historic and heretical pronouncement that curiosity was something that all people had, that nothing could prevent it, and that it increases inevitably with knowledge. Montaigne (1533–92) had explained how to use curiosity in ordinary life, urging his readers to probe 'the mystery of everyday things', and not to be afraid, when travelling, of 'the contagion of an unknown air'. Curiosity, added Thomas Hobbes (1588–1679), is 'lust of the mind', differing from 'the short vehemence of carnal lust' in that the more one indulges it, the more it captures one.

But most people remained wary of curiosity. In the late nineteenth century, even in the USA, 'pure curiosity', in the form of pure scientific research, was regarded as an unnecessary luxury, and businessmen were rarely willing to support anything but applied science, useful to the masses, the rest being 'idle curiosity'. Even in the 1950s American women thought that research scientists were undesirable husbands, not just because they did not care about money, but because they were not 'regular guys' with the same tastes and interests as everyone else.

The history of the spider shows that curiosity is still far from having won. The spider is one of the few creatures existing 250 million years ago to have survived unaltered. But though humans have had so long to get used to its six eyes and eight legs, they still find it difficult to take an interest in the problems that causes. They have done everything to avoid being curious. At first, they preferred to worship the spider: among Africans and among the Incas, the spider has been a god, a creator of the stars, an intermediary between the mortal and the divine. Indians imaginatively made it the symbol of liberty, it being the only creature that can raise itself up by its own bootstrings. In Siberia, Vietnam and Colombia, it carries dead souls to heaven. However, the Jewish prophets turned against it: Job thought its web too flimsy and Isaiah did not like its venom. Contempt turned into panic in the fifteenth century, when Italians became hysterical at the sight of it. Ever since, the majority of humanity has just not wanted to know what spiders do, or how indispensable they are to its survival, controlling pests, unappreciated. Even in the USA, dedicated to freedom, 70 per cent of women have been unable to free themselves of fear at the sight of a spider.

It might have been different had the spider been domesticated, like the bee, as it almost was. One Frenchman spun socks and gloves from its silk; another (the inventor of artificial incubators, Réaumur) started breeding them, but gave up because it was too expensive, their intolerance towards each other requiring that they should each be kept in a separate cage. The female spider has indeed a gift for living without males, since she can store sperm for up to eighteen months, and so tends to eat the male once he has served his purpose. But this did not arouse much curiosity among either male or female humans. Réaumur lived at a time when insects (though spiders are wrongly thought to be insects) were judged unworthy of being mentioned in an encyclopedia, even by the liberal Diderot. When Victor Hugo came to the spider's rescue, it was from generosity, not curiosity: 'I love the spider and the nettle,' he said, 'because they are hated.'

In order for curiosity about animals to be possible, people first needed to get rid of all the frightening myths they had invented about them, to stop the war they had waged on them from the beginning of time, and above all to cease imagining that all that lived existed for the benefit of the human race and it alone. That meant ceasing to find oneself more interesting than anything else. Only then could the idea that humans should not behave cruelly to animals be born. The role of children, and of those who do not forget their childhood, in creating that idea has gone unrecognised, cruel though children can be: an enquiry about the attitude of Parisians towards rats revealed that 80 per cent of adults were frightened of them, and only 6 per cent of children were. So perhaps the time has come for more curiosity to be shown about the rise and fall of curiosity.

Nothing has been more difficult than to be curious about an object or a person, without being obstructed by preconceived ideas. Occasionally the veil is lifted, and the one who lifts it is called a genius. However, the unveiling has happened often enough for it to be possible for ordinary mortals to see how it is done.

One of the first mysteries humans tried to solve was the meaning of the constant beat of their own hearts. That they were not stupid was soon shown, because by 2000 BC they had worked out that it was the pump which circulated blood through their bodies. But this was understood only in China, where the *Medicine Book of the Yellow Emperor* was subtle enough to distinguish between twenty-eight different kinds of pulse. Nevertheless these ideas remained veiled from general curiosity, wrapped up in the complexities of traditional Chinese medicine. Though humans are not stupid, they have usually been obstinately attached to their old ideas, not just from fear of the unfamiliar, but because an old idea is part of a system of thought, which is like a cobweb: every part sustains every other, and once you are in you cannot escape. China's discovery was ignored by most of the rest of the world, which was tied up in other cobwebs and which continued to believe tall stories about what the heart did, right down to the seventeenth century AD. Though a Cairo professor, Al Nafis (who died in AD 1288), passed the discovery on in a book which was translated into Latin, very few were prepared to listen or to look.

Both Christians and Muslims preferred to remain entangled in the cobweb spun by Galen (AD 130–200). Educated in Alexandria, and subsequently physician to a school of gladiators in Asia Minor, he maintained that the heart was not a pump but a sort of fireplace, producing the body's heat. His textbooks were memorised by virtually all doctors in the West and the Middle East for about a thousand years, and nothing they saw in their patients could stir their curiosity towards a different explanation. The reason was that Galen's ideas formed a coherent whole. Having been trained as a philosopher too, he showed doctors not only how to look at the body, and how to advise on nutrition, growth and vitality in a consistent way, but also how to behave as human beings and as physicians. Indeed, he set himself up as the model physician, never demanding a fee from his pupils or his patients, saying he practised from 'love of humanity . . . staying awake the greatest part of my nights not only for the sake of the sick but also for the beauty of study'. He lived modestly, possessing only two garments, two sets of household utensils and two slaves. Of course many doctors did not imitate this selfless dedication, but they cherished and needed the ideal. Their self-respect became bound up with his teachings.

Galen's mistakes survived even though he taught that one should not slavishly believe ancient books. If his disciples had listened to him more carefully, they would not have been paralysed by his ideas. Some people laughed at him, he said, because 'he moved slowly, and so they thought him a most distrustful person', which he was, and which he urged others to be. 'I do not know how it happened, miraculously or by divine inspiration, or in a frenzy, or whatever else you may call it, but from my very youth I despised the opinion of the multitude and longed for truth and knowledge, believing that there was for man no possession more noble or divine. . . . We must be daring and search after Truth; even if we do not succeed in finding her, we shall at least come closer than we are at present.' People who try to think for themselves know that the cobwebs they spin are fragile and incomplete; but those who are content to be disciples, and become entangled in the cobwebs of others, forget that fragility and imagine they have landed on firm, stable ground. Borrowed ideas, which were originally intended to be only gossamer, thus harden and fossilise; ideologies become dogmas, and curiosity, which should blow freely like the wind, suddenly becomes motionless. But it does not need to. Curiosity is not doomed to lose its freedom.

A few scientists did have the curiosity to look at the heart and see what it was doing, but they discovered how difficult it is to escape from a tightly spun cobweb of ideas. Andreas Vesalius of Brussels (1514–64), for example, devoted his life to studying the human anatomy, determined to think for himself. 'I studied without a teacher,' he insisted proudly, though he spent many years in the universities of Louvain, Paris, Bologna and Padua. He used to scour the roadsides for the bodies of executed criminals, their flesh picked half clean by the birds, and under cover of darkness carried them home in pieces to study. In Paris he spent long hours in the Cemetery of the Innocents turning over old bones, 'though gravely imperilled by the many savage dogs'. His students worshipped him, because he dissected corpses himself, instead of getting an assistant to do it; they became so enthusiastic – he called them 'the dear companions of my studies' – that they once kidnapped the body of 'a handsome mistress of a certain monk', snatching it from the tomb immediately after burial, flaying the whole skin to make it unrecognisable, then offering it to their master for dissection. With the corpses hung on pulleys, Vesalius' dissections were great theatrical performances, even though the stench was unbearable. A judge who attended was so impressed that he made available the bodies of convicts, sometimes delaying executions to suit the professor.

Vesalius showed that Galen's descriptions of the human anatomy were often wrong, the latter having based himself on deductions from the bodies of animals, mainly dogs and monkeys. Vesalius mounted two skeletons side by side, one of an ape and one of a human 'articulated from the bones of a French priest', to make the differences clear. And he produced a very beautiful book in seven volumes, *The Structure of the Human Body*, showing it in striking poses drawn by the pupils of Titian. But many people refused to believe him. If Galen seemed to be wrong, they said, it must be due to an error in the translation of his books, or to 'the degradation of the human species' since his time. Vesalius himself did not succeed in escaping Galen's cobweb either. There were still things he could not see, and which he imagined to be there, because Galen had said they were. He could not get rid of the invisible 'sooty vapours' which according to Galen the heart produced, nor did he notice the circulation of the blood. Even William Harvey (1578–1657), who eventually established this, remained entangled in even older Aristotelian notions about 'spirit' in the blood. He worried that some of his ideas 'are so novel and hitherto unmentioned, that in speaking of them, I not only fear that I may suffer from the ill will of a few, but dread lest all men turn against me. To such an extent is it second nature for all to follow accepted usage and teaching which, since its first implanting, has become deep-rooted; to such an extent are men swayed by a pardonable respect for the ancient authors.'

The blinkers which all humans wear do not fall off if they simply become more scientific. Thomas Kuhn has shown how most scientists work to reinforce the systems of thought which dominate their epoch, how they make new facts fit into these systems, or 'paradigms', or cobwebs, which usually take centuries to collapse. And now it is becoming clear that the limits which science places on curiosity have increased in recent times. Laboratories are not merely outposts in the war against ignorance. They are also fortresses in which specialists cut themselves off from the distractions of other forms of knowledge. Few scientists are able to reflect on the assumptions of their work, because the competition to obtain recognition is so intense; without the support of those in power, they cannot continue; and those in power are experts in politicking as much as in research.

The knots into which scientists tie themselves have been described by two anthropologists who visited an endocrinology laboratory and recorded every act as though it was part of the religious ritual of a tribe. They found a scientist claiming that Napoleon's campaigns provided the inspiration for his research methods: rival teams in different countries had been struggling against one another to solve a puzzle for eight years, and he spoke of his 'field' as 'a battlefield strewn with the corpses of competitors'. A physician joined this team, saying he was abandoning his medical practice not just to earn more money but in order to obtain 'a very rare commodity, recognition from my peers; I wanted positive feed-back proving my smartness – patients are not so good for that'. But recognition is hard to get: of the sixty-four learned papers published by the team, only eight were ever cited by anybody in the whole world. The vast majority of scientists devote a large part of their efforts to writing articles which are never read. They enter one field rather than another not just because they find it fascinating, but no less often because it seems to be less crowded, offering more chances of success, or because they find patrons able to help them obtain grants. The more organised, expensive and specialised science is, the more is individual curiosity fettered. Science by itself does not dispel fear.

It is only curiosity that knows no boundaries which can be effective against fear. However, around the beginning of the eighteenth century, that option was abandoned, and the ideal of encyclopedic knowledge was replaced by specialisation. Withdrawal into a fortress of limited knowledge meant one could defend oneself on one's home ground; it gave one self-confidence of a limited kind; but it left one helpless in vast areas of one's life, particularly the emotional part. Now that the silences produced by specialis- ation have become deafening, and now that information fills the air as never before, it is possible to reconsider the choice, to ask whether many people might not be better off if they began looking again for the road which leads beyond specialisation, if they tried seeing the universe as a whole.

The first pondered analysis of curiosity was made by Alexander von Humboldt (1769–1859), whose discoveries in physiology, zoology, botany, anthropology, archaeology, meteorology and geography (of which he is considered one of the founders) are probably unparalleled in range. What he did with his knowledge is even more interesting. Unlike Einstein, who used his genius to fight uncertainty, which terrified him, and unlike Hawking, who, says his wife, can find nothing with which to replace God except him- self – neither of whom have in any way changed the purposes or attitudes of ordinary people – Humboldt tried to extract a new way of life from his researches, abstract though some of them might seem. This is rare, because it conflicts with the rules of specialisation, which require one to keep one's mouth shut on subjects on which one is not a trained expert; and since nobody can be an expert on the art of life, it has become dangerous to speak about it. Intellectuals have therefore increasingly been limit- ing themselves to lamenting the lack of values in modern times. The importance of Humboldt is that he dared to make a link between knowledge and feeling, between what people believe and do in public and what obsesses them in private.

Humboldt was a pioneer of global thinking, without concealing that his purpose was not merely to understand the universe in its entirety, but no less to avoid the pain caused by the tragedies it constantly produces. His *Views of Nature* (1808) is dedicated to 'minds oppressed with care . . . [needing] to escape from the storm of life' adding, 'Let those who are weary of the clash of warring nations turn their attention to the silent life of vegetation . . . and remember that the earth continues to teem with new life.' So the way to get over the tragedies of existence was to view them from a more distant perspective, establish their interconnections and develop a sense of participating in a universal process. As Humboldt's brother Wilhelm, the eminent linguist, remarked, Alexander had 'a horror of the single fact', believing that 'in order to explore any one thing, one needs to approach it from all sides'. The reward, said Alexander, is that one feels 'in touch with the entire earth'. The romantic idealisation of life seemed futile to him, and the view that humans were nature's prisoners unacceptable: his goal was to point the way to freedom. Having a universal perspective meant, for example, rejecting racism, which imprisoned the minds of the majority of educated Westerners until after 1945; but Humboldt wrote firmly, 'There are no inferior races. All are destined equally to attain freedom.' Another unusual opinion he held was that without diversity of opinion the discovery of truth is impossible. And the object of knowledge of the truth should be 'not power but the enjoyment of life'. His research was directed therefore to 'awakening an understanding of all that is lovable'. Discovering harmonies compensates for the tragedies. He convinced himself that 'the single idea emerging from history' is 'the concept of humanisation, the tendency to break down the barriers of prejudice and religion, and the belief in mankind as one large community capable of evolving its inherent capacities'. Humboldt remained a prisoner of the cobweb of his age – naive optimism – failing to see that history could go backwards as well as forwards, but if his thought is extricated from this naivety, it remains substantial and powerful.

He was one of the most admired men of his time – American tourists took home little busts of him – not just because he explored unknown territory (Latin America and Siberia), nor because he came back with new information and thousands of unknown plants, but because he helped people to think about the world in a different way. For example, he made mountains not mere accidents, but explanations of why continents were the shape they were; he did not merely climb volcanoes, but revealed them as being connected along a worldwide geological rift. 'I shall never forget,' said Darwin, 'that my whole course of life is due to having read and reread as a youth his *Personal Narrative*.' Humboldt did not have curiosity simply about facts, but also about how to focus curiosity. The climax of his exploration and research, after fifty years of rumination, was *Cosmos*, a book which described the world in all its physical manifestations, explaining how everything was connected to everything else.

'If I ask myself what gave the first impulse to my ineradicable longing for tropical countries,' he wrote, 'I would have to name: George Forster's descriptions of the South Sea Islands, a painting of the banks of the Ganges by Hodges in the house of Warren Hastings in London, and a colossal dragon tree in an old tower of the botanical gardens in Berlin.' From this personal experience he generalised (in volume two of *Cosmos*) that curiosity has increased in modern times because of the stimulus given to it by the opening up of the imagination through literature and art, descriptive poetry, landscape painting and the cultivation of exotic plants. Every discovery opens up the imagination further, stimulating more discovery: it 'enlarges the sphere of ideas', excites a taste for investigation, while the creation of new instruments of observation increases the intelligence. He carried forward Descartes' idea that curiosity was like an infectious disease, and he showed how it could become an epidemic.

But books and pictures are not enough. Humboldt is important also because his life was dominated neither by his work, nor by his affections, but by an unusually intimate union of both. There was 'not the slightest sympathy' between himself and 'the people who loved me and showed me kindness' in childhood, which he remembered as unhappy; his mother seemed 'a stranger' to him. But he formed very strong attachments with a series of men, most of whom were intellectual colleagues. The intensity of his feelings may be gauged from a letter to one of them: 'For two years I have known no other bliss on earth but your gaiety, your company and the slightest expression of your contentment. My love for you is not just friendship, or brotherly love, it is veneration, childlike gratefulness and devotion to your self as my most exalted law.' The objects of his affection included two of France's most eminent scientists – Gay-Lussac, with whom he shared a room at the École Polytechnique, and François Arago, who reciprocated with these words: 'Outside my family circle, there is none to whom I am more deeply attached than you. You are the only friend upon whom I could count in difficult circumstances.' Whether these were homo-sexual attachments, as some have argued, or not, Humboldt defined his relationship with Gay-Lussac thus: 'He is my best friend and I find his company most consoling and stimulating, and the stimulus seems to be mutual.'

The emphasis on stimulus is the key. Goethe, who met Humboldt several times, wrote, 'His company is exceedingly interesting and stimulating. Within eight days one could not learn as much from books as he imparts in an hour. [His] presence would suffice to render an entire lifetime interesting and to stir up everything that could possibly be exciting in chemistry, physics and physiology. . . . He is like a fountain . . . forever refreshing.' Wilhelm Humboldt dismissed his brother's 'capacity for forming quick attachments . . . and for sacrificing himself for others' as a sign of weakness, lack of self-esteem. It is true that Alexander had occasional bouts of 'melancholy', and his sister-in-law was doubtless right when she said he was an 'incredible mixture of charm, vanity, soft feelings, cold and warmth'; he did say that he thought 'blowing one's own trumpet is part of the job'. But when he was in the jungle, he was delighted to be free from 'the distrac-tions constantly arising in civilised life from social claims. Nature offers unceasingly the most novel and fascinating objects for learn-ing. The only drawbacks of this solitude are the want of infor-mation on the progress of scientific discovery in Europe and the lack of all the advantages arising from an interchange of ideas.'

Being fully alive, fully curious, involved obtaining ideas both 'directly from the world around us' and through 'the medium of an intelligent mind'. Humboldt always needed a human being to sharpen his sensations, to trigger his imagination. That was why he lived for twenty years in Paris, which was then the undisputed intellectual capital of the world. It has been claimed he never experienced fear, except once when he suddenly met a jaguar in a South American forest (but he had the presence of mind to keep walking very quietly as though nothing had happened). It would be truer to say that he had a full measure of human weaknesses, including fear and worry, but his unceasing curiosity enabled him to absorb those fears into the general mystery of the universe, to have a global vision of fears, turning them from personal threats into natural phenomena. He might not have found it so easy to do that if he had been what his parents wanted him to be, an expert on one subject, nor if he had chosen to worry about why his parents did not like him.

At a party Humboldt once met Napoleon, who could think of nothing else to say to him but this: 'You are interested in botany. So is my wife.' And the emperor then moved on. It is no longer satisfactory to be uninterested in what interests one's wife. The limits of curiosity are at the frontiers of despair, and people have always tended to fight what they despair of. I now turn to the problems that fighting causes.

C. D. O'Malley, *Andreas Vesalius of Brussels*, California UP, Berkeley, 1964; Owsei Temkin, *Galenism: Rise and Decline of a Medical Philosophy*, Cornell UP, Ithaca, 1973; A. von Humboldt, *Cosmos: A Sketch of a Physical Description of the Universe*, Eng. translation, 1849; L. Kellner, *Alexander von Humboldt*, Oxford UP, 1963; Helmut de Terra, *Humboldt*, Knopf, NY, 1955; C. C. Gillespie, *Dictionary of Scientific Biography*, 18 vols., Scribners, NY, 1970–90; D. Gorlitz and J. F. Wohlwill, *Curiosity, Imagination and Play*, Erlbaum, NJ, 1987; H. I. Dutton, *The Patent System and Innovative Activity During the Industrial Revolution, 1750–1852*, Manchester UP, 1984; Norbert Wiener, *Invention: The Care and Feeding of Ideas*, MIT Press, 1993; Roger L. Geiger, *To Advance Knowledge: The Growth of American Research Universities, 1900–1940*, Oxford UP, 1986; Paul Von Blum, *Stillborn Education: A Critique of the American Research University*, University of America Press, 1986; Jerome Kagan, *Unstable Ideas: Temperament, Cognition and Self*, Harvard UP, 1989; Bruno Latour and S. Woolgar, *Laboratory Life*, Sage, NY, 1979; Warren O. Hagstrom, *The Scientific Community*, Basic Books, NY, 1965; B. Barber and W. Hirsch, *The Sociology of Science*, Free Press, NY, 1962; J. W. Stigler et al., *Cultural Psychology: Essays on Comparative Human Development*, Cambridge UP, 1990 ('Mathematics Learning in Japanese, Chinese and American classrooms'); J. Céard, *La Curiosité à la Renaissance*, Soc. d'Édition supérieur, 1986; R. E. Thayer, *Biopsychology of Mood and Arousal*, Oxford UP, NY, 1989; Ian I. Mitroff, *The Subjective Side of Science*; I.

F. Clarke, *The Pattern of Expectation, 1644–2001*, Cape, 1979; Judy Chicago, *Through the Flower: My Struggle as a Woman Artist*, Double-day, NY, 1975; Geneviève Fraisse, *La Raison des femmes*, Plon, 1992.

12

Why it has become
increasingly difficult
to destroy one's enemies

Antoinette Fouque's strength is that she knows what she thinks. That is the source of her charisma. But it is also the reason why, to many, she is highly controversial. Everything said about her has been passionately for or against.

What can she do about those who disagree with her? Sometimes she says, 'I have never been preoccupied with what people think about me', and indeed, despite the ups and downs of public opinion, she has remained one of the most persistently active leaders of the women's movement over the past quarter of a century. Sometimes she says all that matters is to express an opinion sincerely held, and that 'convincing people is a form of violence'. But the word 'combat' also occurs frequently in her conversation. She describes herself as a 'militant theorist' engaged in a 'political battle' to free women from discrimination and violence, and to win for them not only equality, but a recognition that their contribution to civilisation is quite different from that of men. The military metaphors are not mere turns of phrase. They imply a view of how one can best get what one wants.

In appearance she is far from military: stricken, since the age of sixteen, by a slowly spreading disease that has made movement increasingly difficult and exhausting; obliged, after a long resistance, to use a wheelchair, she is slight and frail. Her smile is vivacious, even flirtatious. She wears clothes by fashionable designers. Her psychoanalyst's curiosity, the breadth of her reading and the firmness of her convictions are seductive. A considerable number of well-educated women have become not merely admirers, but devoted ones, for she has answers that seem to solve simultaneously their problems as individuals, as women and as citizens. One of her disciples is a rich heiress, so there have been no financial obstacles to her political activity on an international scale and to her becoming a major publisher of women's literature.

It is not because she is physically fearless, and because pain, having become a habitual companion, holds no more terrors for her, that she sees herself as a militant. For all her disclaimers, she is sensitive to criticism, worried that she is misunderstood and misrepresented; when she suspects hostility, nervousness transforms her appearance and she defends herself with a torrent of arguments. Her opponents say she is aggressive. She says they are, and that she is only defending herself. That is how all wars start; but there is more to it than that.

Her basic idea is that women are essentially mothers; that pregnancy is the most important of all social experiences, in which a mother lives at peace with a person inside her who is different from herself; that the link between mother and child is the most precious of all. Men who have rebelled against their mothers have become pathologically narcissistic and violent, terrified of being castrated, though deep down in their unconscious their attachment to the mother survives; the task of women is to get rid of that narcissism and to get men to realise that their own kind of desire, obsessed with dominance, is not the only one: women have a different way of feeling and 'thinking with the body'. The abolition of sexual differences would only mean that women would become pale copies of men. She was shocked that her sister, on becoming engaged, suddenly turned into a football fanatic. Rather, women should try to become aware of the masculine prejudices lodged in their unconscious, and free themselves by understanding better what it means to be a woman, developing a science of women and a language of women, cultivating their difference. Democracy, as it exists at present, is inadequate because it has replaced monarchy – the rule of fathers – by the rule of brothers – fraternal fratricide, she calls it – ignoring the crucial importance of mothers.

The person she has felt safest with was her own mother, a Calabrian, 'who always understood me', a 'strong woman', who went out to work despite her husband's protests, who did not allow him to 'lay down the law', who 'ruminated' all the time, though she was illiterate, and who kept on telling her, 'You must be free.' 'If I were reincarnated, I would like to be my little (female) dog, because I think she is the reincarnation of my mother, very loving and soothing. . . . But anybody who adopts a maternal role – not that I do – is detested.' In her view, what she is defending, when she fights her opponents, is not herself, but the women's movement and her publishing firm 'Des Femmes', as a mother might defend her children, without thought for herself.

Is there something surprising in a psychoanalyst having lived for many years in apartments which had no mirrors, and having hardly ever been photographed in childhood? One of her favourite texts is, 'Give me the courage to look into my heart and at my body without disgust.' She is glad that appearing on television with Catherine Deneuve did not make her 'think that I was less beautiful in her company'. It is this detachment that has enabled her to devote her life to fighting not one war, but several at the same time, to cure the world of its own paralysis.

Her first war is against 99 per cent of the human race, which is her estimate of the proportion of women-haters in it, for most women have been brainwashed into underestimating themselves. Men, as such, are not her enemies; her quarrel is only with the 'imperialism of the phallus'. She likes the company of men, adored her father, fell in love with her husband, had a daughter by him (against doctor's orders), separated but did not divorce, because she values his friendship still. The most influential person in her life has been a man, the psychoanalyst Lacan: 'He loved me and I loved him.' He analysed her for five years, before she moved on to another analyst, the Hungarian Grunberger ('who was Lacan's great enemy, though I did not know it'). 'Working with Lacan meant I was not cut off from men, and so I was saved from the sectarian feminist hatred of men, which means hating oneself.' She denies being a Lacanian, who said various disobliging things about women, but 'he was beautiful; perhaps love is blind; but he understood me, even if we disagreed.' She lives in a household of women disciples, because women need to get away from men to think out their strategy, but she is as opposed to lesbianism as to androgyny. Alas, Lacan really preferred androgynous women. She insists: 'I am not a man, I do not want to be a man. I don't think like a man.'

The second war being waged by this leader of the women's movement is against 'the feminists'. She rejects that appellation for herself, as she rejects the heritage of the suffragettes. Simone de Beauvoir's statement that 'One is not born a woman, one becomes one' she describes as the 'most idiotic pronouncement of the century': the childless, 'mentally frigid' Beauvoir was not a model, her relations with Sartre ('a liar and a hypocrite') were far from admirable; her ideas were pre-war. She has not bothered to read Elisabeth Badinter's latest book because 'Badinter is a disciple of Beauvoir . . . and Beauvoir declared war on me'.

Many feminists, while conceding that Antoinette Fouque is intelligent, accuse her of being megalomaniac, an 'impostor' and a 'cheat', who has attempted to take over the women's movement, by registering the name Mouvement de Libération des Femmes as her own trademark, so that no other group could use it. (She says she saved the movement from extinction.) More recently, she has sued the publishers Plon for using the letter F on the title page of their *History of Women*, because that too is a registered trademark of her own publishing house. (She replies that the feminists want to destroy her, as Sartre and Beauvoir destroyed Camus, and even more to destroy her publishing firm.) Her disciples, say the feminists, have been regimented into a sect like the Moonies, becoming parrots repeating her words: one lapsed follower remembers being told, when she disagreed with the leader's views: 'That is because you do not love us enough.' Both sides suspect the other of bad faith and overweening ambition. Animosity against dissenters and heretics has, historically, been far more vicious than hostility to wholehearted enemies.

Personal vendettas and power struggles have, of course, been endemic in France's intellectual life, in which Antoinette Fouque is an active participant and which she describes again using military metaphors. The first enemy she encountered when she began her bid to become a theorist was Sartre, whom she saw as the leader of the 'ancients', published by Gallimard, 'the Bank of France of the literary establishment'. Against him were ranged the 'moderns', published by Seuil, founded as part of the Resistance during the war. ('Why did Beauvoir not join the Resistance, instead of cycling around the country, having affairs?' she asks.) Each side had its own journals, used as 'veritable machines of war', hurling slanders, 'knives at the ready'. Each side invented formulae, encapsulating their theories, which were like war cries designed to silence their opponents. Lovers of literature and ideas, she says, were fascinated by this 'theoretical war ... even when they did not wholly understand the precious language used, bordering on incomprehensibility'; but she herself found participating in it 'very hard', rejecting accusations that she, too, influenced by the word play of Lacan, became incomprehensible. To her, Sartre and Beauvoir were 'tyrants', but their opponents were 'an oligarchy', exercising a 'terrorism' over those who disagreed with them. Baudelaire had said, 'Art is a battle in which the artist cries with pain before being defeated.' Sollers still says, 'Literature is an art of battle.'

In her youth, Antoinette Fouque started writing a doctoral thesis on the subject of the 'avant-garde in literature', and that is still the place she wants to have, independent of the great masters and of every clan: she attacks women who, she claims, are winning fame in the shadow of men, citing Julia Kristeva, for example, 'the wife of the Pharaoh Sollers' (whose book, *Les Femmes*, she calls 'a manifesto of misogyny'). 'I was penalised by all, even the women, though less so by Hélène Cixous' – but then she was Cixous' publisher and called her the greatest writer in the French language, and indeed in the world. When Elisabeth Badinter wrote on the occasion of Simone de Beauvoir's death, 'Women owe her everything', Antoinette Fouque was contemptuous: she retorted that Simone Signoret had five times more mourners at her funeral. When the philosopher Althusser murdered his wife and was saved from prosecution by his friends, Antoinette Fouque annoyed them by declaring that they should have been more sorry for the wife.

'My opponents did not want to debate with me on the pretext

that I would always end up being right.' The combination of isolation and insult was 'very painful', and in the early 1980s, finding it intolerable, she escaped to California for four years of recuperation by the ocean. But then she returned to fight again. Now she is something of an elder stateswoman, decorated with the Legion of Honour, and awarded a doctorate in political science by the University of Paris 8. Her thesis is like no other, a collection of her speeches, proclamations, interviews and newspaper articles over the years, in five volumes, with only a brief introductory 'synthesis', but even so there is more thought in these unacademic pronouncements than in many learned dissertations. She knows her writings have been important, but she is determined to win public recognition of the fact.

Does she imagine she is a new Cortés, able to conquer a continent with a few horsemen? No, she says, she thinks of herself as a little girl, three and a half years old, as she appears in a rare, treasured photograph, already mature, already what she has remained. Timidity is not what her enemies associate with her, but beneath her audacity and implacability there is a sense of fragility, a fear of being squashed by others: in her refusal to join any party but her own, there is a terror of institutions as 'great intimidators', 'constraining', demanding submission. 'My first reaction is to say yes, and my second one to say yes but. My aggression comes after a period of sympathy.' Having a father who was a Corsican labourer 'made me humble, it was one source of my timidity. . . . I was not dominating when I was an adolescent, I hated originality and did not want to be distinguished; I did not want to go in for competitive examinations; when I went shopping, I let other people go ahead of me in the queue. Simone de Beauvoir did that to me, and I let her pass. . . . I don't like to be defiant. The image I like to have of myself is of a person who is open and friendly. The qualities I appreciate above all are gratitude and courage, not a utopian courage, but the courage to get up in the morning every day.'

Some of her conclusions parallel those of Germaine Greer, who has ended up seeing her ideal in the Indian family, where sons love their mothers above all other women, and who illuminated the meaning of contentment by studying her pregnant cat. The torment of timidity echoes the experience of Gloria Steinem. However, Antoinette Fouque has never published a book, because she is never satisfied with what she writes, 'because I am afraid of writing and because I have been attacked for crimes I did not commit'. Having written her memories and announced their publication, she decided not to publish.

'I am not without hatred. When I am the object of hatred, I see the hatred in me; psychoanalysis enables me to sublimate it into creative or militant activity. There is frustration inside me. What I am most conscious of is jealousy, which lurks under hatred, jealousy of what others will not give you, love for example, or of what others can be and you cannot. If you create something, others will be jealous. The more creative you are, the more enemies you have. The more creative people become, the more tensions there will be. We live in a world of savage narcissism, where everybody wants to be first. We are moving towards a terrible narcissistic war, because the more women succeed, the more there will be a backlash.' So, for all her optimism, she predicts that before the end of the century there will be an apocalyptic 'genocide of women'. The rapes of Yugoslavia are a warning of horrors to come.

Antoinette Fouque is accused of being sectarian and dogmatic, and yet declares that what she hates above all is sectarianism and dogmatism. She says, 'I have a very weak ego. But after every blow, I forget. I do not think I am unlikeable. When I was a schoolteacher, I was much loved by the children. I am a very happy person.' Being surrounded by admirers may explain why she has not given more consistent attention to her public relations: though she has occasionally approached top agencies to help her, they have achieved little. 'Looking back, I was stupid with the press. I did not understand the media. I didn't appreciate its power. Mine was the timidity of a French intellectual. What I was interested in was knowledge.'

Calvin and Luther were both Protestants, but their followers took about four centuries to stop eyeing each other with mistrust. Communists and socialists, though both defenders of the working class, did their best to hasten each other's demise. On the occasion of the bicentenary of the French Revolution, and of the rapes in Yugoslavia, Antoinette Fouque did succeed in rallying many women outside her own persuasion into common protests, but she noticed that some did not greet her, while others did so only secretly. What would she do if, as a psychoanalyst, she was confronted by a fascist? Her answer is that she would try and direct his hatred to a different object, where it did no harm. But how can the personal animosities around her, which have upset her so much, be eliminated? She has no ready answer.

Patching up coalitions with hypocrisy, while continuing to hate one another, is the traditional remedy, not worth inheriting. Hoping that a common enemy can efface the memory of old disputes, among women who genuinely disagree about their vision of the future, is not a remedy at all. My view is that the whole notion of what to do about enemies needs to be rethought.

Antoinette Fouque is not a typical leader of the women's movement – nobody is – but I have chosen her because even with her prodigious psychoanalytical erudition, she is foxed by this problem. She says, 'The adversary is perhaps he who is in me and whom I do not want to recognise in myself.' I want to investigate another approach. The deeper roots of these wars go beyond her psychology or her timidity. She has not been able to rid herself of an ancient, monarchical and military tradition, which the working-class movement also adopted, and in which the women's movement persisted, that the way to get justice is to declare war on your oppressors. The victory of both movements has been only partial for the same reason: never in the whole of history has war been a wholly efficient method of achieving one's aims.

So far, humans have used three strategies to deal with their enemies: fight them, run away or somehow manage to love them. But none of these methods has been particularly successful, and the world is still full of enemies.

The problem with trying to destroy one's enemies has been that it has become increasingly difficult to do so, despite the wonders of high technology. When people believed in witches and could cast spells on each other, there were fairly simple ways of finding the culprit responsible for every misfortune. In 1829, a farmer of Maine-et-Loire called Poirier consulted a diviner to discover why his wife was ill; he was told that if he prayed in front of a jug of water, her enemy would be revealed; he imagined he saw the face of his brother-in-law in the water, and promptly killed him. At the same time, in Carney (Marne), a veteran of the battle of Austerlitz, unable to rid himself of the vermin that infested him, assaulted his neighbours, convinced they had cast a spell on him. In Ardres (Pas de Calais), a man grievously wounded an old lady, because she had unexpectedly greeted him in the street and he feared she meant him harm. Sorcery was a sign that people saw enmity everywhere, even in the glance of a stranger, but it was also a protection. Today, however, the malice of neighbours, the resentments of squabbling families and the envy of colleagues can no longer be eliminated by magic. Misfortune is now the result of elusive social or economic forces, in the face of which the individual feels bewildered. This means that there are ever more potential threats, personal and impersonal, institutions, regulations and mechanical gadgets which backfire.

There was a time when settling private disputes by force was a private matter, but then laws put an end to that kind of liberty. The first industry to be nationalised was that concerned with the destruction of enemies: only kings and nations were henceforth allowed to eliminate those who displeased them. One would have thought that after so many centuries of fighting, they would have purged the universe of enemies, but that did not happen, because war became a way of life. If the great conquerors have any message for private individuals, it is that the more they conquered, the more they were on the look-out for new enemies. As early as the fourth century BC, an Indian treatise on how to govern, the *Kautiliya Arthasastra*, advised kings to make war on all those who were weaker than themselves, and to treat every neighbour as a natural enemy. Machiavelli reinforced this advice when he said, 'Princes should have no other thought but war.' Invariably, the more guns and soldiers they commanded, the more power they felt they needed to tackle the next enemy. Within a few weeks of the Allied victory over Germany in 1945, the CIA was preparing a campaign plan for a possible war against the USSR. Nations have been just as prolific as kings in discovering new enemies. The military historian Quincy Wright has calculated that, in Europe, war has increased on the following scale, taking into account not only its duration but also the size of the fighting forces, the proportion of combatants to the total population and the number of casualties:

12th century:	18
13th:	24
14th:	60
15th:	100
16th:	180
17th:	500
18th:	370
19th:	120
20th, until 1945:	3080

The Second World War, whose horrors were expected to cure humanity of its love of fighting, was followed, between 1945 and 1990, by some 160 armed conflicts in various parts of the globe. Nor is it just tyrants who make war: between its foundation and 1965, the USA had only twenty years when its army or navy was not in active operations somewhere. Britain participated in more wars than any other European country, seventy-five of them between 1480 and 1945. France came next, with 72. Absolutist Spain had 64, Russia 61. In the sixteenth and seventeenth centuries, the major European powers were at war for 65 per cent of the time; that figure fell to 38, 28 and 18 per cent over the next three centuries, but if colonial wars are taken into consideration, fighting hardly ever stopped. Though the number of days devoted to battle has diminished, as well as the proportion of people who have been killed, the percentage of the population involved in hostilities, and of civilians injured, has increased.

In 1898 *The War of the Future*, by Ivan S. Bloch, published in St Petersburg, argued that war had become so costly, so murderous, and so complicated, that it was impossible to win, and war was therefore bound to become obsolete. In 1991 another book, *On Future War* by Martin van Creveld, modified the prediction, saying that it is only nations which have lost the power to win victories: their 'incredible fighting machines and vast armies will crumble into dust' because although they are theoretically able to destroy each other, in practice they dare not use their most lethal weapons and they are incapable of coping with terrorists, who do not abide by the conventions of war. The defeat of colonialism marked the end of an era in the fight against enemies: small movements of rebels were able to defeat large empires. Then the USA found it could not impose its will on Vietnam, nor Russia on Afghanistan. The United Nations could not stop the Yugoslav war. Big wars, backed by high technology, are being replaced by low-intensity conflicts, a multitude of guerrilla wars, a revival of intermittent tribal harassment, which will baffle organised nations. Accumulating weaponry is futile; the future lies with small groups resisting large ones, sniping at them, not destroying them, but wearing them down, making life uncomfortable and dangerous. And in private life, the thunderous battles to bring down great institutions like marriage have been abandoned; instead individual divorces, cohabitations, separations are creating a mosaic of unclassifiable confrontations.

The second difficulty about fighting one's enemies is the growing realisation that fighting often ends up having no purpose other than the fighting itself. Sun Tzu, who is considered to be one of the greatest of all military theorists (his *Art of War*, written in the fifth century BC, is still studied by generals), advised that 'to subdue the enemy without fighting is the supreme excellence'. But for most of history, the world has been ruled by professional soldiers who considered themselves to be engaged in the most noble of all occupations. When the hero of the American Civil War, General Robert E. Lee, confessed, 'It is good that war is so terrible or else we would love it too much', he made it clear that, for soldiers, it has been the fight itself, and the skills it demands and the thrills it offers, which are more important than the ultimate fate of the enemy. They have wanted to kill dissatisfaction with themselves more than their foe. Adventure and honour have been their goals. The knights of medieval chivalry fought in order to win approval from God, from their lords, from their beloved lady, to convince themselves that they were not contemptible. If they made money in the process of fighting, captured booty and acquired more land, that helped, but it was only a means. War has remained prestigious so long as it has been valued as the most dangerous of all excitements.

Even unwilling conscripts, having suffered the terrible torments of the two world wars, have not infrequently looked back on them as the happiest years of their lives, because they found in war what they had searched for in vain in the monotony of their ordinary existence. When they put their lives at risk, they discovered how much they valued life, in its simplest form, much more than the vanities which accumulate around it. When they confronted an enemy they realised how precious a friend was. In the trenches and in moments of danger, comradeship could sometimes achieve the intensity of love, a sense of belonging, a comforting certainty that they would do anything to help those who faced the same perils as themselves, who would do the same for them, oblivious of all risks. Pride in joint achievements, eliminating selfishness and the jealousies of status, was sustained by a determination not to be unworthy of the life-and-death trust that each placed in his colleague. They had no choice but to transcend themselves, to reveal qualities they never suspected they had, to be more heroic, loyal, proud – even egalitarian – than they ever thought possible, to concentrate the mind so that no other worries troubled them but mere survival, not just for themselves but for those who had suddenly become brothers and who not long ago had been strangers. That at any rate was how some remembered war when its horrors ceased to be present, and they consoled themselves for their sacrifices with the belief that they had found a higher meaning in life, defending their nation or their principles. The brave felt bound together like a nobility, hating the cowards and malingerers in safe jobs away from the front line much more than the enemy. Humans have continued to fight wars not merely because they cannot agree, but even more because so many of them have loved the exhilarating sensations it created. Animosity against enemies has been a steadfast substitute for positive goals in life.

'Hatred is holy,' said Zola, who fought the enemies of Dreyfus not just from love of justice, but because he enjoyed fighting, and believed that he did not truly exist unless he was being attacked. So he rejoiced that he had made 'pride and hate my two companions. . . . I have felt younger and more courageous after each one of my revolts against the platitudes of my time. . . . If I am worth anything today, it is because I stand alone and because I know how to hate.' The more the two sides hated each other, the more they had in common, but that was ignored.

Anthropologists have found tribes which never fight, and which praise timidity, but they are not models to imitate, for they are obsessed by the fear of violence. Tribes have also been found which fight all the time, but which have to sustain their aggression by drugs or other stimuli; even cannibals are frightened by their own ferocity, and paradoxically drink the blood of their victims to calm themselves down, to liberate themselves from the feeling that they have become tigers.

Working oneself up into a rage was once almost a form of art: 'Sweeter by far than the honeycomb is wrath,' said Homer. Divine fury used to be admired as heroic. In the last couple of centuries anger has lost its prestige; not that it is diminishing, but people are beginning to be ashamed of their anger. Nevertheless, an Australian enquiry – the only one of its kind – found that anger was experienced five times more frequently than sympathy.

So there has been little progress in the art of confronting enemies. Once an enemy is identified, propaganda now multiplies proofs of his depravity to reinforce the hostility, on the assumption that people find it gratifying to have their opinions confirmed. For example, John Foster Dulles, the US Secretary of State during the Cold War, when confronted with any new information about the USSR, systematically ignored anything that challenged his judgement of his opponent as an implacable and dishonest enemy. There is a firmly established tradition which encourages enemies to blind themselves to each other's point of view. On this basis, war can continue for ever. And what helps it to continue is that private individuals are prisoners of attitudes of mind which make them seek out enemies, just like nations.

If you believe that there are people who despise you or wish you harm, if you have simmering inside you not just a fear of them, but a loathing and a disgust at the very sight of them, if you are convinced that they and you are totally incompatible, then it may be that your deepest roots stretch back into ancient Persia, and that you are an unwitting disciple of the prophet Zarathustra, who lived in the tenth century BC. His recommendations on how to react to enemies are still widely followed, particularly in the West, even if he is remembered only because his priests, the Magi, visited the Infant Jesus. Until Zarathustra had the idea that there was only one true god, and that all other gods were really wicked and hateful demons, enemies were different from what they have become. It used to be silly to imagine that one had implacable enemies, when what happened was believed to depend on the whims of a large number of gods and ancestor spirits, and on one's performing the appropriate rituals to win their favour; there was no need to hate those who harmed one, because magic and sacrifice and prayer were more practical ways of dealing with them. The most ancient divinities were thought of as having the power to be either helpful or nasty, and much depended on how one treated them. Zarathustra replaced that with the belief that life was a perpetual battle, that every individual was surrounded by enemies, ruled over by Satan – the arch-enemy who hated one irrespective of what one did, just for being human.

Zarathustra put all the blame for the opposition which his prophesyings aroused on Satan. Only wickedness could explain that opposition, he thought. Satan refused to understand him, told lies about him: Satan was The Lie. That is how the all-purpose scapegoat was born, and nothing has paralysed intelligence more than the search for scapegoats. There was no need to probe the motives or the difficulties of one's enemies, once one learnt to spot Satan in them, and so to hate them. Several great religions took up the idea that it was a duty to fight Satan, who hid behind the people one disagreed with. Cardinal Newman wrote, 'One must learn to hate before one can learn to love.' Zarathustra explained how to find the right object for one's hatred, whom to blame for every misfortune. In other respects, he was a prophet of generous instincts, whose ideal was that people should be peaceful and neighbourly; but he could not understand those who rejected his ideas; he could not understand disagreement.

Scholars now dispute about how much of his doctrine he invented himself. His religion has officially almost vanished. Today, only the Parsees, of whom there are now less than 60,000, found mainly in Bombay, do him homage, but they have inherited a much modified doctrine; made wealthy by their dedication to education and hard work, they became India's leading industrialists and have quite lost Zarathustra's obsession with enemies. Iran, where they originally flourished, of course became Muslim, and cultured Iranians, in developing a civilisation devoted to refinement and subtlety, were probably the people who altered Zarathustra's simple teachings most thoroughly: their mystics saw Satan as the inventor of poetry; their poets revelled in the consolation that nothing should be accepted as being what it appeared to be, and used ambivalence as a means of protecting themselves against oppressors. The ceremony in the pilgrimage to Mecca, where the faithful throw stones to symbolise their hatred of Satan, represents only one part of what the Koran says about how to treat enemies. Zarathustra lives not in a particular part of the world, but in a temperament.

He is the prophet of those who hate uncertainty above all else, who wish to eliminate doubt, having suffered from it and found it torture. He is the inspiration of those who are not interested in their enemies as individuals, nor care to know that these enemies might not be fully committed to the other side, or as hostile as they seem. His splitting of the world into friends and enemies made it possible for people who were perfectly gentle in private to condemn others to the stake, or wage holy war on them, or terrorise and punish them without ever having met them, without listening to explanations, precisely because they did not want to know their enemy.

It is the silence between enemies that emerges most powerfully from the history of their confrontations. Silences can be broken. Mutual ignorance of each other's fragility prevented enemies from exchanging any emotions other than anger or hatred. Humans have continued to develop hatred in the same way as they have fallen in love, by a visceral reaction, or they have taken it for granted that enmity is the inevitable consequence of the way the world is made. But fabricating enemies is one of the oldest and busiest of human industries, and the raw material may be nothing more than hurt pride and anger, gradually hardening until the manufacturers become prisoners of their hatred. When they do not choose their enemies for themselves, others do so on their behalf.

Hitler wrote, 'The art of all truly great national leaders consists among other things primarily in not dividing the attention of a people, but in concentrating it upon a single foe.' There is no reason why that kind of thinking should last for ever.

When the German Democratic Republic collapsed in 1990, it was discovered that the secret police had files on six million individuals, over a third of its population: close friends, even members of the same family, had been denouncing one another as enemies. That may seem an aberration of a paranoid regime. But other countries might be surprised if they started collecting statistics about who is engaged in a secret war against whom.

What is the alternative? Before I go into that, I must investigate the two other ancient ways of coping with enemies: running away, and enveloping them in love.

John Keegan, *A History of Warfare*, Hutchinson, 1993; Quincy Wright, *A Study of War*, 2nd edn., Chicago UP, 1965; Martin van Creveld, *On Future War*, Brasseys, 1991; Fred McGraw Donner, *Early Islamic Conquests*, Princeton UP, 1981; George L. Mosse, *Fallen Soldiers: Reshaping the Memory of the World Wars*, Oxford UP, NY, 1990; Sam Keen, *Faces of the Enemy: Reflections on the Hostile Imagination*, Harper, San Francisco, 1988; Roderick C. Ogley, *Conflict under the Microscope*, Gower, Avebury, 1991; Jim Forest, *Making Enemies Friends*, 1987; Bob Altemeyer, *Enemies of Freedom*, Jossey Bass, 1988; Leonard W. Doogs, *Panorama of Evil: Insights from the Behavioural Sciences*, Greenwood, Westport, 1978; E. Zola, *My Hatreds/Mes Haines*, Edwin Mellor Press, Lewiston, 1991; A. F. Davies, *Skills, Outlooks, Passions: A Psychoanalytic Contribution to the Study of Politics*, Cambridge UP, 1980; Rosemary Ridd and Helen Callaway, *Caught up in Conflict: Women's Responses to Political Strife*, Macmillan, 1986; Signe Howell and Roy Willis, *Societies at Peace: Anthropological Perspectives*, Routledge, 1989; Carol Z. and P. N. Stearns, *Anger: The Struggle for Emotional Control in America's History*, Chicago UP, 1986; Roy Mottahadeh, *The Mantle of the Prophet: Religion and Politics in Iran*, Chatto, 1985; Howard Schuman, *Racial Attitudes in America*, Harvard UP, 1985; *Sun Tzu's Art of War: The Modern Chinese Interpretation by General Tao Hanzhang*, David and Charles, Newton Abbot, 1987; T. Haas, *The Anthropology of War*, Cambridge UP, 1990; Maurice Keen, *Chivalry*,

Yale UP, 1984; John Ellis, *The Sharp End of War: The Fighting Man in World War II*, Corgi, 1980; John Burton, *Conflict: Resolution and Provention* [*sic*], Macmillan, 1990; Joseph V. Montville, *Conflict and Peacemaking in Multiethnic Societies*, D. C. Heath, Lexington, 1991; David Binns, *Beyond the Sociology of Conflict*, Macmillan, 1977.

13

How the art of escaping from
one's troubles has developed,
but not the art of
knowing where to escape to

This is not an ordinary success story, of ambition triumphing over obstacles. My tale is of a brother and sister who rose from poverty to fame, not because that was what they were seeking, but because they could see no other way of escaping from the intolerable, and each time they escaped, they still found themselves in an intolerable place.

They remember the block of flats in which they were raised as 'hell', 'a curse', a great din of arguments, fights and drunken lovemaking, with a smell, which never went away, of alcohol and sex. Their father was a factory worker, who did try to escape himself, by going to evening classes and becoming a bookkeeper and eventually a warden of a children's home, but he lost his chance by buying a house, the mortgage for which made him poorer than ever. Observing him, Gérard Colé swore at the age of ten that when he grew up, he would never be short of money; and he did not wait to grow up. Leaving school with no certificates of any sort, he went to the local abattoir and bought the horses condemned to death – some of them thoroughbreds whose only crime was that they had not won enough races. Sensing that there were many people like himself, fascinated by riding, but feeling they did not belong to the social circles of the horsy, he set up a poor person's riding stable, eventually with seventeen horses. That was his first escape, from poverty.

But then he thought he needed to escape also from a life which made little use of his intelligence. The *New York Times* had never employed a Frenchman: he persuaded its Paris correspondent to take him on as the office boy; and so he became a journalist and a broadcaster, and eventually a successful public relations man. But then, he says, 'I saw wealth was a dead-end street. It's terrible to have no money; it's good to wear nice clothes and eat well, but I wouldn't give up a single day of my life for money.' So he escaped from the world of money to the world of power.

He decided that François Mitterrand, who, at the time, was in the wilderness losing elections, would eventually become the most powerful man in the country: so he became his public relations adviser, in charge of his 'image', giving up his luxury 350 square metre flat for one of 40 square metres, selling his Jaguar for a tiny car. He played a very significant part in getting Mitterrand elected president, and then re-elected. But the skills he had used in these victories, he believed, came from his having remained, very profoundly, 'viscerally', a 'man of the people'; he never belonged to the élite and did not want to, because he felt sorry for those who did, for their bloated sense of their own importance and their inability to see they had no contact with the electorate, that no one was listening to them any more. The social life of the powerful did not attract him.

So he has escaped once more, to become the chief organiser of escape in the country, as head of the national gambling company, the Française des Jeux. His aim is to make it the world's biggest escape factory, serving those billions of humans who cannot imagine escaping from their ordinary lives except on the back of chance. He has already expanded to twenty other countries, from China and Kazakhstan to Germany and Senegal.

Meanwhile his sister, Michèle Blondel, has become famous as a sculptor. It may seem that they have gone their separate ways, but she too is devoting her life to investigating the art of escape, in more intimate detail, discovering what a woman in particular needs to escape from. At seventeen she left home; at nineteen she was a mother; her son has given her great satisfactions, but, even though they remain close friends, at eighteen he also left home, to lead a life of his own as a designer in Milan, and, in any case, a child could not protect her from the way people looked at her. 'Nothing is more cruel than a look.' The air is full not only of the smell of alcohol and sex, but of glances which are like knives puncturing confidence and leaving scars that are hard to heal. As a child, Michèle Blondel had believed she was unattractive. She cut her hair short like a boy's, wore trousers, fell in love with a male homosexual, and moved in a circle which treated her as a boy: she relished their desire but was safe from harassment. Her best men friends are still homosexuals, though she does not believe they quite understand her; only women have been able to see deep into her. As a result the subject of her art became not what things or people look like, but what it feels like to be looked at.

What she hates most about that stare from outer space is that it immediately puts her in a box. 'I am a woman: that is a box. I am a mother: that is a box.' 'No,' she protested, 'I am invisible': people could not see what she was really like. Her first paintings were all in white, varying shades of white, which contains all colours, but they are colours that cannot be seen. The paintings were of herself, and she thought she was successfully concealing herself from view by this method, until she decided they were revealing her more naked than ever and that to reveal herself was useless; nothing irritates her more than for people to believe what she tells them about herself, which imprisons her in a box of her own making. So she started making sculptures out of crystal, supremely invisible, unfathomable, but throwing out a thousand hints, never the same, of colours that are the result of the light and the look that falls on it. Having her crystal made specially for her at Baccarat, by new methods, to be like no other crystal, urging the technicians to mix the ingredients so that it should come out purer than any they have ever made, staying overnight in the factory to see it born, not knowing what it will look like, then smashing the rock into individual boulders, arranging these in suggestive poses, she is retracing the whole process of how people come to appear as they do and still remain elusive. Every time a crystal is broken by the misadventures of life, it becomes different: that is an affirmation of hope against those who would have one always recognisable, for ever imprisoned in the same box.

It is as a creator of fountains, as a magician with water – transparent but impossible to grab hold of, always escaping – that she has become best known. Every railway commuter at the Gare de l'Est in Paris has passed by her fountain in its front courtyard. She likes them to pause and stare at it. Sometimes she comes and watches them do it. The very first man she saw using it as she wanted it to be used was a photographer with a polaroid, who offered to take a picture of a girl he had never met before standing in its spray, and before long he was embracing the girl and going off with her. That is her ambition, that her sculpture should induce people not just to look at each other but to meet and to talk, to mix like water, to love.

It is unlikely, however, that the photographer lived happily ever after. How can men learn to love women without trying to possess them? That is the refrain that her fountains sing. As a poor child in a school for the rich, unable to afford posh clothes, she refused to be frustrated because she was different, and valued her independence. But most of the men she has known have been prisoners of their frustrations, trying to prove over and over again that they are attractive, but never convincing themselves for long, using violence or domination or contempt as their tools, always really more interested in themselves than in her. They have told her that she is impossible to understand, like a Breton menhir, a mysterious Stone Age monument of rock whose message no one can decipher. And it is true that she does not like to express herself too directly; she cannot tell her lover who she is, it is for him to find out. That is what love is, the discovery of another. She wants her lover to discover that she is a creative artist, and not to be frightened because there is a man as well as a woman inside her. Of course, when she is loved, she turns into liquid, and that is a most satisfying state to be in, completely shameless, until she realises that she is perceived simply as an object of desire, and not a person like no other. 'To be put into a category is to be put in a coffin.' Even to be perceived as a sculptor is to be categorised. She wrote a book and made a film to escape, but the world does not appreciate people jumping from one box to another. No lover has lasted more than six years in her life: when they do not demand dependence, they become dependent. 'I am a bar of soap,' she says. The bubbles in her fountains are women like her, impossible to capture.

At times, she despairs of the possibility of communicating: perhaps humans are no more than ships in the sea, separated by fog, who occasionally glimpse each other's lights dimly in the distance, and exchange brief salutes as they pass. But the loneliness is intolerable and she never ceases trying to escape from it. Her resistance takes the form of turning what depresses her upside down. If her body is not beautiful enough, if it is falling apart, she escapes the pain by recognising it and transforming it into pleasure, or at least the pleasure of honesty and discovery. If she feels megalomaniac, as an artist must inevitably, for brief moments, she tries to transform the sensation into generosity. If she feels alone, she thinks that at least makes her more open to new people. If people do not understand, she becomes provocative and aggressive: at least they will not react with the blank stare of indifference: she would rather they hated her work than not notice it is there.

So she loves to make her sculptures inside factories, among workers who have never been able to make sense of contemporary art. For example, she has used the waste products of a pottery, seizing the warm earth as it came out of the machines, to transform it into images of sexual organs. At first the workers were horrified; they had expected her to make little figurines with two arms and two legs; but gradually they understood that she was trying to show how life came out of earth. Then they began asking questions; one worker finally said she would now look at the world with different eyes. When commissioned to redecorate churches, she has added crystal sculptures of liquid semen, in an attempt to reconcile Catholics with eroticism, to celebrate the place of sexuality within love.

Now she has a studio in New York, too, where she spends part of the year, sculpting, exhibiting, being provocative and resisting whoever tries to put a label on her work. Many of France's greatest artists since the nineteenth century have found it easier to win recognition in America than at home; but Michèle Blondel feels particular closeness with one French woman artist, Louise Bourgeois (a pupil of Léger, a friend of Le Corbusier and once a lodger in Isadora Duncan's house) who emigrated to America in 1938, and who was there able to express more freely her obsession with the theme of women imprisoned in a box, or in houses or in social situations, from which they are endlessly trying to escape. Michèle Blondel should have gone to America at least a decade ago, says her brother. A famous art dealer promised, 'If you aren't burned as a witch, I shall do a giant exhibition of your work.' But usually artists have to wait to be burned before they are recognised to be as interesting as gangsters or murderers.

One of those men on motorbikes who go round Paris clearing the streets of dog-shit told her he too found joy in his work, which suited his personality, because he loved the independence and hated dogs. She has not found as simple a solution to her problems, but in her workshop she feels that she does succeed in escaping temporarily from the most intolerable human weaknesses. In her workshop, she juggles with chance, like her brother, never knowing what will come out of her experiments, or how a smashed crystal will look. The great consolation is to know that one is not repeating oneself, that one is moving on.

All humans are, by origin, escapists. All are descended from ancestors who migrated from Africa and Asia. Every religion has been an escape from the sordidness of real life, a withdrawal from the aching body into the safe haven of the soul. When religions have become too conventional and superficial, there have always been escapes into mysticism and fundamentalism, away from reality. Industrial society began as an escape from poverty. Now it has become an escape from work to leisure, hobbies and sport. But even a civilisation dedicated to leisure contains enemies, so the art of escape has been refined still further: detachment, humour and parody have been cultivated to prevent any worry from being taken too seriously. The escape route from marriage to divorce and back again is a motorway to which new lanes are being added all the time. More humans have run away from their enemies than have fought them. Escape has been an unrecognised art, because the many forms it takes have never been seen as a unified response to life.

A survey of Americans born between 1946 and 1966 found only 10 per cent who were aggressive 'competitors' in the way they tackled their daily problems. Far more built their lives around the art of escape: 25 per cent were 'pleasure seekers', 15 per cent felt they were 'trapped' and did not know how to escape; 28 per cent were 'content' because they had 'survived'. Only 20 per cent were fully satisfied because they had a balanced life. According to the Harvard Business School professor responsible for these findings, if his respondents had been more honest, even more would have admitted to being trapped. At any rate, a case study of an American firm of accountants, published in 1992, found that only 6 per cent of private grievances among its members produced accusations and protests; 21 per cent preferred to suffer in silence. The majority ran away from their enemies: 31 per cent 'temporarily avoided' them, 14 per cent adopted an attitude of 'strategic alienation', 8 per cent sought counselling. In Britain, only 18 per cent of the population admit to having made a verbal complaint in a shop, and only 2 per cent to having taken part in a demonstration or boycott.

Running away has its philosophy and its devotees, no less interesting than those of war or rebellion. The most vocal modern advocate of escape as a way of life has been Henri Laborit, the scientist who invented one of the most widely used of modern tranquillisers, Largactil: that is the great escape route from pain and worry. Ever since, he has been using the royalties from his chemical discoveries to finance research into alternatives to aggression. In Alain Resnais' film *Mon Oncle d'Amérique*, he told the world that science confirmed the opinion of the ancient sages for whom to run away was the truest wisdom.

Avoid confrontation, says Laborit, because its only result is to establish an order of dominance, such as monkeys seek when they fight to decide who may mate with whom. Once you are caught in the competition for superiority, you lose your independence. The purpose of life – he speaks as a biologist – is to survive, which requires one to keep calm, to avoid stress. He has measured the stress of rats he tormented in his laboratory: those he allowed to run away into another compartment had a normal blood pressure one week after the experiments. However, when he prevented the rats from escaping, even a month later their blood pressure was still high; when they can find no way out, they develop ulcers and lose weight, and hope, so that when the cage is opened, they are too frightened to escape. A third set of rats were put in cages in pairs, and while prevented from escaping, were allowed to fight each other; after being subjected to the same torments, their blood pressure remained normal. Fighting and running away, he concludes, are alternative ways of countering stress; but fighting, if successful, becomes addictive and draws you into the stress of a competitive life; moreover, there will be occasions when it is impossible to fight your competitors, so you will fight against yourself, and that will produce stress. It is better, he insists, to run away.

And when circumstances do not permit you to escape physically, you can do so in your thoughts. The imagination is the only part of you which nobody and no group can touch. You may be powerless, but in your imagination you can transform the world. The best escapologists are artists who abstract themselves from the realities of daily life and from the constraints of hierarchy; they create worlds of their own, expressing their independence and originality. Laborit does not advocate running away in order to avoid emotion, for that would leave you colourless and indifferent. Every artist must be a dissatisfied person, even an anxious one, but being an artist means that one is dedicated to discovering ways of making that anxiety fruitful and beautiful.

Of course, the employees of a mean and nasty boss cannot always escape, because they would simply be condemning themselves to unemployment. They cannot always fight, because there is a hierarchy organised to paralyse them. So they become 'inhibited'. People who do not have enough information to decide how to react to a threat also become inhibited, as they do when they have too much information. Inhibitions engrave themselves in the memory, which, mindful of past failures, discourages action and so encourages more failures. Laborit has found a chemical which inhibits inhibition, though new problems then arose. Meanwhile he urges people to escape from their inhibitions in every way possible, by talking, or writing, or getting angry, or insulting those who annoy them. Otherwise the inhibitions will destroy their health, by inhibiting their immune system, and they will develop psychosomatic illnesses, which is a way of punishing oneself when others fail to understand one.

However, he does not claim that his solution is a formula for happiness. Reflecting on his own life, he admits he has had trouble escaping. 'I have a pathological need to be mothered, and to be protected in my work,' he says. His work has been a long, unceasing battle with his superiors, who thought him quarrelsome and insubordinate. He protested that promotion was given for licking the boots of the establishment and conforming to its values, not for professional achievements. 'I should have liked to have lived in the age of chivalry,' he says, but the nearest he got to it was being a surgeon in the French navy, where he claims the chivalrous spirit still lives. However, his employers did not appreciate him enough, offering him early retirement, refusing him the promotion to which his seniority entitled him (he never became Surgeon General, as he believes he should have). He managed to escape into a research job in a military hospital (though he does not like soldiers, nor merchants, nor various other categories). His temperament, as he himself confesses, is aggressive, and has been since childhood. It is no accident that he has devoted himself to studying mood-altering drugs and painkillers, and that his other inventions include anaesthesis by artificial hibernation. He still deeply resents the French medical establishment's failure to recognise the importance of his work; only foreign countries have honoured him with prizes, including the Lasker prize, which in forty-five cases has been followed by the award of the Nobel prize, but he did not get that.

Making friends was not what he was best at; all his friendships, he says, date from his childhood; since then, he has known only competitors. That is not strictly true, for a circle of admirers has grown around him: Alain Resnais, for example, cured himself of some of his gloom by reading Laborit's books, which was why he made his film. And Laborit, in his seventies, fell in love with a much younger woman, with whom he and his wife now form a trio, which is 'neither completely satisfying nor frustrating'. The most serious failing he recognises in himself is that he is not a woman, for women know how to use both hemispheres of the brain and thus achieve a more balanced life. At any rate, there are things he cannot escape from; it is not a perfect solution by his own admission, but it offers respite.

Humans have long experience of escaping into their imaginations. Everybody's original enemy was hunger, and when it could not be defeated with food, the escape was by drugs. The most common European drug in the Middle Ages was probably poppy seed (vast areas were devoted to its cultivation, on an industrial scale) and it was used to make bread, as was hemp seed, spiced with coriander, aniseed, cumin and sesame. With such help, the poor escaped into a dreamlike condition, often pursued by goblins and vampires and terrifying visions, but at least their fears were not as debilitating as hunger. Restless children were given infusions of poppy to keep them quiet. When hunger turned into famine, 'insect men' reminiscent of swarms of locusts ate anything that could be swallowed, raiding refuse dumps, feeding even on excrement, and slowly withdrew into dazed stupefaction, oscillating between narcosis and neurosis, but dreaming that they were eating.

Escape into altered states of consciousness, into sedation or excitement, has been a constant ambition everywhere, in all centuries. There has been no civilisation which has not tried to escape from normality with the help of alcohol, tobacco, tea, coffee and plants of all sorts. The Aztecs, who had a particularly gloomy view of life, used to have 400 gods of drink and drunkenness, called the 400 Rabbits, to help them escape. Where to? When they were sober, they freely discussed their hallucinations. They saw themselves being devoured by wild beasts, taken prisoner in battle, convicted of adultery and undergoing the penalty, which was having their heads crushed, or they imagined they were rich and masters of many slaves. Drink thus did not liberate them from habitual preoccupations, but rather allowed them to contemplate them like a horror or fantasy film watched over and over again. The cacti and mushrooms they ate gave them 'horrifying or comic visions' for several days, but they persisted: 'It filled them with courage to fear neither battle nor thirst nor hunger, and they say it preserves them from all danger.' Strongly conscious of the risks of intoxicants, they punished a drunkard, for the first offence, with the shaving of his head in public, to the jeers of the crowd; and if he persisted, he was sentenced to death; officials and priests were executed on the first offence, while nobles had the privilege of being strangled in private. But when really serious trouble came, nothing could stop them, 25 million of them, from drinking and drugging themselves to death, until there were only one million left. This was the greatest mass suicide in history, doubtless hastened by the import of European diseases, but it was the loss of morale caused by the Spanish conquest, which declared their whole civilisation bankrupt – rather as Communism has been declared bankrupt – that was devastating.

In Europe, opium became the fashionable way to escape from pain and boredom after the Swiss doctor Paracelsus (1493–1541) mixed it with alcohol to produce laudanum. The leading English physician of the eighteenth century put opium into a cocktail flavoured with cinnamon, cloves and saffron and called it 'one of the most valued medicines in the world'. In 1854 a standard English medical textbook repeated that opium was 'undoubtedly the most important and valued remedy in the whole *materia medica* to be used for maladies of everyday occurrence', though smoking it was judged 'dangerous for the physical and moral character, especially of the lower classes'.

The United States was remarkable for the suddenness and versatility with which it moved from drug to drug, from one escape to another. In the early nineteenth century, Americans doubled their consumption of alcohol, identifying it as the symbol of their egalitarianism – all were equal before the bottle and none could refuse: 'They drink to be free.' The man on the tavern floor who shouted, 'I am as independent as the USA', though he could not stand up, was expressing a sentiment that even the temperance movement dared not deny, for it allowed its members to get drunk on the fourth of July. An abrupt change came in 1830–50: alcohol consumption halved and henceforth dependence on the bottle became the speciality of a minority. But then a craze developed for patent medicines with a high opiate content, which reached its peak in 1900, quadrupling the import of opium per inhabitant. At that point, doctors began to have doubts. So during the next decade, smoking opium became the fashion, till 1909, when the import of opium was banned. So cigarettes took over, much boosted by the First World War. When prohibition of alcohol spread through the country, starting in the south and west, cola drinks, laced with cocaine, stepped in as the saviour. Cocaine was declared to be the best way to escape from addiction to alcohol, opium and morphine, and to be an excellent general tonic too. William Hammond, Surgeon General of the Army, proudly announced that he took a wine glass of cocaine with each meal. Cocaine was the official remedy of the Hay Fever Association. Bars served it in whiskey; shops sold it in the form of cordials; it was distributed to miners and construction workers. American doctors vied to praise it. The professor of medicine at the University of Pennsylvania, Dr George Wood, president of the American Philosophical Society, even recommended it as an aid to those who wished to be true and religious Americans, for it offered 'an exaltation of our better mental qualities, a warmer glow of benevolence, a disposition to do great things but nobly and beneficiently, a higher devotional spirit and withal a stronger self-reliance and consciousness of power. It seems to make the individual, for the time, a better and greater man.' The search for escape from mediocrity and monotony has been relentless.

The French have for long held the world record as drinkers of alcohol. The reason is not that they produce the best wines, for they have now added to their laurels the world record for the consumption of tranquillisers and sleeping pills. It is more plausible to suggest that theirs is a civilisation which has esteemed artificiality as well as art; to trim a garden hedge into strange shapes, to wear clothes which make each individual even more unique, to speak in precious prose, these are all part of a similar attitude, a belief that humans are perfectible and should be sharpened and polished rather than left as they are. Their escape has not been into oblivion but into a condition which enables them to approach more nearly their ideal of a sociable and interesting being, able to cope with the hazards of existence. Out of every ten of them, three depend on coffee (particularly middle-aged married women), two on both coffee and tobacco (mainly the young), another two on tranquillisers only (particularly the old and the less well off), one on tea, with occasional smoking and alcohol (mainly highly educated young people), one on tea and nothing else (mainly highly educated young women), one on wine or beer, but without smoking; only five in a hundred are heavy smokers and heavy drinkers of alcohol and coffee all at the same time. Only 1.5 per cent are total abstainers from alcohol, tobacco, coffee, tea, tranquillisers and sleeping pills, but they arouse only mistrust or puzzlement. The noble savage was seldom sober.

The problem with escape is knowing where to escape to. Laborit says you cannot escape from misery into happiness, because that is an unattainable goal. The ancient Greek Heraclitus (552–487 BC), who renounced his hereditary claim to the kingship of Ephesus, in what is now Turkey, said that because the universe was constantly changing, there was nowhere to escape to: 'Human nature has no set purpose'; the whole universe was engaged in running away from the way it was. The Taoists, who date from the second century BC and who were China's specialists in escaping from ambition and the normal concerns of ordinary beings, were original in saying that women as well as men, young and old, could learn to escape from misery, not quite to escape death, but rather to escape dying the wrong way; with the help of drugs which increased sexual potency but gradually destroyed the body, they tried to wither away with a smile, drunk, dancing to the rhythms of nature; but their dream of equality, of men and women 'mixing their mingled breath', being spontaneous and natural, got lost in magic formulae, because they sought a short cut to bliss through alchemy.

When the Japanese writer Mishima committed suicide, he was carrying escape to its ultimate conclusion. He said he was inspired by the example of Oshio Hehachiro, a hereditary police inspector of Osaka. At the age of thirty-seven this man gave up his job because he got tired of fighting corruption; he decided that success in life was less important than making a heroic gesture: one should escape from mere pettiness, and that was within the capacity of anybody: even an illiterate peasant woman toiling in the fields all day could thus become a sage. 'It is the journey, not the arrival, that matters', and one should embark on it 'like a madman'. ('*Il faut toujours être ivre*,' wrote Baudelaire at about the same time.)

Sacrificing oneself voluntarily made one impervious to others. So Hehachiro organised a rebellion of 'the wretched of the earth', starting it off by burning down his own house, and when the rebellion flopped, he cut his own throat. He was found guilty posthumously of 'criticising the government' and his pickled body was ordered to be publicly crucified. He became a Japanese hero nonetheless, as a noble failure, and Mishima copied this suicide because even though it achieved nothing it was 'an act of sincerity', an escape from hypocrisy. It was escape for its own sake. The enemy survives.

People who wish to escape from the grasp of the institutions of their time, and the opinions of the crowd, and indeed from ordinary life, are not misfits in modern society: their roots go back into furthest antiquity, as far as those of warriors; they were singing songs like these in ancient China:

> *I arrive all alone, I sit down all alone.*
> *I have no regrets that people today do not know me.*
> *Only the spirit of the old tree, in the south of the city*
> *Knows for certain that I am an Immortal passing by.*

To ask what the practical results of escape might be is to miss the point of escape, which includes escape from purpose. Those who want a purpose must look beyond escape.

Henri Laborit, *Éloge de la fuite*, Laffont, 1976; Henri Laborit, *La Vie antérieure*, Grasset, 1990; Dorothy Rabinowitz, *New Lives: Survivors of the Holocaust Living in America*, Knopf, 1976; Vladimir Bukovsky, *To Build a Castle: My Life as a Dissenter*, Viking, NY, 1977; M. Magdelaine and Rudolf von Thadden, *Le Refuge huguenot*, A. Colin, 1985; Dino Cinel, *The National Integration of Italian Return Migration, 1870–1929*, Cambridge UP, 1991; M. M. Ktitz et al., *International Migration Systems*, Oxford UP, 1992; A. Hourani and N. Shehadi, *The Lebanese in the World: A Century of Emigration*, Tauris, 1992; Jerome L. Singer, *Daydreaming and Fantasy*, Oxford UP, 1981.

Roderick Phillips, *Putting Asunder: A History of Divorce in Western Society*, Cambridge UP, 1988; Roderick Phillips, *Family Breakdown in Late Eighteenth-Century France: Divorces in Rouen, 1792–1803*,

Oxford UP, 1982; Dominique Dessertine, *Divorcer à Lyon sous la Révolution et l'Empire*, Lyon UP, 1981; H. Kent Geiger, *The Family in Soviet Russia*, Harvard UP, 1968 (on the very high divorce rates after the Revolution); G. B. Spanier and Linda Thompson, *Parting: The Aftermath of Separation and Divorce*, Sage, 1987; Elizabeth Martin, *Second Time Round: How Divorce Affects Future Relationships*, Macdonald, 1989; Sandra S. Kahn, *The Ex-Wife Syndrome*, Ebury, 1990; Mavis Maclean, *Surviving Divorce*, Macmillan, 1991.

Jerome David Levin, *Alcoholism*, Hemisphere, NY, 1990; G. E. Valiant, *A Natural History of Alcoholism*, Harvard, 1993; David F. Musto, *The American Disease: Origins of Narcotic Control*, Yale, 1973; Q. J. Rorabaugh, *The Alcoholic Republic: An American Tradition*, Oxford UP, NY, 1979; Herbert Fingarette, *Heavy Drinking: The Myth of Alcoholism as a Disease*, California UP, 1988; H. B. and H. C. Jones, *Sensual Drugs: Deprivation and Rehabilitation of the Mind*, Cambridge UP, 1977; Timothy Leary, *Flashbacks: An Autobiography*, Heinemann, 1983; John Rosecrance, *Gambling Without Guilt: The Legitimation of an American Pastime*, Cole, Pacific Grove, California, 1980.

14

Why compassion has flowered
even in stony ground

She has had misgivings about how people treat each other since the age of twelve. The first sign she gave was when she suddenly stopped speaking to her parents for several months, and began thinking about the fragility of adults, their tendency to fall to pieces. Their world seemed to be held together with a glue made of hypocrisy: her Alsatian father felt bound to obey his employer – 'a detestable man who used him as a whipping boy'. The first decision of her life was that he had chosen the wrong road: injustice should not be allowed to do as it pleased. But who would be her allies against it? She could not understand why her father did not rebel. Nor did her elder sister rebel. In the slums where she now works, she still cannot understand why some rebel against their humiliation, but others do not.

Marie-Thérèse Gaab was a solitary child, seeking refuge in the music of her violin. 'Nobody could understand me. My parents nicknamed me Cohn-Bendit and sent me to a boarding school, where I played truant, taking my violin with me. When I saw boys fighting, I tried to stop them.' She felt some affinity with her German mother, whose strong character she admired, but communication with adults was difficult. No teacher at school, no professor at the university of Strasbourg held out a hand to her.

A striking beauty, tall and blonde, she married Michel Krieger, whose body had been devastated at the age of three by a rheumatic disease and who, after eight years in hospital, lived in a wheelchair, radiating intelligence, sensitivity and triumph over suffering. He is the very opposite of submissive. Together, they have been replying to injustice, each in their own way.

As an artist, he has now attracted international attention, by reflecting the cold glare cast on the handicapped back to the supposedly normal public. He paints, for example, giant walls, with a tiny door. When you look carefully, you are not sure whether the door lets you out or lets others in. He reveals humans to be like molluscs, constructing shells around themselves, seeking protection from dangers they prefer not to know too much about. The boundaries between them become, in his work, puzzles rather than obstacles, interrogations of prejudice which are at once haunting and serene. Sometimes he feels that having married and had children and received recognition, he has melted the frozen looks that his appearance at first produces. But only sometimes.

She says he has come nearer to curing himself of the sense of being excluded than she has. For her work is with children for whom all doors seem to be immovably shut, in a school where nine-tenths of the pupils are from homes below the official poverty line. The high-rise blocks stand like scattered trees in a desert, from which there is no escape; there is nothing to do, nowhere to go, not even a cinema, no other signs of civilisation apart from a police station and the security men guarding the supermarket. The high culture of Strasbourg evaporates over the wastelands which separate it from this suburb of Le Neuhof, where half the population survives on benefits from the social services. 'It's a miracle that the children manage to come to school with their satchel each morning; they need courage to do so. I can't do much for them, there's no way of overcoming the obstacles in their path. When a girl whose brothers are drug addicts and whose parents are penniless asks me for help, I can't change her family; and after I've talked to her, in the evening, she returns to her alcoholic father and her depressive mother. The pupils leave school without qualifications and can't get work. But the school is the only effective institution here; the pupils would be utterly lost without it; the summer holidays they spend getting into trouble in the wealthy parts of town. Immigrant unemployed fathers, humiliated by their inability to support their families, recover their prestige through religion, become the representatives of God, and demand obedience from their wives and daughters. But the daughters don't want to be like their mothers. They work much harder than the boys at school, and reach out for their freedom at eighteen.'

What can Marie-Thérèse do for the boy who has been abandoned by his mother, and whose father is an alcoholic living with four others in a two-roomed flat? If she had a larger apartment, she would have taken him in herself: 'He comes to see me at least five times a day at school, just for a word from me. He didn't dare tell his teacher he had no money to pay for the swimming lessons; he has his pride; he is constantly violent; I'm the only one he will speak to. I accompanied him to court so that he could find his mother, but she said she doesn't want to see any of her children again.'

What can Marie-Thérèse do for a gypsy girl whose father is in prison, who is the eldest of eight children, whose mother keeps her at home to look after the rest? 'I'm revolted by how parents break their children because they find life too difficult. With different parents, some children would have succeeded brilliantly.'

Marie-Thérèse never stops being deeply moved by her encounters with such children; even talking about them brings tears to her eyes; she persists in fighting for their rights though she feels she cannot do very much for them. Still, she is someone they can confide in, for often they dare not speak to their parents. 'If after talking to a boy over a whole year, I can get him to accept his school, that's something. It's hard for me, because I feel isolated; the teachers say I'm too lenient, that I listen too much. I can't do anything about the economic, social, political problems; I'm only a catalyst; it's the children who have to win their freedom themselves. I remain a rebel; I can make things change from the inside, I change the way they see things: to that extent I believe I am doing well.'

To renew the battle every day, when the battle as a whole can never be won, takes enormous energy. She gets that strength from Michel. When she returns from work, to find him painting peacefully in his studio – a back room in their flat – 'my worries disappear and I'm in the clouds'.

I have crossed the Atlantic and come to a street whose tarmac is potholed and littered with trash: few cars pass this way. The buildings are crumbling, boarded up. Empty spaces are the gravestones of houses that have vanished. The shell of a factory suggests there were once workers here, taking wages back to their families, but the people in this street do not work: they just stand, or sit, waiting. Sometimes they look like scarecrows, their clothes hanging over bent, swollen, raw bodies. I am in the richest country in the world, which has built its marvels out of hope, but the eyes which observe my intrusion have no hint of hope in them. Or are they vultures waiting for prey? No need to be scared: they are victims, not vultures. They are waiting for the shelter for the homeless to reopen for the night.

The Pine Street Inn is not one of the tourist attractions of Boston, because this once beautiful building, with its elegant Italianate tower, is cordoned off by desolation. Inside it has become a huge garage, except that instead of cars, there are beds in long rows, several hundred in each dormitory, open plan. During the day, the guests, as they are called, have to go out into the world; many of them stand outside in the street, waiting to be readmitted and to have their clothes fumigated while they sleep. Only the sick are allowed to stay, and they sit clasping their blotched heads in rough hands, like so many Van Goghs, studies in still life, except for the occasional angry mumbling to themselves.

For seventeen years Barbara McInnis has been a nurse here, trying to make the pain go away, with no more success than people in hot countries who shoo away the flies, only to see them come back again. When she first came, she was living with a plumber who could see only the violence in the world: he always carried a gun, ever scared of being robbed. The Wild West has become urbanised here, an inseparable part of a city full of beauty and culture. The contrast became unbearable for Barbara, spending the day amid kindness in Pine Street and then going back to her lover's violence at home. She left him. With a dozen friends, she rented a cheap house in Jamaica Plain, bought cheap furniture and lived the simple life. They made Stone Soup (you beg a carrot from one person, a potato from another ...) and distributed it in a soup van; usually, they also had about five homeless people living with them. Now she is helping hundreds of them twenty-four hours a day.

How can she help them? She used to be a nurse employed by the city's department of health, where nurses have to do what doctors tell them to. But in Pine Street she was able to invent her own way. When the shelter first opened, the organisers made plans, but nothing turned out as expected. Barbara makes no plans; it was pointless telling the guests how to behave: 'I am a friend to them. That is what I do. My longest goal is to hope, if I see a man in the afternoon, that he will come back in the evening for another dose of medicine. Working here has meant a constant shortening of goals, to preserve my own sanity. Band Aid therapy is what I call my work. I have no dream of a different society. I never think about that. I'm busy surviving, like the guests. I can't see a new way of doing things, because I'm too busy surviving. When we started, we had 240 guests. Now they are 780. We are over-whelmed. My work is done minute by minute. I don't have any hopes that people will change.'

She used to hope. In the early 1960s, as a radical, she thought Ted Kennedy was a hypocrite, 'so I worked for his opponent, who later revealed himself to be no better'. That was the only time she got involved in politics. Never again. 'I feel helpless about who gets into office. Besides, making changes to society is not what I am good at. I am intuitive. I try not to make judgements. I don't force my beliefs. I don't proselytise. Everybody has the right to choose what they believe. But I have no tolerance for cults which destroy people's wills, though I don't do anything to change that.' She says she admires radicals who still have hope, or who are trying to get a working-class woman elected as governor of Massachusetts, but adds, 'Somebody should tell her that she won't be able to raise the money, that she hasn't a chance.'

Feminism once seemed a solution. She once tried to get some local feminists to help at the shelter as volunteers, 'but they wouldn't have anything to do with us. I had no experience of homeless women, so I asked the women's clinic to help us; they made many appointments but never turned up. Feminism has got its place, but I wouldn't call myself a feminist; I am not a militant one, though I believe in the rights of women.' Planning the future would be like writing fairy tales.

The down-and-outs who end up in Pine Street do not come to be lectured about morals or anything else. 'Some have no morals, they talk about stealing as though talking about their job.' And what does she talk to them about? 'I ask them how much they have begged.' She herself does not give money to beggars except on Sunday. For the people she moves among, 'money is not a big value'. Some need money because they are alcoholics, but though they ask for money, a conversation is as good for their loneliness as money. The world which is concerned with money is a different, outside world, which, she says, 'has a right to its beliefs: if I was in their place, I might do the same. I don't feel horror at the consumer society.'

Where does she get her energy to continue? From the need of her guests. And from St Francis of Assisi, whom she has studied. In her free time, she attends meetings of his followers, the Young Catholics, mostly in poorly paid jobs. 'Spiritually, I feel I belong with those people.' After Mass, they have a discussion, a drink in a bar, go dancing. But belonging has become more difficult because for security reasons they now lock the doors, and her work stops her being able to get in on time.

Though Barbara lives from minute to minute, dismissing all long-term plans, she is nevertheless waiting, as everybody in the world is waiting. 'I am still living in the 1970s. I need to change or I won't survive.' The only new thing she has learnt in twenty years is how to use a computer. Her mother is sick, and takes up a lot of her energy. 'When my mother dies I will be able to change.' Though she needs more time (like everybody else) to be able to work out how to change, she believes she will change only when 'things are not certain for me'. What she is doing at the moment is so pressing, so necessary, that she is very certain she ought to be doing it. It will need a great spiritual trauma to shake her out of her routine. Meanwhile the world goes on as it has, and she continues to put dressings on the sores of vagabonds who were indistinguishable not so long ago from the city's ordinary residents: one is a former lawyer, whose father was a Supreme Court judge; another is a doctor destroyed by drink; others have lost their skills, or their jobs, or their families, or their homes, as people lose the keys to their front door. They have slipped on banana skins, and been catapulted into a lonely planet.

In Pine Street, Boston, I also found the descendant of an Alsatian family which had emigrated to the USA, and prospered, and sent one of its children to Harvard and then to Oxford, an education which culminated in a brilliant doctoral thesis on socialism in nineteenth-century France (which is of course all about hope). But the author of it decided that rather than pursue a conventional academic career, and rather than fight all the forces that have made the world what it is, he preferred to mitigate humanity's sufferings on a daily basis; and now he lives frugally and humbly, caring for these homeless people. He has learnt too much to believe in panaceas: when it comes to making arrangements for the future, he is baffled, his brow creases; he hesitates even to make appointments a few days in advance. But when he moves among his broken, hurting, crumbling guests, he seems to feel that here at last all illusions and hypocrisies are stripped away; he does not have to pretend to be better than he is or to wonder what he is worth; there is a soft, gentle, cheerful look on his face, the look of one who can see dignity beneath the suffering, and whose reward is being able to say a kind word to a human in despair.

In 1944 the mother of the poet Yevtushenko travelled from Siberia to Moscow, where she witnessed a procession of 20,000 German prisoners of war marching through the streets. The generals strutted at their head, oozing contempt, determined to show that they still considered themselves superior. 'The bastards smell of perfume,' someone shouted. The crowd yelled its hatred. The women waved their clenched fists in anger, and the police had great difficulty in holding them back. But when the Russians saw how pitifully thin and ragged the ordinary German soldiers were, dirty, battered and completely miserable, many of them hobbling on crutches, the street became silent. Suddenly, an elderly woman broke through the cordon and held out a crust of bread to one of the soldiers. Then from every side, other women copied her, giving food, cigarettes, whatever they had with them. 'The soldiers were no longer enemies. They were people.' But such spontaneous outbursts of compassion have seldom been more than rainbows in the sky; they have not changed the climate; they have not so far stimulated a desire to listen to what enemies have to say.

Since the world began, compassion has been the most frustrated of emotions, more so than sex. Individual humans have normally felt attracted by only a small minority of the opposite sex, but when they have recognised suffering in almost anybody, they have frequently been deeply moved. Nevertheless, they have gone to great lengths to prevent compassion disturbing their other priorities. All sorts of philosophies and prejudices have been like chastity belts, keeping it firmly under control. Impulses of generosity by tribes, nations or groups have repeatedly withered away; but then suddenly they reappear, usually through the example of an eccentric individual, who breaks the convention that people should concentrate their compassion on their own families and beware of strangers. The obstacles that humans have created to prevent themselves from feeling compassion have accumulated over the centuries, reinforcing a persistent reluctance to get to know an enemy or a stranger personally or intimately.

The first obstacle has been a taboo against feeling compassion for the wrong person. Confucius (551–479 BC) drew a series of circles of compassion around the individual, of diminishing intensity, suggesting that one should love one's father most warmly, then one's family, then others in lesser degree according to their distance from the core: filial piety teaches 'how to love and how to hate people'. In other civilisations, too, 'good breeding' has meant knowing when to give and when to restrict one's compassion. Confucius' attitude was widely shared in every continent, because there seemed to be no wholly satisfactory alternative. Mo-Tzu (479–389 BC), who in ancient times was as famous as Confucius, asked, 'If everyone in the world practised all-embracing love (loving everyone as they loved themselves) ... would there be thieves or robbers? Would clans contend among themselves? Would states attack each other?' But there have been few humans who have loved everybody. Compassion has only been a truly powerful force when it has been felt for a particular individual, by an individual. All efforts to organise it as a system, in which everyone is treated equally, have spread it so thinly as to render it invisible. Mo-Tzu, and the many utopians who have repeated his ideas, wanted compassion to be not an emotion but a choice, a duty, a recognition of how things ought to be, denying that it needed to involve personal affection. He distrusted the emotions because they seemed too unreliable a basis for justice. The welfare state has agreed with him. So today compassion is carefully rationed.

There are no statistics to say how often perfect strangers have come to the aid of the sick and the sad, expecting no reward, simply because they were moved by suffering, and because grief is humanity's common enemy; if there were, many civilisations would probably seem less magnificent than their monuments suggest. But it is possible, to a limited extent, to observe compassion rise and fall and change its shape by watching the way the sick have been nursed at different times. Just as there have been waves of sexual promiscuity and of puritanism, so there have been periods when hospitals flourished, treating their patients with more or less respect, and others when they could not care less.

Hospitals for the sick have not always existed. In 1800, the USA had only two, in 1873 only 178. That country began erecting its temples of health in significant numbers only a century ago – by 1923 it had 4,978. The reason was that nursing the sick was originally the responsibility of families. The second, man-made obstacle in the way of compassion for strangers was fear of disease, deformity and every kind of handicap. Hospitals in ancient times were for the poor and the orphaned, excluding the sick, the insane, epileptics, the incurable and those with 'humiliating' or sexual diseases. The Assyrians – and virtually all civilisations since – spread the rumour that illness was a punishment for sin, and could only be cured by repentance, or magic. There was therefore little respect for those who attended to the physical needs of the sick, a task normally left to widows, fallen women, unemployed peasants: nurses were often unpaid, given only shelter and food, and treated as servants. The Code of the Emperor Theodosius

(AD 438) forbade nurses to go to the theatre because of their 'shamelessness, coarseness and violence'.

On rare occasions, compassion was deliberately practised as a virtue by rebels protesting against the cruelties of the world. In Rome, Fabiola, a patrician woman twice divorced, cured her unhappiness by becoming a Christian and establishing a hospital, working in it herself, gathering sick patients from the streets. Basil the Great, the genial and playful bishop of Caesarea (300–79), built a whole suburb where he could care for misfortune of all sorts, kissing lepers to show his support for them, personally attending to their needs. Such people seemed to be trying to turn the world's habits upside down. The motive behind their efforts was self-sacrifice. But that became the third obstacle erected in the way of the spread of compassion: most people did not want to be martyrs, nor monks or nuns, for whom souls mattered more than bodies.

In 1633 the Sisters of Charity were founded in France and became the archetype, in both Europe and America, of the lay woman nurse, generous and open-hearted. They did not live in convents, nor seek holiness in meditation, but wearing their own regional dress, travelled all over France, and later abroad, giving practical help and solace to the poor and the sick. Yet even they treated their work as a form of penitence and martyrdom: 'No one could regard them as other than holy victims who from the excess of love and charity for their neighbours willingly hastened to the death which they courted amidst the stenches and infections.'

The founders of this nursing order were an astonishing pair of saints, united in platonic love. Vincent de Paul (1581–1660) was a peasant by origin, kidnapped by pirates, a slave in Tunis for a year until he escaped, while Louise de Marillac (1591–1660) was the illegitimate offspring of an aristocrat and a servant, brought up 'like a man and also like a woman', trained in both philosophy and painting, married to a royal secretary and tormented by the thought that she ought to leave her husband and do something more useful. They believed that every pauper was another Christ on earth and that every sick person re-enacted the Crucifixion, and therefore had to be served with humility. To achieve humility, a nurse, they said, must work in an unfamiliar region – 'it is essential to be a stranger' – accepting that 'nobody is happy in foreign lands'. Happiness for themselves was not their goal. Instead they taught the Sisters to spread cheerfulness and cordiality among strangers, gaiety in the face of adversity: St Louise said that not a single day of her own life had passed without pain. All the tensions which afflicted nursing in subsequent centuries were foreseen by this extraordinary couple, who were both deeply practical and intensely idealistic. They were determined that there should be no struggles for power, no pride or rancour among the nurses, who rotated offices, with none being superior. The model for the selfless nurse of the future was set by them.

But it was not a model without snags. Nursing used to be done by both men and women, each for their own sex. In the seventeenth and eighteenth century it became an exclusively female profession. That opened up enormous opportunities to women, but with unexpected and unfortunate emotional results. People came to believe that only women were suited to nursing, that it was a task comparable to being a housewife and should be subordinated to male control. The Staff Surgeon of New York Hospital wrote in 1860, 'Men, even if they be of the best quality, cannot meet the demands of the sick. They have not the instinct for it.' Gentleness (though monks had cultivated it too) was considered a feminine monopoly. So a fourth artificial obstacle was raised in the path of compassion: the stereotype male was left out.

It was religious enthusiasm that created the great medieval nursing order of St John of Jerusalem and Malta, with its white cross; the red cross comes from another crusading order, the Knights Templar, while the black cross was flown by the Teutonic Knights Hospitallers: they all confusedly reconciled war with care for the sick. In the sixteenth century, the Knights of St John built one of the world's most magnificent hospitals in Malta, for 700 patients, in a ward 500 feet long, 34 feet wide and 30 feet high, each bed in a bell tent; the sheets were changed, when they needed to be, several times a day, and the food was as outstanding as the personal attention: rice and vermicelli and herbs, minced meat, fowl, beef and veal, fresh eggs, almonds and sweet biscuits; the knights had double portions. But in 1786 when the English reformer John

Howard visited it, he was horrified by the filth and the stench, concealed with perfume, and by the nurses whom he described as 'the most dirty, ragged, unfeeling and inhuman persons I ever saw. I once found eight or nine of them hugely entertained with a delirious dying patient.' The horses in the stables received more care. This has happened again and again to hospitals. The interests of the institution eventually override those of the patients. That became the fifth barrier in the way of compassion.

In the past, hospitals seldom employed physicians, so long as they were essentially alms-houses. An apprentice surgeon might examine new entrants, but only to exclude those who were unsuitable, because too sick. The nurses concentrated on feeding the patients, for that was what the poor seemed most in need of. However, in the late eighteenth century doctors protested that overeating was not necessarily the way to recover strength, and it was from this time that they gradually began taking over control of hospitals, transforming them into facilities for medical research, concerned with the technical cure of diseases, rather than with the spiritual needs of the patient. In the end, hospitals became institutions which had to be above all financially viable. When technology became more prestigious than bedside care, administrators got the upper hand. Compassion did not disappear, but it was subordinated to efficiency.

Florence Nightingale had said, 'I look forward to the abolition of all hospitals.' Her ideal was that nursing should be done in the home, and she was wary of nurses becoming hardened by too much medical learning: 'You cannot be a good nurse without being a good woman.' Now the whole world admires the almost superhuman kindness of nurses, and yet they are probably more frustrated than they have ever been. A recent enquiry into their morale by the University of Edinburgh found that less than one-fifth of British nurses were wholly satisfied with their work, while one-quarter were positively dissatisfied. That degree of discontent is far greater than exists among comparable workers. It is due not so much to low pay as to the feeling nurses have that they are prevented from giving their patients the quality of care they would like to give; there is a conflict between the values of the hospital system and their own. An Australian study reveals them to be troubled by the emotional detachment they are professionally required to show, by the inability of outsiders to appreciate how stressful it is to violate the taboos which keep sex and excrement and death under a veil, and how their superficial jollity conceals continuous tension. When nurses were elevated into a trained profession, when their numbers multiplied – in England, there were only 1,000 hospital nurses in 1861, but 56,000 in 1921 – many believed the solution was to fight for a larger share of power, against doctors, administrators and the state. But nursing could never be a job like any other.

That so many people are determined to be nurses, even when the conditions of employment are oppressive, is proof that compassion is like a fire that cannot be extinguished, however much cold water is thrown upon it. From time to time, nevertheless, the embers die down, at least on the surface. The great ancient Indian hospitals closed, one by one, at the end of the first millenium AD. In 1160, a visitor to Persia found sixty well-organised hospitals in a single city; in the fifteenth century, it was Florence which provided the most attentive hospital care in Europe; but all institutions eventually lose their enthusiasm. In the eighteenth century it was France, with over 2,000 hospitals, that had the largest hospital population in Europe, sixty times more numerous than England's. Every country has passed through alternating periods of charity and callousness.

The most insidious impediment to compassion expressing itself has been a cynical or despairing view of humanity. This can be illustrated by the experience of the USA. By origin it is a collection of complete strangers, who, as the first Governor of Massachusetts, John Winthrop, said, had to learn to love strangers – 'we must delight in each other' – but they have only half succeeded in doing so. Today 45 per cent of all adults engage in voluntary work, helping others for at least five hours a week. But more of them (54 per cent) believe that people generally bring suffering upon themselves, and that charity is not an answer, only a temporary 'Band Aid'. It is the opinion of four out of five Americans that people have to find solutions to their troubles for themselves.

Americans remain a nation of strangers, even when they are generous.

Though 42 per cent say 'I want to give of myself for the benefit of others', when the sentence is rephrased – 'I want to sacrifice myself for the benefit of others' – only 15 per cent are ready to do so. Albert Schweitzer insisted that compassion requires sacrifice not only of time and energy but also of the usual joys of life: it means one is forever haunted by the anguished faces of the poor and the cries of the sick. 'Anyone who experiences the woes of the world within his heart can never again feel the surface happiness that human nature desires.' But few Americans are willing to go that far.

The vast number of voluntary organisations have been shown, in a recent Princeton study, to have had the effect of placing limits on charitable feelings, assigning clear roles to be fulfilled, and so protecting people from being made unbearably miserable by the misery they see. Two-thirds of Americans say it is important not to get too involved in the problems of others: you have to take care of yourself first and if you have any energy left, then help others. Once they have done their good deeds, they switch off, and are not necessarily either warm or compassionate in their ordinary relationships. Church-goers have been found to be no more compassionate than non-church-goers; they do not stop to help a broken-down car, or look after elderly relatives more often. Only one-quarter of Americans believe that there is genuine concern for the needy in their country. Many volunteers confess that they help others because of the rewards, rather than what it does to those they help. Some say they enjoy 'the ego trip' of being perceived as generous, or heroic, and admit to being adventurers who happen to do things which are caring; it is the adventure they like. Those who say charity allows them to get to know people outside their own class often do not pursue that acquaintanceship very far. In the old days, Americans tried to be compassionate in obedience to God's commands. Now they more often use the language of therapy to explain their motives: it makes them feel good, it improves their self-esteem. But helping others comes at the very bottom of the list of things which improve their self-esteem. All this self-criticism does not cancel out the vast amount of open-handed philanthropy which has distinguished the USA, but it does show that many Americans have difficulty in presenting or perceiving themselves as compassionate, assuming that they must denigrate themselves to be convincing. The sixth obstacle in the way of compassion is the idea people have of what a human being is really like. That is no more unchangeable than any of the other obstacles.

In 1977 the country with the longest expectation of life was Sweden. Since then it has been Japan. They have each developed different forms of compassion to help them achieve their longevity. Sweden democratised compassion, more thoroughly than anybody else, providing every sort of care to everybody, more or less free, from the cradle to the grave. Japan, however, chose to combine modern with ancient methods, Western with Eastern practices, to surround itself with a veritable pot-pourri of compassions. The average stay in Japanese hospitals when it won its longevity record (WHO figures for 1977) was 42.9 days, compared to 8.1 days in the USA, 16.7 in West Germany and 12 or 13 in Scandinavia, Britain and Italy. While receiving modern treatment from as many doctors and nurses as anywhere else, Japanese patients are additionally looked after by relatives, who take up residence in the hospital and even cook for them, supplementing the three meals the hospitals provide. Hospital is not just a medical experience, but a social one, too; all one's relatives and acquaintances visit, each bearing a gift, to provide reassurance that one is still esteemed by the world despite one's misfortunes – 114 visitors was the count for one patient. The most popular gift after food is nightwear, because, whereas American hospitals treat patients as medical cases and wrap them up hygienically in uniform sterile garments, the Japanese see hospital as a holiday from conformity and the rigours of ordinary life, and they emphasise their normally restrained individuality by wearing their own nightclothes. Modern medicine, which treats diseases, is not enough for them; they indulge themselves with traditional Chinese medicine too, which treats each individual differently and as a whole person. They often also add religious medicine, even having their cars 'purified' at holy shrines. In this way they try to collect different kinds of compassion, which they savour, openly discussing all their medical symptoms, however trivial, with relish. Eighty-eight per cent of the Japanese claim to be suffering from some kind of illness.

After sexual liberation, it may be the turn of compassion to be liberated. But it is naive to imagine that a primitive, basic goodness is simply waiting to be set free and that everybody would then reveal the tenderness of a self-sacrificing mother. Emotions have always had to compete against each other.

Four thousand years ago, a Babylonian book of advice on how to behave said:

> *Do not do evil to one who has a dispute with you;*
> *Return good to one who does evil to you;*
> *Maintain justice to one who is bad to you;*
> *Be pleasant to your enemy.*
> *Do not utter slander; speak well of people;*
> *Do not say nasty things; speak favourably.*

Virtually all religions have embroidered on the same message, but without any more success than they have had persuading people to be chaste. 'Be free from animosity to all creatures' is the command of the Hindu god Krishna. Buddha and Christ showed in their own lives that it is possible to be compassionate even to those who harm one. Judaism contains the message that revenge is not wholly satisfying: 'Whoever is merciful to his fellow man is certainly of the children of Abraham.' The God of Islam is Merciful and Compassionate.

These sublime exhortations have had only a limited effect because the majority of the faithful have obstinately narrowed them down to make the saving of their own souls their first priority, showing more interest in winning a reward in the next world in return for being compassionate, than in the act of compassion itself. They have seldom seen it as part of their religion to appreciate infidels; and even the personal relations between donors and recipients of charity have always been ambiguous.

Today, however, different sorts of relationships between humans are becoming possible as a result of three innovations which are like three new sorts of emotional adhesive. First, interest in psychology has given additional meaning to the old injunction 'Forgive them, for they know not what they do'. Aggressive enemies now seem to be inspired as much by fear as by malice, imagining that they are acting in self-defence, battling with themselves as much as with their opponents. The assumption that everybody is basically selfish seems over-simple now that the confusions of the mind are being laid bare; enemies have been revealed as concealing shared passions which happen to have put them in the service of opposing goals; fanatics fighting on opposite sides have much in common; and the enemies of fanatics, though on opposite sides, have a lot to say to each other. For a whole century, humanity has been finding reasons for showing compassion to those who commit stupid or even horrible crimes; the law is everywhere becoming more merciful. So hitherto inconceivable attachments between the most unlikely people have come to seem quite reasonable.

Secondly, interest in the processes of communication has suggested new options, beyond the romantic dream of two souls understanding each other so well that they fuse into one, in blissful, total intimacy, sharing the same thoughts, experiencing compassion in the most literal sense; the dream seemed sweet to poets desperate to escape from their loneliness, but in ordinary life perfect harmony has usually been found to be suffocating. The Theory of Relevance has destroyed the belief that communication is simply a matter of encoding and decoding messages; it has revealed people as interpreting what they observe in the light of their own past experience, always being more or less approximate translators, and never wholly certain. There is no key to enable one to enter fully into another mind. Each individual's dignity and mystery remain intact. Communication is a fluid and flexible adhesive.

Thirdly, the world now contains a vast number of educated women, who have made old-style relationships impossible. The preachers who said 'Love your enemies' were never thought to be referring to the war of the sexes, but it is precisely in the intimate relations of men and women, which can no longer be based on dominance and dependency, that compassion is acquiring a new significance. Instead of being a mere gift of nature, which some have and others have not, it becomes the essential quality which the partners have to develop to make the partnership work. When the ideal relationship becomes one of equality and exchange, compassion has to be deliberately kept alight. Couples trying to be gentler with each other naturally change their attitudes to strangers, too, and even enemies, because personal experience, rather than the commands of authority, is increasingly the inspiration of public behaviour.

So though most of what newspapers report suggests that humans can still be as cruel and inconsiderate as they ever have been, there is no doubt that the revulsion against all forms of cruelty has greatly increased. Compassion is a rising star, even if it is from time to time obscured in the skies. But it will only rise if people push it, and the way they can do that is by deciding whether they are content with the old style of compassion, which meant helping others in order to clear one's conscience (and there was no need to speak to them, writing a cheque would do), or the new style, which means discovering others as individuals, exchanging understanding with them. For the only acceptable compassion in a world which regards all as of equal dignity is that all must feel they are contributing something, both sides must listen, and if the meeting is without a trace of compassion, it is incomplete, a wasted meeting.

The unemployment queue was not invented with this second possibility in mind: it was a necessary invention to free those in distress from the arrogance of those who monopolised the world's goods, but the anonymity it gave was only half a victory. It then became necessary to construct new affinities, new ways in which people could meet. However, for a long time individuals have been more preoccupied by the desire to be left alone, by the hope that mutual toleration would end their conflicts.

It is time to investigate why toleration has not achieved as much as was hoped from it.

Karl F. Morrison, *I Am You: The Hermeneutics of Empathy in Western Literature, Theology and Art*, Princeton UP, 1988; M. A. Nutting and L. A. Dock, *A History of Nursing*, 4 vols., Putnam, NY, 1907; Celia Davies, *Rewriting Nursing History*, Croom Helm, 1980; Christopher Maggs, *Nursing History: The State of the Art*, Croom Helm, 1987; Colin Jones, *The Charitable Imperative: Hospitals and Nursing in Ancien Régime and Revolutionary France*, Routledge, 1989; Monica E. Baly, *Florence Nightingale and the Nursing Legacy*, Croom Helm, 1986; Edinburgh University Nursing Studies Association, *Nursing Morale*, 1988; Jocelyn Lawler, *Behind the Screens: Nursing, Somology and the Problem of the Body*, Churchill Livingstone, Melbourne, 1991; Lindsay Granshaw and Roy Porter, *The Hospital in History*, Routledge, 1989; Charles E. Rosenberg, *The Care of Strangers: The Rise of America's Hospital System*, Basic Books, NY, 1987; Edwina A. McConnell, *Burnout in the Nursing Profession*, St Louis, Mosby, 1982; Jeffrey Blustein, *Care and Commitment: Taking the Personal Point of View*, Oxford UP, NY, 1991; Richard Waite and Rosemary Hutt, *Attitudes, Jobs and Mobility of Qualified Nurses*, University of Sussex, Brighton, 1987; Madeleine Leininger, *Transcultural Nursing*, Wiley, NY, 1978; Robert Wuthnow, *Acts of Compassion: Caring for Others and Helping Ourselves*, Princeton UP, 1991; Paul Adam, *Charité et assistance en Alsace au moyen âge*, Istra, Strasbourg, 1982; Fondation Jean Rodhain, *La Charité à l'épreuve des cultures*, SOS, 1985; J. Calvet, *Louise de Marillac par elle-même*, Aubier, 1958; Robert Sabatier, *Les Plus belles lettres de Saint Vincent de Paul*, Calmann Levy, 1961; Henri Lavedan, *Monsieur Vincent*, Plon, 1928; M. A. and L. Wallach, *Psychology's Sanction for Selfishness*, W. H. Freeman, San Francisco, 1983; 'Philanthropy, Patronage, Politics', *Daedalus: Journal of the American Academy of Arts and Sciences*, vol. 116, no. 1, Cambridge, Mass., 1987; David Thomas, *The Experience of Handicap*, Methuen, 1982; Cyril Elgood, *A Medical History of Persia*, Philo, Amsterdam, 1951; Jeffrey Hopkins, *Compassion in Tibetan Buddhism*, Rider, 1980; Emiko Obnuki-Tierney, *Illness and Culture in Contemporary Japan*, Cambridge UP, 1984; Annie Cheetham and M. C. Powell, *Women's Values and the Future*, New Society, Philadelphia, 1986; Carol Gilligan, *Mapping the Moral Domain*, Harvard, 1988; Jo Campling, *Images of Ourselves: Women with Disabilities Talking*, Routledge, 1981; Mary Adelaide Mendelson, *Tender Loving Greed*, Vintage, 1973; S. P. and P. M. Oliver, *The Altruistic Personality: Rescuers of Jews in Nazi Europe*, Free Press, NY, 1988; Jean L. Briggs, *Never in Anger: Portrait of* an *Eskimo Family*, Harvard, 1970; John H. Ehrenreich, *The Altruistic Imagination: A History of Social Work and Social Policy in the US*, Cornell UP, 1985; Anne Hudson Jones, *Images of Nursing*, Pennsylvania UP, Philadelphia, 1988; C. Hardyment, *Home Comfort*, Viking, 1992; Lauren Wispe, *The Psychology of Sympathy*, Plenum, NY, 1991; Robert H. Frank, *Passions Within Reason: The Strategic Role of the Emotions*, Norton, NY, 1988; Nancy Eisenberg, *The Caring Child*, Harvard, 1992; Seth Koven and Sonya Michel, *Mothers of a New World: Maternalist Politics and the Origins of Welfare States*, Routledge, NY, 1993.

15

Why toleration
has never been enough

When Sue is in the north of England, she feels at home, and relapses into the local accent. That is where she comes from. When she is in the south of Italy, she feels at home too, and is cross if the natives say she is a foreigner. That is where her husband comes from. How is it that she can appreciate two places which are so different, when most people have such difficulty in tolerating foreign ways? The roots of toleration are to be found in attitudes to compatriots as much as to foreigners.

Sue's father was a pillar of Tory respectability, a bank manager, son of a Baptist minister. 'Do what you promise to do,' he used to say, 'but don't rock the whole boat.' Nevertheless, supporting the establishment did not mean he was a member of it, nor complacent about it, nor satisfied with himself. He had been prevented from getting a university education; he had not received the promotion he felt he should have had; these resentments lay hidden under his bowler hat. Sue made fun of that hat, and disagreed with his politics, and protested that he would not give reasons for his point of view when she argued with him. She grew up an independent only child: being alone is a fact of life, like death, her mother used to say, discussing loneliness with her at the age of twelve.

There is no telling where a Protestant English background,

made of both tradition and dissent, can lead: in Mrs Thatcher's day, the leaders of all three main political parties were Nonconformists. In Sue's case, dissent expressed itself in an enthusiasm for things foreign. At university she changed from studying English to French when a charismatic left-wing lecturer captured her admiration with his preaching of existentialism, and Third World cultures, and the ordering of everything into systems of thought. She made a name for herself protesting. A year in Paris, 1967–8, revealed, however, that it is not so easy to be admitted into the fraternity of Francophiles. And yet Sue spoke the language so perfectly that no one could tell she was not French.

They did not welcome her on the barricades, or rather when they discovered her nationality, they sent her to the Foreigners' Committee. The stones she longed to hurl at the authorities who had been too pompous with her never got thrown. The rhetoric of Cohn-Bendit did not move her; the way the crowds were inflamed, their alternation between infatuation and scepticism puzzled her and made her feel that she did not belong. France jilted her. The lesson she drew was that the left ought to be more cynical, more foxy, there was no point in being idealistic; and she personally could not be a revolutionary, she did not know how to throw stones. To be a revolutionary, she believed, dogmatism was needed, and she had none. That problem has remained with her ever since. With the years, she has become increasingly fascinated by people whose views she does not share.

Without knowing any Italian she went to Italy and there found a foreign country to her taste. Can one explain this as one explains love? Should one accept that she likes the Italians for all the qualities they are admired for, their conviviality, their strong family life? Did they admit her into their society more generously than the French, whom she calls arrogant? In the same way that, as a northern provincial, she does not fit into London any more than Paris, so in Italy she is wholly comfortable only among the unpretentious, in some cities more than in others; even her husband considers himself almost a foreigner in its cooler climates. Their friends are Scots and southern Italians. She does not want family life to be an oasis. Her husband dreams of returning to his rural birthplace, so that he can bring up his son as he was, free to play in the streets, but it is a dream, like his dream to become a doctor in alternative medicine, his real interest. Wherever one finds oneself, the soil is never suitable for every crop. If one has only one's family roots to give one meaning, then roots can be like shoes that are too tight. Sue is lucky that she has in-laws she adores, admiring their resilience and the full flavour of their characters.

Here is someone fluent in three languages, who is so far from being a European that she says she dislikes international people. Her friends are all foreigners, the guests at her wedding were from seven nationalities, but she abhors the idea of being an expatriate. She is bringing up her child to be an Italian, he will have his English side added later. Her roots are now in Italy, she says, she needs roots. But she adds that the hectic colours of the Mediterranean do not really suit her.

How exactly did she come to be an expatriate? At university, she was training to be a teacher of French and history (English history, mainly) to English children. Practising to teach in a Manchester comprehensive school immediately brought out the same independence, or discomfort. She got on much better with the children who were shunned or mocked than with the conventional ones. 'I don't like the people one is supposed to like.' Her popularity with the older children, whom she treated as adults, disturbed the headmistress. 'If you try to make people think and get inside themselves, you almost automatically get up their nose – unless you find like-minded people, which you occasionally do.' But only occasionally. The clash with the headmistress did not end in a fight. Sue decided not to fight. She abandoned her training and went to Naples: she thought she could always come back. And for many years after, from her Italian exile, she struggled obstinately writing a doctoral thesis for her English university, never cutting off the possibility of return. Eventually she won her accolade. That makes it less necessary to return. Yet the vast amount of research she did had prevented her from playing the role of an Italian mother.

Her hesitations are the result of a discovery she made in her first year at university, that philosophers are divided about everything, none can claim to be right; that was 'a liberation', wonderful for her self-confidence, because it meant she could have opinions that were worth listening to; but it was disturbing because there has been no end to her questioning. And secondly, she has been infected by the intellectual virus that education can spread in some, the desire to read and read, and force apparent chaos into order, and argue and argue about how things fit together. But then she has doubts: though often excited by her discoveries, sometimes she decides she is not a great thinker: 'Others have always had a higher opinion of my intellectual abilities than I have.'

It may seem from her effervescent manner, her peals of laughter, her warmth, the astuteness of her judgements about people and the sensitivity with which she caresses their vulnerable spots, that she is above all a sociable person. 'People think I am friendlier than I am. I am intimate with only a very few people. I get very quiet if I am worried or upset.' At a first meeting she appears volcanically effusive, but she says, 'I don't like effusive people.' Behind the charm, she is a loner. She is attracted, she says, by thoughtful people, 'working inside themselves': that is a self-description.

Crossing the bridge between herself and other people has always been her obsession. As a child she used to follow strangers in the street, imitating their walk, practising being someone else. The theatre is her sesame. She has been directing plays since her teens: acting, for her, is the supreme method for getting under other people's skins. 'Acting gives you the ability to see other people's point of view, to understand other ways of looking and feeling. In acting you have to tap something in common. In actors I look for fluid people, who are not attached to one definition of themselves.' When her Italian students act English plays, she forces them to discover hidden parts of themselves. One of the attractions of Italy is perhaps that, in daily life too, it allows everyone to act in grander gestures than England does. In her lecturing, at any rate, she enjoys the performance; she pours all her energy into it, is delighted by her 'amazing results', loves the students, falls out of love with them, decides she must do research, then rediscovers that she loves teaching most of all: there is unending, exhausting passion in her search for contact. And yet she regrets that she has never found a director to push her to her limits, as she does to others.

Sue is not sure where she will go next. The ideologies of her youth have lost their sparkle. Is she still a rebel, was she ever a true rebel? 'I grate against people, I irritate people in the hierarchy, without meaning to.' Bringing up a child leaves much less energy for protest. 'The search for purpose has gone; instead, I now want quietness.' If only her husband would come out firmly in favour of a withdrawal into rural bliss, she would be tempted. But he is an intellectual too, who needs the stimulation of cities, imperfect though they are.

Sue is a double person, one of whom is always trying sympathetically to get under the skin of other people, while the second resists, unable to feel comfortable in certain kinds of company, or in certain relationships. She inhabits the very frontiers of toleration, where minute nuances both attract and repel, on the borders of a country known as Alienation. If such a gentle person has so much difficulty being a European, or even a citizen of one country, or one city, what new kind of passport needs to be invented for her?

Will people stop irritating, hating or fighting each other when the spirit of racial, political, religious toleration gradually spreads over the world? Only those with short memories can believe it. Toleration has always been like a summer season, followed by cold and storms. Its sunshine has come and gone through the centuries, warming the earth during the empire of Haroun al Rashid, for example, and even through the war clouds of Genghis Khan, only to disappear as though it had never been. Toleration is the glow which radiates in periods of euphoria and prosperity, when everything seems to be going well, so that there is no need of scapegoats. Or else it is autumn, deceptively calm, an accompaniment to the exhaustion that comes after periods of blood and strife. It has never been a permanent cure for hatred or contempt. Preaching toleration to the humiliated, the angry, or the desperate has always been useless. But it is possible to break out of the cycle of toleration and persecution; there is a third way, more exciting than both. Toleration has been only a preparation for it.

Today, only 39 per cent of the French regard toleration as one of their important values (33 per cent on the right wing, 45 per cent on the left). Exactly the same percentage of the British have said that it is impossible for people of different colour to live in the same area without problems. Toleration has failed to capture the popular imagination, because it is not a passion: the reluctant acceptance of a burden, putting up with what one cannot avoid, is not exciting enough. Nor has education been of much help, as was seen when in 1933 the most educated nation in Europe suddenly became the most intolerant that had ever existed. The educated have as poor a record for tolerance as the ignorant, because it is as easy to be infected by intolerance as by the common cold. Though prosperity has repeatedly been the reward of the countries and cities which have been the most open to strangers, those who have longed to be rich have always been tempted by intolerance, fearing envy. In the USA, over the last two decades, 29 per cent of blacks and 16 per cent of whites say that they have become less tolerant of people of different colour.

Toleration is not the modern medicine it is made out to be, but an old folk remedy, with only short-term effects. Though some civilisations have enabled different races to live peaceably side by side, anger against foreigners and minorities has flared up again and again, often with a suddenness which catches them unawares. And even after so many centuries of experience, the tolerant are still liable to be held in suspicion, or accused of lax morals. Little has changed since the Roman Emperor Honorius, in the fourth century AD, condemned over-tolerant dandies to banishment for the crime of wearing trousers, which was the garb of his German enemies. Roman soldiers wore miniskirts: that was the limit of toleration.

Proclaiming the Rights of Man did not automatically make France a tolerant country. In the early nineteenth century, when there was competition for work in the new textile factories, riots broke out against English and German immigrant workers. Then Belgian immigrants became the target, when unemployment struck in towns like Roubaix, where they made up half of the population. The foreigners in Paris, who doubled in number while Haussmann was rebuilding the city, should not have been surprised that welcome turned into animosity as soon as the boom ended. The attacks on Algerians today echo the 'Marseilles Vespers' of 1881, when a crowd of 10,000 went on a rampage through that city, attacking Italians and their property.

However, this does not mean that humanity is powerless in the face of fundamentalism and dogmatism. The taste for toleration has deep roots, but it is not necessarily from one's ancestors that one acquires it. The roots stretch back to ancient India, which has the longest tradition of toleration in the world. There the major religions, and many others, co-existed more or less harmoniously for over a thousand years. And yet India succumbed in 1948 to a horrific outburst of fanatical intolerance, in which more than a million people were killed. This was not inevitable, but the story of how it happened has echoes in everybody's biography, however far they live from the Ganges River, and whatever they may think of its waters, which are supposed to purify humans of their sins.

Hinduism is essentially undogmatic: it is possible to be a good Hindu whether one believes that there is only one God, several, or none at all. The Hindus' ancient hymn book, the *Rig-Veda*, stands out among religious works for its tolerance towards doubt:

> *Whence all creation had its origin,*
> *He, whether he fashioned it or whether he did not,*
> *He who surveys it all from highest heaven,*
> *He knows – or maybe even he does not know.*

Some Hindus worship Vishnu and some Siva, each looking upon their god as supreme, but each accepts that the other's god is worthy of worship too, and that ultimately, perhaps both sides are right. Hinduism started 5,000 years ago with the premise that things are more complex than they seem, that reason is not enough to discover the truth, and that how near one gets to the truth depends not just on one's knowledge but on the sort of person one is, on how morally one lives. The attitude of Hindus to new doctrines has been to approach them with the thought that most probably have a trace of truth in them, worth absorbing, but that in the process it may be modified, as food is when it is cooked and eaten. If the result of combining other people's ideas is contradictory, they take the view that the contradictions of life have to be accepted, even though this can produce disconcerting results, in politics for example, where everybody seems to disagree, and then they vote unanimously for a motion they have argued against.

The most influential of all India's philosophers, the Buddha (who died around 483 BC), did not claim to be a god and was not interested in being the head of a sect. In northern India, he was absorbed into religious life by being looked on as the ninth of the ten incarnations of Vishnu, and most ordinary laymen regarded his teachings as one of many faiths, not mutually exclusive, all being worthy of respect. Buddha did not attack other religions, nor was he persecuted. Goodwill, compassion and friendliness to all living beings was his message: 'Just as a mother, as long as she lives, cares for her only child, so should a man feel all-embracing love to all human beings. He should feel boundless love to all the world, above, below and across, unrestrained, without enmity. Standing, walking, sitting or lying down . . . he should be firm in the mindfulness of love. For this is what men call the Sublime Mood.' Buddha's respect for disagreement was such that he did not require a vow of obedience even from monks, who had to be unanimous before taking a decision.

Free speculation about heaven and earth has remained a widespread passion in India; 300 million gods and countless gurus offered innumerable models of behaviour and belief. Jainism, which has survived since the sixth century BC and which is an atheist religion, is based on the 'doctrine of maybe', holding that it is impossible to know or describe the world accurately: it preaches non-violence towards everything that exists, for even stones and insects deserve respect.

This almost limitless toleration of belief existed side by side with the world's most long-lived intolerance of free social contact, the caste system. Toleration means leaving people alone. Caste means keeping away from people who might pollute one, not eating with them, not marrying them. The two ideas are related. Toleration does not necessarily mean that minorities cease to live as isolated communities marrying only among themselves, interested only in themselves. That isolation is a major reason why they have often suddenly found themselves friendless and no longer tolerated.

The first ruler, anywhere in the world, to attempt to stimulate the spirit of religious toleration was the emperor Asoka, who reigned over most of India between 264 and 228 BC, and who was not, as so many monarchs have been, just another bundle of human weaknesses trying to do his duty. Asoka ascended the throne at the age of twenty-four, after arranging the murder of his elder brother and many other rivals: a typical potentate with a huge harem, slaughtering hundreds of animals and birds each day for his table, he tortured his subjects in a terrible prison he built for the purpose. Then, according to legend, he encountered a young Buddhist monk who was seemingly immune to the sufferings inflicted on him. Asoka was so impressed that he became a Buddhist and rethought the whole purpose of government. Abandoning war, he devoted himself to increasing 'safety, self-control, justice and happiness'. His laws, preaching humanitarianism, vegetarianism and compassion to all living creatures, are engraved on cliff faces and on huge rocks which still survive, written in the form of a personal confession, speaking of his remorse for his cruelty and blindness in the past, and of his desire to dedicate himself to improving the way people treated each other. He established a new class of official, the 'officer of righteousness', to help them behave more decently. Originally known as Asoka the Fierce, because of his malicious and impetuous nature, he became Asoka Pleasant-to-Behold, even though he was apparently ugly and had been hated by his father for that reason. Withdrawing from the royal sport of hunting, he urged his subjects to entertain themselves instead by travelling on pilgrimages; he ordered that rest-houses be built every half mile on the roads, and hospitals and herb gardens for both humans and beasts. Though a Buddhist, he also encouraged rival sects: his Edict Number 12 proclaims that people of all faiths should be respected. The ultimate purpose was to set an example so that all the kings of the world would change their ways too; and he seems to have convinced five petty monarchs in Greece and the Middle East to adopt his principles. One legend claims that his mother was a Hellenic-Syrian princess; at any rate one of his laws, in Kandahar, is engraved bilingually in Greek and Aramaic. Of course, in real life there were limits to Asoka's generosity and pacifism: he did not disband his army, nor abolish the death penalty. But the significant fact is that the Indians have chosen to remember him as the greatest of all their rulers, and one of his monuments features on the seal of the present Indian republic.

When Islam spread to India, and Muslims became the ruling class (from 1021 to 1858), and when Christians arrived as merchants, missionaries and finally rulers, the Hindus avoided confrontation, preferring an ambiguous exchange of beliefs and cults. The three religions mingled their traditions, borrowing from each other. St Francis and St Thomas were admitted into the Hindu pantheon as healers, while Hindu divinities were transformed into Muslim or Christian martyrs. Christians joined Hindus in ceremonial processions while Hindu rulers sponsored Christian shrines. The Indians who converted to Islam saw it as an additional form of divine protection, as well as an impressive military power, but they did not throw away their old gods. Muslims were treated as an additional, new caste. This incoherent mixing of beliefs is absolutely normal in the early stages of conversion and conquest, found not only in Melanesia and Africa, but in Europe too, where Christian saints were adapted to perpetuate pagan healing rites. It is absolutely normal for those who are not interested by theological niceties to pay lip-service to official doctrines, while interpreting them in their own way.

Living with contradictions is an art, involving, like all art, both spontaneity and cunning thought. If it is simply woolly thinking, it leads nowhere. To be fruitful, it has to be a deliberate choice, an imaginative creation. But there are always those who find it difficult to cope with contradictions, or with art which is not straightforwardly simple, and in the nineteenth century they became more influential in India (and elsewhere too). Its three major religions lived in remarkable harmony until leaders arose demanding an end to ambiguities and contradictions, and a clean break between doctrines they considered to be incompatible. Hostility between the religions was generated by learned theologians, who studied the ancient books, recreating what they believed to be a purer doctrine, and urging obedience to it as a cure for all discontents. They might have gone unheeded, except that economic crises provided opportunities for inter-communal animosities to be aroused;

poverty was blamed on moral corruption, and disorder on the neglect of religious duties.

When the British became masters of India, dislodging its Muslim rulers, they used Hindus to run the administration, emphasising the divisions between religions (bureaucracies specialise in putting people into pigeon-holes); then they played the religions off against each other. The Hindus were sucked into English schools and spat out as minor officials, while the Muslims remained isolated in Islamic schools, listening to their own teachers, who as an alternative to government jobs offered their pupils salvation and respectability through religious orthodoxy. The old cosmopolitan rules of gentlemanly conduct (originating in Persia and known as *adab* or decency), in which Muslims used to delight when they felt in control of the world, were now replaced by a new model of behaviour, involving strict obedience to holy law (*shariah*), grounded in suspicion of other religions. The masses were attracted, because they were given a new dignity: whereas they used to be despised by the élites, as ignorant and incapable of refined behaviour, now anybody could, through piety, become an ideal Muslim. The elimination of foreign influences became a thrilling new passion, a purpose in life, in the pattern that fundamentalisms everywhere have followed. Many of these intellectuals were brilliant scholars and brave men, with high ideals, who never foresaw the bloodshed that would result from their pointing out of the distinctions between religions; but some thought blood was worth shedding in the name of truth.

Is any answer possible to intolerance resulting from people defending themselves by emphasising a separate identity? The most interesting experiments to find out were carried out by M. K. Gandhi (1869–1948), and they are significant far beyond their Indian context. He believed that in order to be tolerant an ordinary person needed first to acquire enough personal strength, and that meant fearlessness. As a child he had lived in terror of the dark, of thieves, ghosts and snakes; as a youth he was timid; when taken by a friend to a brothel, he was speechless and immobile. All his courage was the result of a determination to overcome his weakness, which he worried about ceaselessly. Peace inside himself was his goal; and he thought he could attain it by enabling others to find it too: harmony between individuals and communities could come only from a cure for inner anxieties. Such a solution meant transforming politics into a psychological adventure, depending not on a leader, but on the efforts of every citizen; instead of blaming others for one's discontents, one should change one's own private behaviour; example was the best way of exerting influence on public events.

So he spoke openly about his own private life, admitting his personal difficulties, discussing his wife's unhappiness with his rejection of ordinary domestic comforts, his insistence that they should live as simply as possible, that all wealth, beyond what was necessary for basic essentials, was a trust, to be used for the general welfare. His sons resented his neglecting them, which he did not deny, insisting that one's love should not be confined to one's kin and should be offered rather to everyone with whom one could develop an affinity: he counted no less than 150 individuals as his 'relatives'. The spread of 'fellow feeling', of personal friendship, seemed to him to be the way to overcome the barriers of religion, nationality and class. Love should be expressed above all by service, forgetting one's selfish interests, devoting oneself to others.

In experimental villages, he attempted to put these ideas into practice. To the great scandal of many Hindus, he included untouchables in them, the caste whom everybody avoided. He too had been brought up to keep them at a distance, to leave unpleasant work like cleaning lavatories to them, but here he insisted on cleaning lavatories himself, and on his wife doing likewise. Each day he devoted an hour to helping with menial tasks in a hospital. Only thus, he believed, could old barriers be eliminated.

However, being friends with everyone was not easy, particularly since Gandhi was not a sociable person in the conventional sense. He did not have friends who were his equals: Nehru was like a son to him, Gokhale like a father. It was among the mature women who helped him in his work that he found the broader emotional sustenance he needed. That is significant, and not just anecdotal. Though his views on women were old-fashioned, though he thought of the ideal woman as essentially a faithful wife, in these friendships he discovered how much more they could be, without however seeing that friendship between the two sexes might become the basis of a new kind of society. One of his women friends he used to call Idiot, and she called him Tyrant; he listened to her nonetheless, but he heard only part of her message.

Gandhi did not aim to abolish all traditional distinctions. It was not his hope that either Christians or Muslims would ultimately see that Hinduism was a superior religion. For him, all religions had both defects and virtues. Preaching them did not improve the world, because most people did not practise their religion properly. Far from wanting to convert others to his beliefs, he urged that everybody should strive to fulfil the teachings of charity of the religion into which they were born. Truth was many-sided: there was no need to simplify it into a single creed. But this meant that he did nothing to counter the zealots who treated their separate religions as blinkers.

He demonstrated that one individual can change the behaviour of 600 million people, momentarily, that something near to a miracle can happen. When in 1947 Muslims fleeing to Pakistan were butchered 'by the trainload' – his comment was 'we have turned well-nigh into beasts' – when Muslims retaliated, when Calcutta rioted and news of atrocities brought about new atrocities, Gandhi the Hindu took up residence in the Muslim quarter in a Muslim's house, unprotected by police, a symbolic gesture of fearlessness and reconciliation. Within hours, Muslims and Hindus were embracing each other, even praying together at each other's temples and mosques. Then the rioting resumed. Gandhi went on a fast, announcing he would continue until sanity returned. Once more the violence stopped, and people began handing in their arms. The Viceroy Mountbatten said, 'He has achieved by moral persuasion what four divisions would have been hard-pressed to have accomplished by force.' But he achieved it only momentarily. The crowds, despite their prejudices, were deeply moved by his willingness to sacrifice his life in the cause of peace. However, it was not long before the hatred resurfaced.

So Gandhi both succeeded and failed. He succeeded in that he showed that animosities could be overcome. But his success, though real, was only temporary, because the deep roots of animosity were left untouched. He failed because he never got to know the Muslims well enough. Humans, he said, were all the same, the soul of each was part of a universal soul; toleration for everybody seemed natural to him. However, his experience showed that general goodwill and 'fellow feeling' towards the whole of humanity could be drowned by sudden waves of resentment. In his experimental villages, bickering and incomprehension were the rule. Had he lived 120 years, as he thought he could do with his frugal and sometimes bizarre diet – he once thought he could survive on fruit alone – he might have placed less emphasis on the spread of general goodwill and more on the building of individual relationships, on becoming personally interested in the beliefs and ways of thinking of other individuals. He himself was always showing concern for the health and families of the people he met, but he devoted less time to penetrating their mentalities.

He was convinced, rather, that it was possible to shame people into behaving more generously. He allowed the British army to beat him and imprison him, offering no resistance, simply going on a fast, defying them to kill him. His willingness to accept martyrdom certainly demoralised the British. But he was wrong to believe that soldiers would not beat innocent people: many, on the contrary, thought that they were giving the Indians 'what they deserved'. The Viceroy Field Marshal Wavell, who believed in force and traditional politics, did not possess enough imagination to be interested in changing the rules of politics or war, as Gandhi was proposing: 'This malevolent old politician,' he wrote, 'for all his sanctimonious talk, has, I am sure, very little softness in his composition.' However, Gandhi did not understand his opponents either. That people should enjoy being nasty was incomprehensible and unbelievable to him. Convinced that quarrels are ultimately due to misunderstandings, and that friendly discussion could resolve difficulties, aspiring to be an 'interpreter', he never had a genuinely friendly conversation with the leader of India's Muslims, M. A. Jinnah (1876–1948), the creator of Islamic Pakistan, 'the Land of the Pure'.

The two men began as political rivals in the same Congress Party. Gandhi never understood how his own triumphs were experienced as humiliating defeats by Jinnah, who developed an ever-growing personal dislike for him. It was only after twenty years of frustration that Jinnah abandoned his aim of Muslim–Hindu collaboration in an independent India; he had wanted the Hindu majority to solve its minority problems by making concessions, arguing that a minority, feeling oppressed and excluded from power, had nothing to concede; but he never succeeded in obtaining the gestures of generosity he was seeking, and his proposal of a constitution in which elected representatives would be obliged to obtain the votes of both the minority and majority religions was rejected. Far from being a fanatical Muslim, Jinnah was an urbane, cosmopolitan, extremely rich lawyer, married to an independent young woman of non-Muslim origin (the daughter of a Parsee or Zoroastrian millionaire industrialist), and their daughter married a Parsee convert to Christianity. It was only in public that he concealed his taste for ham sandwiches and pork sausages. His private ambition was 'to play the role of Romeo at the Old Vic', and his relaxation from politics was reading Shakespeare aloud. When his Indian career was at its nadir, he took up the practice of law in London and thought of becoming a British Member of Parliament. Gandhi failed to spot the links, the qualities and attitudes they had in common, which might have prevented Jinnah from turning increasingly inwards to his own Muslim community. Jinnah did that only out of despair with what he considered to be the arrogance and insensitivity of the Hindus, for whom Muslims were, as he put it, outcastes. On proclaiming Pakistan independent, he promised it would be a country without discrimination of any sort. This relationship illustrates how personal animosity becomes a fog which separates individuals searching for the same goals and makes them crash in a public disaster.

In Gandhi's life, one can see the limitations of toleration. He shared with other Hindu sages a mistrust of admiring followers, who might turn his doctrines into a fossilised catechism; only 'fellow scientists and co-experimenters' were welcome, keen to struggle with themselves, side by side with him, in search of something better. Nostalgia for ancient rural bliss comforted him, but he really wanted to change the world and create a completely different one. 'In the secret of my soul I am in perpetual quarrel with God that He should allow such things to go on.' Ultimately, his weakness was that he was a utopian: his extreme asceticism, his denial of the most elementary pleasures, like family, food and sex, meant that he was a saint too remote from ordinary people. Praising poverty as spiritually cleansing, condemning cities as demoralising, could not be convincing to the starving and unemployed. He demanded too much heroism, and sometimes got it, but not for long, not for a lifetime. For him, release from his anxieties came only during his self-imposed torture of his body: he was wrong to assume others felt the same way. A saint can be an inspiration, but not a model: it is not possible to educate, persuade or force people to be tolerant for very long. Even Gandhi, with all his charisma, did not 'melt the hearts' of his oppressors, as he had hoped. After softening, hearts harden again. Asoka too was wrong to think that he was changing the course of history, and that his righteousness would last 'as long as the sun and the moon'.

As a youth studying law in London, Gandhi had learnt to dress as an English gentleman, but as he became more and more involved with his 'experiments with truth', he took to wearing only a loin-cloth, as a symbol of his rejection of Western civilisation. In fact, he was a great synthesiser of Western and Eastern traditions, as much influenced by the Sermon on the Mount, Tolstoy and Ruskin, and by his battles against racial discrimination in South Africa, where he lived from 1893 to 1914, as by Buddhism and Jainism. Though he used non-violence and passive resistance, which were ancient Hindu devices for resisting oppression, and developed them with unprecedented skill, he condemned Hindu society as corrupt and in need of complete rejuvenation, almost (but not quite) as decadent as the West's urban culture.

Having won independence for India, he looked on that achievement as his greatest failure, because what emerged was nothing like the country he dreamed of, purged of intolerance, devoted to spiritual self-improvement and rejecting violence. That the Muslims should break away to form the separate state of Pakistan was intolerable to him, because it denied India's pluralist talent for embracing the apparently incompatible. To avoid it he proposed to make his Muslim enemy Jinnah president of Hindu India, but only he could contemplate such an ingenious solution: his followers were horrified, and unwilling to sacrifice their own ambitions. For appearing to be too tolerant towards Muslims, he was assassinated by a Hindu fanatic: even Hinduism could spawn intolerance. Gandhi's life confirms that toleration is an insufficient remedy even when practised by a very exceptional man, and even in a country where tradition is in its favour.

The West has for most of its history been intolerant, in varying degrees, and only began thinking seriously about becoming tolerant during the Reformation and the wars of religion, when governments found that they could not get all their subjects to think alike, however severe the penalties for dissent. Persecution eventually exhausted the persecutors. It became impossible for them to be certain that they monopolised the truth, or that the truth could ever be known for certain. So toleration was adopted for largely negative reasons, not out of respect for other people's views, not out of deep knowledge of what they believed, but in despair of finding certainty. It meant closing one's eyes to what other people believed.

That is no longer satisfactory. The tolerated are increasingly demanding to be appreciated, not ignored, and becoming more sensitive to suggestions of contempt lurking behind the condescension. They do not want to be told that differences do not matter, that they can think what they like provided they keep to themselves, out of the way of the majority. Besides, there is no longer a majority to tolerate the minority, because majorities are disintegrating into more and more minorities. Mere toleration would end up with general indifference.

The ideal of toleration can now be seen to be not a goal, but a stepping stone. Understanding others is the great adventure that lies beyond it – more ambitious than the ancient obsession with conquest. Exploring the mystery of other people's thoughts and feelings is the new spiritual quest. Finding empathy is the new reward of intimacy. A pathway for these aspirations has been opened up for the first time in history, now that the two sexes are trying to understand each other on a basis of equality, as they have never done before. It is wrong therefore to assert that the world has lost its sense of purpose, and can never regain it, and that there is nowhere to go but backwards. A new adventure has begun, even though old habits survive.

John S. Strong, *The Legend of King Asoka*, Princeton UP, 1983; Susan Bayly, *Saints, Goddesses and Kings: Muslims and Christians in South India 1700–1900*, Cambridge UP, 1989; L. S. May, *The Evolution of Indo-Muslim Thought from 1857 to the Present*, Uppal, New Delhi, 1987; Katherine P. Ewing, *Shariat and Ambiguity in South Asian Islam*, California UP, 1988; Asim Roy, *The Islamic Synthetic Tradition in Bengal*, Princeton UP, 1983; Barbara Daly Metcalfe, *Islamic Revival in British India: Deoband 1860–1900*, Princeton UP, 1982; Barbara Daly Metcalfe, *Moral Conduct and Authority: Adab in South Asian Islam*, California UP, 1984; A. A. Engineer, *Communalism and Communal Violence in India*, Ajanta, New Delhi, 1989; Judith M. Brown, *Gandhi*, Yale UP, 1989; Bhikhu Parekh, *Gandhi's Political Philosophy*, Macmillan, 1989; Stanley Wolpert, *Jinnah of Pakistan*, Oxford UP, NY, 1984; Henry Orenstein, *Gaon: Conflict and Cohesion in an Indian Village*, Princeton, 1965; Helmuth von Glasenapp, *The Image of India*, India Council for Cultural Relations, New Delhi, 1973 (on Europe's discovery of India's culture in the eighteenth century); A. Leslie Willson, *A Mythical Image: The Ideal of India in German Romanticism*, Duke UP, Durham, 1964; A. L. Basham, *The Wonder that was India*, 3rd edn., Sidgwick and Jackson, 1967; Jean Herbert, *Spiritualité hindoue*, Albin Michel, 1972; R. C. Zaehner, *Hinduism*, Oxford UP, 1966; Preston King, *Toleration*, Allen and Unwin, 1976; R. Dahrendorf, *The Modern Social Conflict*, Weidenfeld, 1988; T. Raychaudhuri, *Europe Reconsidered: Perceptions of the West in Nineteenth-Century Bengal*, Oxford UP, 1988 (illuminating far beyond its apparently limited scope); Herbert McClosky and Alida Brill, *Dimensions of Tolerance: What Americans Believe about Civil Liberties*, Sage, 1983; Colin Holmes, *A Tolerant Country?: Immigrants, Refugees and Minorities in Britain*, Faber, 1991; A. D. Falconer, *Reconciling Memories* (in Ireland), Columba, Dublin, 1988; Michel Wieviorka, *La France raciste*, Seuil, 1992; Barry Troyna and R. Hatcher, *Racism in Children's Lives: A Study of Mainly White Primary Schools*, Routledge, 1992; Philip Cohen, *Multi-Racial Britain*, Macmillan, 1988; Robert L. Holmes, *Non-Violence in Theory and Practice*, Wadsworth, Belmont, California, 1990; Louis Dumont, *Homo hierarchicus: Essai sur le système de caste*, Gallimard, 1966; Joelle Affichard and J. B. de Foucauld, *Justice sociale et inégalités*, Esprit, 1992.

16

Why even the privileged are often somewhat gloomy about life, even when they can have anything the consumer society offers, and even after sexual liberation

A painter commissioned to do a portrait of Annick Geille would accept with alacrity, thinking there would be no problem. The world never tires of admiring a chic Parisienne. But he would soon be in trouble. Should he show her with an air of assurance which proclaims 'I belong here, with the best of them', or breathless and dishevelled, in clothes which say 'I'm not wearing anything that will distract you from seeing me as a person?' Should he paint her with the smile which transforms her face and makes her look an adolescent, shy behind her fringe, disarmed by a compliment? But those who know the insides of the French media recall that she has been a formidable power in the press, with a determination reminiscent of the founder of *Elle*, Hélène Lazareff. How to paint at the same time her feather-light handshake, the fragility she is so conscious of? When the business moguls headhunted her for her last job, she was struck above all by how large the men were, how small she was.

Annick Geille has spent her life showing her readers what it means to be an independent woman, on the basis of experiments on herself. Whereas Simone de Beauvoir, the last generation's philosopher of women's independence, tried to explain away the unhappiness Sartre's infidelities caused her, and insisted that the relations of men and women could ultimately be put on to a satisfactory basis, Annick Geille doubts it: for her there is no solution: life is inevitably tragic.

Perhaps the truest portrait would be a series of surrealist studies of her nose, which, even more expressive than that of an animal on the alert, constantly changes shape, betraying enchantment, curiosity, awe, pain, disapproval, boredom. What has made her professionally successful is her talent for smelling 'l'air du temps', not the mood of the season as fashion experts do, but longer and deeper trends in emotion: she is a barometer of the climate which subtly modifies the way women respond to men, a thermometer of the temperature at which they like to meet. The successive stages of her career have been a commentary on the evolution of attitudes among educated, ambitious women. The four magazines she has edited each represent a phase in the emotional history of France and, even more, a search for a way out of the dilemma which that history has not solved: when you get what you desire, what do you do next, and why does it remain desirable so briefly?

In 1973, still in her early twenties, selected to launch the tricolour version of *Playboy*, she at once gave the permissive age a twist. The photographs of naked girls, she insisted, were not enough to satisfy intelligent Frenchmen. She persuaded the most respected intellectuals to contribute articles explaining what she called 'refined hedonism'. Her coup was to spot the young philosopher Bernard-Henri Lévy as a rising star: she commissioned the first major press interview he had. But when she sent it for approval to Hugh Hefner, he complained 'Who is this man?' She telexed back, begging to be allowed to publish the fifteen pages of diatribe by an unknown: 'Trust me; you will soon hear of him.' If it flopped, she would be sacked. Three weeks later, *Time* magazine imitated her, and did a cover story on Lévy and the New Philosophers. Never again was her editorial choice questioned.

Her nose had sensed that the moment had come for a new rebellion to replace that of the exhausted students of 1968. She reported one saying, 'I no longer believe in politics. I belong to a lost generation, without hope, lost not only to this or that cause, but to every cause that might ever come along. We know from the start that life is a lost cause.' Even human warmth frightened him.

The only thing he believed in, he said, was the need to tell the truth.

Ever since, Annick Geille has treated delight in innovation and disenchantment with human frailty like balls that have to be kept in the air at the same time. The current which wants the left to be more right wing and the right to be more left wing pleases her, refusing to be either. Her principle is not ambiguity but obstinacy in the face of contradictions. She applied it above all to the raging war between women and men. 'When I am with men, I feel a woman,' she says, 'and when I am with women, I feel a man. Among people of the left, I am of the right. When there is a majority, I am part of the minority. So it is with the feminists: I don't feel part of them. That is liberty.' Annick Geille's Bible as an adolescent had been Simone de Beauvoir's *Second Sex*, but she published Nelson Algren's merciless ridiculing of the philosopher who thought that she was having a romantic affair with him. The feminists were of course vitriolic about *Playboy*. Why does she go out on a limb? 'If I knew, I would be happier.'

As soon as *Playboy* ceased to be avant-garde, Annick Geille became editor of *F Magazine*, turning it into a 'post-feminist' companion for women who were tired of the war of the sexes and who thought that 'life is not much fun without men'. A companion, not a guide, because she claimed only to be reporting what the 'new woman' felt, 'leading the life of a man, while wanting to show off your luck in being a woman'. With Christine Ockrent on her cover, she discussed the new art of attraction, based on couples having contrasting qualities. Once again, she brought in fashionable philosophers, like Alain Finkielkraut, to explain that paying compliments was no longer enough: what was needed now was the 'talent to make oneself original. One has got to be surprising, different, amusing.' So there is no single formula. She published a poll claiming that both men and women rated intelligence as the supreme quality in the other sex; 'sex appeal' came bottom of the list. Make way for women, said her headlines; We want men's jobs, with good salaries and responsibilities; How to obtain a rise. But then followed articles saying, 'Active and dynamic, you manage your career with an iron hand. But in the evening, why not change the décor? Wearing a girdle and silk stockings, become once again a femme fatale, a child-woman, a female woman.' Just as she admired the art of being both left and right, now she saw women wanting to alternate their charm and their strength.

However, when women won increasing success in reaching the top of their profession, she sensed there would be something missing. Without 'culture' they were incomplete. Jack Lang was her prophet. She founded a glossy magazine called *Femme*, paid for by expensive high-fashion advertisements, but with the text written by the most admired and cultured authors. Literature, psychology and art became as essential as contraceptives. At the same time, *Femme* was a celebration of hedonism and luxury, of the pleasure of satisfying desires in the most refined way, with no expense spared.

Then it was time to move again. Men were dragging their feet. Divorcees of her generation were finding it increasingly hard to meet the right sort of partner. In 1990, she started another magazine, but instead of naked girls, she offered men the same recipe she had offered women: culture. It was men's turn to become cultured, she thought; men's magazines were needed to broaden their sensibilities, to bridge the cultural abyss between them and women, and to set new standards in what it means for a man to be chic.

All this suggests she is an optimist. Her magazines always hold up heroes and heroines as models to her readers. But the snag is that real people never quite resemble those models, who in any case are wholly admirable only at a distance. Though she moves in high society and lives in the most fashionable part of Paris, married to a rich businessman, this world does not satisfy every side of her. Back home from her plush office, she sat at her table late into the night and wrote novels. These contain her more intimate reflections on her experience. In her magazines she seemed to be saying, isn't life wonderful, with all these amazing people doing marvellous things; you too. . . . But her novels reveal what she considers to be the missing link. Their subject is always frustrated love. Her work as a journalist enables her to collect material about how people live. Unlike her magazines, her novels are sad. The novelist, she says, does not agree with the journalist.

There is a scene in her *Voyageuse du soir* in which the heroine comes home to her husband in raptures, because she has just won the top accolade in her profession, and the husband says he is too busy to celebrate, he cannot see what the fuss is about and he has never heard of the great names who have honoured her. That expresses the dilemma of Annick Geille's own life. Work is her passion, her greatest love affair. She often works a twelve-hour day. Many men seem to resent her success. Another of her characters complains, 'Men who appreciate a woman professionally get bored with her very quickly in private'; her books are peopled with successful professional women who 'despite the high salaries they receive have no one to talk to'. At work she has the voice of an editor, quite different from the soft, tentative tones she uses in her feminine mode, when she treats people as though they are made of the most delicate porcelain. Power in the office weakens a woman, she says; at home she wants to be able to collapse, to leave the man to make the decisions; but men are also tired when they come home from the office. She needs to be a child sometimes, and to have someone listen admiringly to her boasting about her triumphs. In 1978 she published a book announcing the arrival of the New Man, who understood this, who combined all the qualities a liberated woman wanted and who made no demands on her. Now she regrets it; the vision was a mirage.

'A woman feels she has failed if a man does not desire her, and prefer her to every other. Women believe they are liberated and strong, but they always need to be loved, to be the chosen one.' However, though love is something she would like to believe in, she always concludes that, alas, she cannot find it in the idyllic, impossible form she wants it. For her, falling in love has always been like a dream – she wakes up and it vanishes. And not only does it not last, but it is destructive while it does, killing friendship, bringing not peace, but conflict, because it is inherently unequal. One partner is always more deeply in love than the other, and so at a disadvantage, for the other, who allows him or herself to be loved, can break off unhurt. Men are constitutionally unfaithful, she believes. Infidelity in all its forms is a constant theme in her books. To accept this tragic incompatibility of the sexes, and 'the impossible destiny of women', means to stop fooling oneself and others. 'I do not believe in love,' she says defiantly, as she might say

'I do not believe in God', but love continues to inhabit her. Sometimes she suggests love is the search for a soul-mate, but unfortunately 'those who love you do not understand you and those who understand you do not love you.' Sometimes she ends up in a vicious circle: one loves the person who loves one; but then one feels one's lover must be stupid to love one; too much intimacy can become an invasion of privacy, a threat; too much honesty can ruin love; one must be wary of falling in love; when one of her characters says, 'I love you', she adds, 'The word has been used too often.'

The problem almost disappears if one contents oneself with friendship, which of course she cannot, but even so she often feels that friendship is what gives meaning to her life, or at least a sense of being safe and protected. One of her early love affairs was a constant argument; now that it has been transformed into a friendship, with no right to make demands, the relationship is perfect. Of course, there is an element of idealisation in this praise of friendship, because a busy woman has little time; she cannot see her friends as often as she would like; and occasionally her friends have no time for her. When one of her books was being published she asked a famous author, to whom she is close, to write a few lines about it. He was too busy, he replied, very sorry, she would understand he could not break off from his absorption in his own book. Angry silence is eventually followed by forgiveness. In friendship, you cannot hold your friends' faults against them; you have to accept them. But despite the value she places on friends, she concludes: 'I have many people around me, but I remain alone . . . Life is composed of solitary worlds which do not meet.'

Sex, which Simone de Beauvoir approved of as a method for achieving rapid and immediate communication is, for the former editor of *Playboy*, another of nature's frauds, a stillborn pleasure. There are 'caresses which few women can resist' and 'masculine odours' which are 'intoxicating', and 'without men a woman is in danger of ceasing to exist'. Her characters jump in and out of bed, and her male ones need several women to give them the illusion that they are alive. But she finds it pathetic that passion should be no more than a quest for those few moments. The dream of the permissive age has not come true. Sex brings together only people who are already together. 'Everybody lives alone.'

Annick Geille was born and brought up in Brittany. Her mother was a banker's daughter and her father a sailor, a scholarship boy who became a naval engineer. That has left her torn between two worlds, loving luxury on the one hand, and on the other feeling that she really belongs to the world of the underprivileged. She resented that she was expected to do the domestic chores simply because she was a girl, and forbidden to go to cafés. It was from the local bookseller that she discovered the kind of affection she was seeking; twenty years her senior, he was her first friend, who inflamed her passion for reading which remains a true passion: she rarely goes anywhere without a book; there are few works of modern literature she has not read. She was further encouraged by a childless woman teacher who half adopted her, and who tried to convince her parents that they were too severe with her, that they were failing to recognise her talent. Annick Geille thought that all her mother wanted was for her girl to get happily married and have children. In Lorient she felt a prisoner of provincial bourgeois rituals. Through books she had glimpsed another life, presided over by the Parisian authors whom she worshipped as her gods. The Balzacian local journalists who dreamed of becoming Parisian celebrities, and the frustrated marginals of the town, unsuitable husbands, were her accomplices in rebellion. Her best friends have always been marginals by origin, or by temperament, or by sexual orientation. She became a journalist because that was the way to meet her gods face to face. Most of Paris's leading literary figures are now in her address book. She treats them with awe, respect and affection, as though a book is not only an idea but also a hug, and deserves a hug back; it is the best proof that total strangers can, despite all that separates them, touch each other's hearts.

Sometimes the chic Parisienne appears at media parties accompanied by her teenage son, as well as her briefcase. In her youth she had laid it down that a woman must not sacrifice herself for her children. She still believes that. Having a child made her discover that what her parents had said to her – 'We did everything for you' – was misleading: a parent gets pleasure from doing everything for a child. 'The child owes you nothing; it is you who owe him the pleasure of living, thanks to him, who educates you, who forces you to improve yourself all the time.' Her love for her son is very strong; there is no worse loss she can imagine than that of a child, and women who have lost a child arouse in her a terrified, almost superstitious curiosity. The one illusion she insists on believing in is that love of children is the only kind that lasts for ever. And yet her *Portrait of a Guilty Love*, which won France's First Novel prize, is a long cry of pain in front of a wall of incomprehension between a mother and a daughter.

Sometimes she wishes that work did not prevent her from seeing more of her husband, but sometimes she likes to be alone. Suddenly social contact makes her go cold and she cuts herself off. Her nose changes its shape, like shutters being closed. Do not assume, simply because she says that women like to know that men lust after them, that you will ever discover all the thoughts in her mind, any more than all the ingredients of the perfume she wears.

'The world belongs to those who know where they are going,' she has written, but the signposts she looks for are within herself; she decides where she will go by how she feels, and since desire is neither constant nor satiable, she does not expect ever to reach a goal: 'There will always be something missing.' There is no danger of her being called naive. 'Women must remain strangers, foreigners, keep their mystery, be unpredictable': that way they are never totally defeated. But now she has finished another novel, her most important one, whose ending is not sad. It is as though she has smelt a whiff of hope in the air, as though she has sensed that ultimately people tire of being struck more by the sadness than the possibilities of life.

It is not the first time that women have been liberated. The women of today who are on the scent of a more interesting life have precursors, for example, in tenth-century Japan.

I am talking of its aristocracy, of course, but then the liberated women of today are an aristocracy too, a small minority. It just happens that the women of Kyoto, in the days when it was the residence of the Japanese emperor and known as 'the capital of peace' (Heiankyo), made a record of what they felt, illuminating human emotion as though with a floodlight which suddenly reveals all that is normally dark, and is as suddenly extinguished. While men wrote learned texts on the usual subjects of war, law and religion, in the language ordinary people could not understand (Chinese, the Japanese scholars' equivalent of the Europeans' Latin), women started writing novels in the everyday Japanese language, and in the process invented Japanese literature. For about a hundred years novels were written only by women, who discussed the subjects which interested them, namely the emotions. The world's first psychological novel is the *Tale of Genji*, written between AD 1002 and 1022, by a widow in her twenties, whose husband, twice her age, had died after only a few years of marriage. Those who say Japan can only imitate will be astounded by this extraordinarily readable and intelligent precursor of Proust, Murasaki Shikibu.

A novelist, she said, is one who is so moved by her experiences that she cannot bear them to disappear into oblivion; and she was moved by the confusion of the love lives she observed at the Imperial Court, reacting first with hope, and then with despair. Her book has 430 characters, excluding servants, as though she wanted to be sure that no additional evidence could be found to contradict her conclusions about the art of existing.

Murasaki's hero was the ideal man, like Annick Geille's New Man. He is wonderfully handsome, of course, but he is ideal not because of that, but because he is, above all else, sensitive. He is a prince, but Murasaki is not interested in his power, nor his skill as a swordsman, nor his physical strength, and there is indeed not a word in the book about his public career. He has one love affair after another but, unlike Don Juan, he never forgets or abandons a woman, treating them all as individuals, responding to the unique temperament of each one: 'He is the great noble when he makes love to Lady Rokujo, the demon lover when with Yugao, tender and unassertive with the timid Lady of Akashi, fatherly to Tamakazura.' He finds an ideal wife for himself, but the author does not allow him to forsake all other women, as though she wished to illustrate the skills of lovemaking in all its forms. Conquering women is not his ambition, because he is an artist in love. His mistresses are content to receive only a share of his affection, because the pleasure he gives is unforgettable, and because he never abandons them entirely, neither when they cease to be young, nor even when he discovers that he has made a mistake, and that they are not as admirable as he thought they would be: he never runs away, but looks after his old flames as tenderly as if he still truly loved them.

In this period, it was shameful for an aristocratic woman to be dependent financially on her husband. She did not move in to live with him on marriage; each kept their own home. The husband paid her visits only from time to time, so she was tyrannised neither by him nor by his mother, but was free to enjoy her leisure and the benefits of her education. This was the only time in history that daughters were more valued than sons. All was not perfect for them, however; they lived secluded in their homes, waiting for male visitors, always waiting. Sheltered behind screens (their alternative to veils), their education did not enable them to use all their energies, being confined to the arts, excluding history, philosophy and law, which it was shameful for them to know about (though Murasaki knew a great deal she ought not to have known).

Theirs was a domestic world; so they had more leisure than they could think what to do with, and talked of 'suffering leisure' or 'relieving their leisure'. However, they had both the ability and the time to reflect on their relations with men, which were unusual also in that there were virtually no restrictions on sexual intercourse. Long before the 1960s, here was a society (or rather a class) completely tolerant of promiscuity, holding that to have as many affairs as possible, with the married or the unmarried, was a source of prestige. Men could have many wives (some went up to ten at a time) and even more concubines. That was one reason why daughters were desirable: they had far better chances than sons of contracting a good marriage, or at least of consorting with the powerful. Wives were encouraged to have all the lovers they could attract, and virgins were thought to be blemished, possessed by evil spirits. Wearing clothes whose colours did not harmonise was more reprehensible than having too many lovers. What mattered was good taste.

This meant that these people became obsessed with conforming rather than with being free. Taste required them to obey unwritten rules. Men seemed often to sleep in a different bed every night, but they had to creep away unseen at dawn, and immediately send the lady a morning-after poem. 'A woman's attachment to a man depends largely on the elegance of his leave-taking.' Love affairs often became a ritual, devoid of feeling. Men did not admire women for their bodies: 'Unforgettably horrible is the naked body,' says Murasaki. 'It really does not have the slightest charm.' Even naked teeth were horrible, and were dyed black to conceal them. Long hair was the only physical attribute worth having. Women were beautiful by what they did, by their skill in choosing and wearing clothes, by their command of the arts, by their ability to create beauty, to invent lovely perfumes, or to play sweet music. A man might fall in love with them simply from seeing a fold of their sleeve, or a sample of their handwriting, calligraphy being a major art; so too was the choosing of the right paper to suit the mood and the weather of a particular day. Courting for marriage was a ritual which involved sending a thirty-one-syllable poem to a girl one may never have met, and then deducing her character from her reply; if the results were promising, they had to make up their minds after three experimental nights were spent together. It was letters and poems which triggered sexual desire. Women were mystified by their capacity to cause emotion, and men were ashamed when they felt passion, which meant they had lost self-control. Even promiscuity has rules, all the more difficult because they are made of subtle nuances, which one cannot so easily pretend to adhere to, as one can with marriage laws.

Nobody expected a partner, either short or long term, to be faithful. A wife, indeed, believed that if her husband had many mistresses, she was more likely to have exciting and affectionate relations with him, provided she was the woman he preferred; that was a constant challenge. But this system became a nightmare because these wonderfully elegant people could not stand the uncertainty. Both men and women were morbidly jealous, even though jealousy was regarded as a breach of good manners. They all pined for security, though they were bored by it. They worried about losing a love, and about committing themselves, and about their future and the future of their children, and about what the gossips might say.

So despite the privileges and the beauty with which they surrounded themselves, they were often miserable, or at least gently melancholic ('aware' is the Japanese word, used over a thousand times in this novel, meaning the feeling that life is both wonderful and awful). Murasaki suddenly kills off her hero two-thirds of the way through her book: 'Genji was dead,' she says abruptly, and moves on to look at less perfect people, and at an endless succession of failures in mutual understanding. Her Buddhist side reasserts itself, in which the ultimate aim is to extinguish desire, which is the cause of all suffering. Though she was for long tempted by the grandeur of unlimited love, in the end she, and all her characters, were overwhelmed by everything they valued being so ephemeral, youth, love, power, social position. Their only solution was to say that it was very sad, and that it was very elegant to be able to recognise the beauty and the sadness of the world at the same time.

Murasaki's individuality surrendered to the traditional Japanese solution for despair, which is to turn it into an aesthetic experience, to find beauty only in what is impermanent, to insist that unless beauty and love were fragile and perishable, they would not be beautiful. As Kenko wrote in his *Essays on Idleness* (1330–2), 'If man were never to fade away like the dews of Adashino [a famous graveyard], never to vanish like the smoke over Toribeyama [a crematorium], but lingered on for ever in the world, how things would lose their power to move us. The most precious thing in life is its uncertainty.' This fascination with uncertainty has a modern ring, and helps to explain why Japanese art, which reflected it, became an inspiration for modern European art. But the desire to be moved at the cost of irremediable suffering is not modern, but very ancient. Helplessness in the face of the cruelty of the world is not peculiarly Japanese at all, but part of the sense of cosmic gloom on which virtually every civilisation has been built. Annick Geille's sombre Breton horizon is a fragment of a global one.

The women of tenth-century Japan were thus unable to get beyond waiting. They waited for the ideal man who would love them, even though they half knew that they would be disappointed and that love does not last. The wait has continued for ten centuries. However much people have modified their sexual arrangements, they have continued to look on the world as a house haunted by a ghost, which is the prospect of failure, decay and disappointment. Protesting against the unfairness of life, or making fun of the absurdities it leads to, does not change a great deal. *The Lady Who Loved Insects*, a novel by another of these extraordinary young women of Heian Japan, shows an early bluestocking refusing to blacken her teeth or pluck her eyebrows, insisting that she is interested only in 'enquiring into everything that exists and finding out how it began'. And of course no man will love her. The ending is unhappy, as endings inevitably have to be for all people who believe in the cosmic gloom.

Even *The Pillow Book* of Sei Shonagon (born 965), which experts say is the wittiest book in all Japanese literature and a stylistic masterpiece, written by another of the Empress's bright young ladies in waiting, cannot escape from the conclusion that 'If I do not come first in people's affections, I had just as soon not be loved at all; in fact I would rather be hated or maltreated. It is better to be dead than to be loved in second or third place. Yes. I must be first.' But that was not enough: 'Nothing is so delightful as to be loved by every one.' (No chance of that: Murasaki, who knew her, says she 'has the most extraordinary air of self-satisfaction'). Besides, 'A lover's visit is the most delightful thing in the world.' Unfortunately, 'I am the sort of person who approves of what others abhor and detests the things they like.' Even worse, 'I realise it is very sinful of me, but I cannot help being pleased when someone I dislike has had a bad experience.' Neither wit nor delicate sensitivity can give her what she most deeply wants.

Several centuries later, Japan's new class of merchants attempted to enjoy the pleasures of the aristocracy, but without the snags. They wanted to invent their own way of behaving, 'the way of the townspeople' (*chonin-do*), seeking individual happiness directly, not by the circuitous routes of tradition and ritual. Not for them the way of the samurai warriors, who spurned love as effeminate and who yearned for glory after death. The inhabitants of Osaka in particular made it their purpose to live for the present, devoting themselves first of all to making money. Coins having been introduced in 1601, they were fascinated by their novelty, delighting sensually in the feel and sight of them. 'Nothing in this world is as interesting as money,' said the novelist Ihara Saikaku (1642–93). 'It is the one thing that is essential in the present world.' There was no need to bother about who one's parents were if one had it; and it enabled one to acquire exotic luxuries.

But the merchants decided the best thing to spend money on was sex, so as to blot out the 'sad world' with a 'floating world' of pleasure, to gratify all types of sensuality, to create a 'nightless' city which would be 'the universal theatre of pleasures and diversions'. There were more Japanese able to read now (40 per cent) than in any other nation, except England and Holland, and they used their knowledge to read books about sex. Courtesans became the centre of the social life of the city, not infrequently daughters of samurai, and extremely expensive: there were soon stories of millionaires who spent huge fortunes on them. Love was dangerous; adultery was punished by death; such obstacles only added to the excitement. Sex was what this precursor of the consumer society immediately focused on. There were limits to what business people could hope for, in terms of social prestige, but, said Saikaku, 'The pleasures of the flesh have no limits.'

He was their Balzac or Dickens, the first Japanese writer to find heroes and heroines among ordinary commoners, with precisely such sensual obsessions. His *Five Women Who Loved Love* shows the search for pleasure being led by women, who do not timidly wait to be wooed, but make the advances and the decisions; the bolder they are, the more they are admired; they often lead to disaster and death, but even so they dominate the execution in the final scene. Sometimes they even have to tear the men away from the boys: in his *Mirror of Manly Love*, Saikaku shows homosexuality competing with heterosexuality; in the theatre, handsome young men playing female roles are the stars most admired.

But though Saikaku had enormous fun describing the frenzy of desire, and the minutiae of lust, expressing his delight in 'the foolish things of this world', he too grew sadder with the years. He began worrying about those who had no money, about the difficulty of making money if one does not have it to begin with, about the delinquency of the young, about courtesans who had had ninety-five abortions, addicted to 'fleshly pleasures' even when these gave no pleasure. Despite his fascination with women, he says, 'All men in the world are beautiful, but among women beauties are rare . . . one must end by discarding women in favour of men.' Those of his characters who try to raise loyalty into a heroic virtue end up feeling that they cannot control their destiny, that individual merit does not bring the rewards it ideally ought to, that they need religious faith to sustain them. But they choose the one which interferes least with their pleasures, the popular Amida version of Buddhism, which does not punish them for their misdeeds, nor demand that they behave themselves, but promises redemption in return for a few prayers.

It may seem that during the thousand years which separate Murasaki and Annick Geille there has been little progress in ridding the world of its sense of gloom: it keeps on coming back. In which case the conclusion would be that humans are condemned to be dissatisfied, and to feel, even in their moments of glory, that 'there is something missing', so that they have no option but to surrender to the superstitious belief that desire is inevitably a source of torment as much as of pleasure.

However, I see the history of desire leading in a different direction. How people have felt about pleasure has depended on what sorts of pleasure they imagined to be possible, what exactly they considered to be missing and how far their horizons stretched beyond their own personal problems. Instead of assuming that there is a basic flaw in the universe, one can look at desire from another angle. The flaws in the universe, like those in Oriental carpets, may not be flaws. Observing the universe, science – having emerged from its adolescence, when it believed that it would conquer ignorance – has learnt to derive its satisfactions from knowing only that every discovery contains an invitation to make further discoveries, and that a failed experiment means only that the wrong question is being asked, not that there are no answers. Most people, without realising it, still perpetuate the habits of mind they have inherited from the days when it was expected that the world would come to an end very soon, and have not learned to view it as having infinite possibilities. Pessimism and optimism thus emerge as being a dispute, to a great extent, about how far one is prepared to look, about focal distances.

There is no need to deny that individuals have been born with a particular temperament, and that their chemistry almost compelled them to see the world in a different shade of pink or grey. That does not mean they were trapped in their bodies. Aristotle said that the liver was the seat of the emotions, and indeed it has been confirmed that the amount of sugar it makes available does modify mood. It has been established that there are physical differences between those who wake up fully only in the evening and those whose liveliness fades gradually in the course of the day; since the invention of electric light, the world has probably been divided into owls and larks as profoundly as by the ideologies of left and right. Drugs can alter attitudes, and every body does have its own internal clock, causing varying receptivity at different times of day. Rats unharmed by a drink of alcohol in the evening have been killed with the same dose taken in the morning. People who lack adrenal hormones (a condition culminating in Addison's disease) have been shown to be many times (occasionally 150 times) more sensitive than others, to the point of hearing sounds which the normal ear cannot, and they suffer excruciatingly because it is not easy to cope with sensitivity. On the other hand, there have always been individuals who have triumphed over the obstacles put in their way by their bodies. No history of the world can be complete which does not mention Mary Helen Keller (1880–1968), whose overcoming of her blindness and deafness were arguably victories more important than those of Alexander the Great, because they have implications still for every living person. Private triumphs over temperamental and physical impediments have, however, seldom entered the history books, and humans are still like car drivers who do not know much about how the internal combustion engine works. But every human is in some way an exception.

Anybody who values freedom needs to remember how people with no particular constitutional predisposition to cheerfulness have cultivated hope nonetheless. The most important method they have used has been to widen their horizons. The most famous of all optimists, Leibnitz (1646–1716), who is quoted derisively as saying that all is well and that this is the best of all possible worlds, was very far from being unaware of the cruelties of life, for he spent much effort urging the kings of Europe to mend their ways and investigating how the different religions could stop fighting each other. What was special about him was that the range of his interests was exceptionally wide, extending through history, geography, philosophy, mathematics, politics, theology and law, in which he obtained a doctorate at the age of twenty-one. The reason he was hopeful was not just that he believed a good God must have allowed evil into the world for good purposes, but that he himself saw the world in the way that scientists see it today, as composed of an infinite number of particles; for him, there were no limits to the wonders of nature and the ingenuity of reason; all that was needed was not to allow the readiness to make new discoveries go numb, as it does in most adults. Leibnitz's ambition, as he himself said, was 'to awaken in us all the sleeping child within us', to see in every person somebody different, as complex as a garden full of plants and a lake full of fishes, and in each plant and each fish, yet another garden and another lake. He believed liberty was possible because he saw beyond the present into the infinite distance. Harlequin was his hero, multiple, crafty, always in search of something else. Leibnitz invented differential calculus, but also an Academy of Pleasures. He was more than he appeared to be, which is how intelligent optimism should be understood, discarding its absurd exaggerations: it is not a belief that everything is perfect, but a willingness to admit that there is more than the eye can see, good or bad; there is always a glimmer of light, however dark it may seem, because life is inconceivable without hope. Optimism is awareness that despite nastiness and stupidity, there is something else too. Pessimism is resignation, an inability to find a way out.

The eternal see-saw between optimism and pessimism has been weighted in favour of the latter by writers whose view of the world has been formed by an examination of their own navel, rather than by the undertaking of new adventures. The last happy writer, according to Roland Barthes, was Voltaire (1694–1778), but he was also the first of the post-moderns, defending himself against unhappiness with acerbic wit, unable to get over being unloved in his childhood, thinking of himself as an eternal orphan, trying to escape from himself, even imagining that he was a bastard, ever insecure, seeking comfort from the company of women and the praise of princes, dazzling them, but uncomfortable in society all the same, insisting that he was incorrigibly a 'doubter and not a doctor'. Voltaire, the great and brave defender of human rights, both liked and disliked the taste of optimism because he was so critical. He became the model of the literary intellectual, for whom thinking means above all being critical, in the sense of being more forcibly affected by what is wrong than by what is right, having one eye more powerful than the other. Cosmic gloom is blurred vision. It is exacerbated by the narrowing of the horizon, and becomes paralysing when a generation mocks its predecessor so fiercely that it cannot learn from its mistakes. Thus Jean-Paul Sartre, for example, a more recent model of the thinking person, narrowed his horizon by excluding all 'swine', and plants also, for he hated the countryside, and even uncooked food, anything he could not control; nor could he appreciate the desire for physical comfort, nor that essential part of existence which is dreaming and day-dreaming. Stuffing himself with drugs to keep reality in harness, no wonder the journey through life looked grim. But to say just that would be to narrow the horizon oneself: Sartre inspired many hundreds of thousands of people all over the world to prefer generosity to selfishness and prudence. To condemn him because of choices which turned out to be misjudgements or contradictions, while ignoring the fact that his vision forms part of the never-ending search for liberty, would be to repeat the same mistake of exclusion. Condemnation is lack of imagination, when one cannot suggest something better.

Gloom is thus not like the blackness of the sky, which nothing much can be done about. Ordinary people have indeed seldom committed themselves wholeheartedly to gloom. When they have despaired of public affairs, they have turned to private pleasures, and vice versa. The escape from gloom into personal amusements has been the world's secret history, since the ancient Egyptians refused to accept that they would die, and built tombs with the same delight as houses, and since they refused to accept the diagnoses of their doctors, consulting specialist after specialist – they had doctors for every part of the body, down to 'shepherds of the anus' – inserting into their medical textbooks the precept that a patient needs above all to talk about himself: 'He likes the doctor to pay attention to what he says, even more than to be cured.' Gloom only seemed to be a divine punishment when the Babylonians decided that humans and gods could not fight on equal terms, that perhaps the gods were fair rather than arbitrary, and that humans should blame themselves for their misfortunes, asking how they had sinned. With the centuries, gloom has even acquired a dusky charm for some and been raised to the status of a familiar companion. Annick Geille insists that literature must always be about gloom of one sort or another, on the principle that there is nothing interesting to be said about happy people.

So gloom has depended a great deal on whether people believe they know all there is to know about how the world works, or whether they are more impressed by their ignorance and the possibility of new discoveries. The most elaborate remedy for gloom so far devised is the consumer society, but it has not succeeded in eliminating it, nor boredom either. The reason is that it is still very young, uncertain about its future, attractive to many but tormented by doubts as to whether it is more ugly than beautiful. It still offers only a limited range of luxuries. Luxury has so far meant lavish spending, lasciviousness, the acquisition of unessential possessions, the full enjoyment of all the senses, and often ostentation too, the wish to impress other people, without really caring about them. These are mere beginnings. The experiments with luxury by the merchants of seventeenth-century Japan were beginnings too, remaining private to themselves, because the majority of their compatriots were peasants devoted to frugality. Only in eighteenth-century England did the mass consumer society of today begin to take shape. Until then sumptuary laws – in every country – forbade people to consume as they pleased, compelling them to wear only the clothes of their own profession or class and to behave according to their station in life. Thrift was what everybody believed in until the Industrial Revolution gave the masses cheap goods and the money to buy them, by offering factory work to women and children, thus creating the two-three-four-wage-earning family with cash to spend.

Entering the consumer age was the equivalent of changing one's religion. The English took to believing in 'Improvement' instead of tradition, though to conceal their apostasy they cultivated nostalgia as a hobby. The 'extravagance of the lower and middle classes', with everyone trying to imitate those slightly richer, was what immediately struck foreign visitors. The national ambition became, as the agronomist Arthur Young said in 1771, 'Universal Luxury', or 'the opulence of all classes'. Newspaper advertising was invented at the same time to stimulate 'a passion for new styles and fashions', and was denounced for causing 'constant titillation' and 'an epidemical madness'. But money alone cannot transform old habits. People do not change their religion without miracles. They believed anything was possible only when they saw new breeds of animals, which God had not created in that shape, when they themselves took to crossing bulldogs and greyhounds, and bought seeds which grew into plants their parents had never known, when children received microscopes as presents to play with, and when knowledge expanded as never before. It was then that they decided that desire deserved to be encouraged rather than controlled. They relearned what they wanted, and how to get it, in the shops, which have been as influential as schools.

The consumer society lost its sense of direction when it adopted two myths to guide it. The first was that private vices are the source of public prosperity. Avarice, pride, envy and greed, rather than friendliness and kindness, are the necessary bases of a successful economy, said Bernard Mandeville (1670–1733), a doctor of nervous diseases, born in the Netherlands (possibly of French origin) and practising in England, who wrote a best-selling *Fable of the Bees*, showing what disasters would occur if people were nice to each other, or if they tried to have any ambition other than self-interest. He was the first person to study luxury, but as will be seen, he did not study it enough: his other works, on hysteria, hypochondria and in defence of public brothels, show that he was concerned with only a limited range of human talents. The science of economics has ever since based itself on this rather narrow view of consumers acting rationally and predictably in their own interest.

The second myth was invented by America's first major science fiction writer, L. Frank Baum (1856–1919), creator of the Land of Oz, where anything is possible and where fantasies can be turned into realities. In 1897 he founded *The Shop Window*, the first magazine devoted to window dressing: seeing an object in a shop window, he said, should 'arouse in the observer a cupidity and longing to possess the goods'. That is what he himself enjoyed most, the feeling of desire. His father having made a fortune in oil, his mother being a feminist, he wanted a different kind of life. Hard work and good causes bored him: he preferred to write fairy tales, preaching the creed of eat, drink and be merry, for tomorrow we die. Thrift he despised as heartily as religion (though he was interested in spiritualism), all taboos in the way of desire irked him. The theatre, photography, motion pictures were his passions. Travel was his supreme pleasure, moving from hotel to hotel, which he thought of as outposts of paradise, the true fairy land.

John Wanamaker, owner of the New Kind of Store in Philadelphia, put Baum's dream into a formula in 1906: America, he said, was the Land of Desire. Previously it had been only the land of comfort. Consumption was redefined as the satisfaction of desire, rather than of need. The worship of pleasure ceased to be thought of as un-American, and soon Christian preachers were trying to convert the USA to a new kind of Christianity, in which the satisfaction of sexual desire was holy.

It was originally believed that if individuals satisfied their desires, of whatever kind, the world would automatically become a better place. However, the consumer society was so successful because it gave people more than luxury; it expressed more than fascination with technical progress or with a system which had to make consumers ever wealthier so that they kept on purchasing what was produced. By buying and spending, people obtained a sense of belonging almost comparable to that which religion gave. When Richard W. Sears reduced prices so that anyone could afford the goods in his catalogue (which offered 35,000 different items by 1928), that meant not only that getting a bargain was a victory against the profiteers. His customers responded above all to his invitation to 'buy the object which everybody has', because each object they acquired diminished their sense of being excluded from the good things of life. Consumption was a long celebration of victory over poverty, a way of making the home into a private fortress safe from the humiliations of the workplace, a home filled with objects of one's own choosing, even if they were cheap imitations and even if there was perpetual frustration because one could not afford everything one wanted.

However, since the consumer society was set up, four changes have completely altered the circumstances in which it operates. First, whereas only 15 per cent of the working population originally earned its living other than by industry and agriculture, today, in developed countries, two-thirds do so by supplying services rather than by making objects. This means, secondly, that what consumers now buy, and desire, more than objects, are services, that is to say personal contact, help and advice. Thirdly, the cost of goods has diminished but that of services has not; for a while it seemed as though self-service would be the answer, but then time became the most precious of all commodities. Ownership of goods did not free time; rather it added new commitments and opportunities for using up time. Fourthly, the idea of luxury has expanded so that the most desirable luxuries are those that money cannot buy, involving not so much possessions as human relations.

To consume has so far been taken to mean using an object until it becomes useless. But it has become self-defeating to use people so that they cease to be people: a tycoon who turns his workforce into automatons was once the ideal, but he is now unquestionably doomed. It is consummation rather than consumption that people increasingly seek, taking pleasure to its ultimate degree of intensity, and not just having free time. Filling time with the most profound experiences, in most cases, involves other people. The advice from the first generation of the Green Movement was to consume as little as possible, but that was a leftover from the ancient tradition of asceticism, which the world has repeatedly rejected, and will doubtless continue to reject as long as there is inequality, as long as there are people struggling to come out of poverty who have not tasted satiety. The Green Movement could not become a major political force so long as it concerned itself primarily with natural resources rather than with the full range of human desires. Its setbacks are yet another instance of idealism being unable to get off the ground because it has not looked broadly enough at human aspirations in their entirety.

Desire for a loved one, and for luxury, are part of a whole bouquet of desires which contains also the desire for culture and the desire to be useful, as well as the desire for excitement, or the substitute for excitement that drugs give. It needs to be remembered that at the same time as the British invented the consumer society, they also took to drink as never before, and the USA followed suit, but over the last generation there has been a radical change in attitudes to alcohol, with a move to moderate drinking of high-quality wines rather than mass consumption of anything fermented. Watching television may become addictive but gradually discrimination is learned.

Consumption is thus always evolving into something more complex. Greed and altruism were once enemies, desire and abstinence were once the only alternatives. But desire can find its consummation in generosity, and greed in curiosity. It is a question of knowing what one values most. Humans have never been simply passive victims of their desires, but have played with them, polished them. Playing with gloom has produced some fine art, but otherwise it has been a waste of energy.

The range of possibilities for dealing with restlessness has not been exhausted. Leisure is not the only goal society can set itself. So I now turn to other paths that have been explored in search of more satisfying links between humans. First, I deal with travel, which has become the most popular method of freeing oneself from the sense that the world is a dangerous and unfriendly place, irrevocably ruled by selfishness. Then I investigate the feeling that all such efforts are doomed to failure, either because men and women are too different ever to put an end to the war of the sexes (chapter 18), or because people cannot change (chapter 19, on 'destiny'), or because they are becoming increasingly stressed by trying to do too many things and going to too many places (chapter 20, on the lack of time), or because the family is collapsing (chapters 21 and 22). But I show that the history of these apparent obstacles contains positive elements which open up many new options. My final three chapters therefore suggest that the human imagination has not exhausted itself, and that the human adventure has barely begun.

Murasaki Shikibu, *The Tale of Genji*, translated by E. G. Seidensticker, Secker & Warburg, 1976; Sei Shonagon, *The Pillow Book*, translated by Ivan Morris, Oxford UP, 1967; Ihara Saikaku, *Five Women Who Loved Love*, translated by W. T. de Bary, Tuttle, Tokyo, 1956; Ihara Saikaku, *The Life of an Amorous Woman*, translated by Ivan Morris, Chapman and Hall, 1963; Ihara Saikaku, *Some Final Words of Advice*, translated by P. Nosco, Tuttle, Tokyo, 1980; Ivan Morris, *The World of the Shining Prince: Court Life in Ancient Japan*, Oxford UP, 1964; Richard Bowring, *The Tale of Genji*, Cambridge UP, 1988; Andrew Pekarik, *Ukifune: Love in the Tale of Genji*, Columbia UP, NY, 1982; Norma Field, *The Splendor of Longing in the Tale of Genji*, Princeton UP, 1987; A. Martins Janeira, *Japanese and Western Literature: A Comparative Study*, Tuttle, Tokyo, 1970; Sokichi Tsuda, *An Inquiry into the Japanese Mind*, Ministry of Education, Tokyo, 1970.

Attitudes to suffering

Ronald D. Mann, *The History of the Management of Pain*, Parthenon, Carnforth, and Park Ridge, New Jersey, 1988; P. D. Wall and R. Melzac, *A Textbook of Pain*, 2nd edn., Churchill Livingston, Edinburgh, 1989; Anthony Diller, 'Cross-Cultural Pain Semantics', in *Pain*, the journal of the International Association for the Study of Pain, Elsevier, Amsterdam, vol. 9, 1980, pp. 9–26; Maurice Sendrail, *Histoire culturelle de la maladie*, Privat, Toulouse, 1980; L. Romanucci-Ross, *The Anthropology of Medicine*, 2nd edn., Bergin and Garvey, NY, 1991.

Consumer society

Gary Cross, *Time and Money, The Making of Consumer Culture*, Routledge, 1993; Neil McKendrick, *The Birth of a Consumer Society: The Commercialisation of Eighteenth-Century England*, Hutchinson, 1982; Simon J. Bronner, *Consuming Visions: Accumulation and Display of Goods in America 1880–1920*, Norton, NY, 1989; T. Scitovsky, *The Joyless Economy*, Oxford UP, NY, 1985; Peter Gardella, *Innocent Ecstasy: How Christianity Gave America an Ethic of Sexual Pleasure*,

Oxford UP, NY, 1985; Ian Proctor, *Service Sector Workers in a Manufacturing City*, Avebury, Aldershot, 1988; G. Paolo Prandstraller, *Le Nuove professioni nel terziaro: Ricerca sul professionalismo degli anni 80*, Franco Angeli, Milan, 1990; Hans Peter Blossfeld, *Expansion of the Tertiary Sector and Social Inequality: Is There a New Service Proletariat Emerging in the Federal Republic of Germany?* Florence European University Institute working paper 91/8, April 1991; G. Akehurst, *The Economics of Services*, Frank Cass, 1987 (discussing Gershuny's theory of self-service); R. G. Streets, *The Impact of Service Industries on Underemployment in Metropolitan Economics*, Lexington, Mass., 1987; Paulette Carrive, *La Philosophie des passions chez Bernard Mandeville*, Paris I thesis, 1979, printed Lille, 1983; Paulette Carrive, *Bernard Mandeville*, Vrin, 1980; Hector Monro, *The Ambivalence of Bernard Mandeville*, Clarendon Press, Oxford, 1975; Thomas A. Horne, *The Social Thought of B. Mandeville: Virtue and Commerce in Eighteenth-Century England*, 1978; Colin Johnson, *The Green Dictionary*, Macdonald, 1991; Luc Ferry, *Le Nouvel ordre écologique*, Grasset, 1992.

17

How travellers are becoming the largest nation in the world, and how they have learned not to see only what they are looking for

The great attraction of a job in the railways used to be security, and the guaranteed heaven was a pension at fifty-five. Both Vivienne's parents came from railway families, three generations back on her father's side. And what effect does that have?

For her father, the world remains full of danger and uncertainty despite all the guarantees. He lost an eye at work and had to give up engine-driving to become a clerk. He was one of eight children and had five himself: life has always been hard, and the fear of poverty has never ceased to haunt him. The great achievement each month was to be able to put a little money aside, in case of worse calamities. 'A penny is a penny' is his philosophy. 'He does not live,' says his daughter. 'He is very down to earth. Even though entitled to free travel on the railways, he does not know his own country.' Now retired, he spends three months each year by the sea, but always in the same place, camping, playing boules and cards. What does he think about? 'I don't know,' says Vivienne. 'I never asked.'

Her mother has never been out of France either, but 'in her dreams, she is adventurous'. Why is that? Because she reads a lot. Adventure starts in the imagination. The mother is responsible for all five children being the opposite of their father. All except Vivienne have married foreigners: a German, an African, an American, a Russian. Her father was deeply upset, especially by the African, refusing to speak to his daughter when she ignored his objections. Having been sent to a German concentration camp during the war, he did not like the German either: 'It was very difficult, but gradually he accepted him.' Now seventy-five, the father remains 'chauvinistic'. He does not enjoy changing.

Vivienne, however, lives by curiosity. 'I like to dream. I was born that way.' Fishing is her hobby, 'because I like to be alone, which allows me to think about the future. I am a dreamer.' Her favourite author is Proust, but she likes many other kinds of books, particularly stories that are 'precise'. When she watches a film, her mind wanders if it does not interest her and she plans for the time when she will be able to do what she likes. At the moment, she works in an old people's home and her ambition is to become its director, or deputy director. But that would mean she would have to live in the home, which she does not want to do: 'I like my independence.' So she is hesitating. Her husband is also employed by the municipality, in the maintenance of the sports ground and swimming pool. They are poorer than her parents, and live in an unpretentious council flat in the unfashionable part of the city. She has security. But her mind is elsewhere.

'I like travelling,' she says. She has not been able to afford to venture any further than Spain and Africa. 'But when I retire, I shall travel to discover everything I do not know. I am more interested in the world than in France.' She would not mind living abroad, but her husband refuses, his taste of abroad having been building work in the Gulf. Meanwhile, her holidays are spent camping in a caravan, which enables her to meet foreigners, for even though 'I cannot speak any languages, I am very communicative. I need to exchange ideas.' Brought up to camping since childhood, 'Hotels do not interest me, because people do not communicate in them.' By contrast, on the beach, or doing the washing up, campers speak to each other.

The children have left home and are doing their military service; one has a certificate in international commerce. She is free to plan for the future. 'I know I'll do what I dream of doing.'

Ask professional people whether they like their work and they will in most cases say yes, even if they have complaints. But ask if they like what their work has done to them, as they have got older, how it has altered them, and they will take longer to answer.

Caroline regretted that five years after getting her degree in information technology, her knowledge was not making her a better person. Pointing her nose at a computer screen gave her a sense only of having learned enough about her machines. She changed jobs, but that proved to be a mistake. Her third job, with a firm of engineers, promised more travel and human contact. Then a chance opportunity, there being no one else available at the time, put her in charge of building some ultra-modern automated warehouses. So here she is, without any training as a civil engineer, raised to the eminence of a project manager, faced with making important decisions and contracts, running meetings of people twice her age.

She is coping very well. When an engineer representing her client protested, 'It's unbelievable that a bit of skirt should be running the project', she thought he was being ridiculous and ignored his comment. 'If your technical armour is sound,' she says, 'you can justify every move.' When, without consulting her, he changed the agenda of a meeting she had fixed, she walked out of the room. She believes women have some qualities when it comes to handling people that men do not have. Conscious that she has other skills to learn, she has applied for a management course; but she gets on perfectly well with most clients and colleagues; she is growing in confidence; having selected a good team to support her, she finds the work is progressing without difficulty, and learning to perform new tasks has been exhilarating.

However, her work involves long hours and completely takes over her weekends. There are times when she thinks she ought to move to somewhere with less pressure, but she remembers that when she was doing routine computing, on a seven-hour day, she got bored, and she concludes she would rather do ten hours to avoid that. 'I give all my life to my work. I do nothing else. I even dream about work. And I am always stressed, because I have never done as much work as I should have, and I cannot say that to the client. I don't know whether I could last ten years; three years perhaps. At forty or forty-five, will I ask myself, Why have I given my life up to this job? What is the point of getting more and more power and money, just to be able to say that I have climbed up so many rungs in the ladder? It's mad; my friends say it's mad; it's wrong to work so much. But we can't change it. We accept it, and play the game, and work weekends, because we are forced to by our clients who are always on tight schedules.' It seems all the more absurd when she thinks about the unemployed, whose plight she finds heart-rending; she feels guilty: 'Perhaps we should share our work with them, accept lower salaries, work shorter hours, live more sensibly: but how can that be done in practice?'

Caroline is not the only one in her office who is restless. Most of the young people move on after a few years. But she does not share their ambition. A brilliant career is not her aim: 'I did not know what I wanted to do when I left school; I feel I have done well to have reached such a high position. I don't wish to be managing director; I don't have the background. If I were a graduate of a top university, I would want to justify it. But I am satisfied with my post.'

Unfortunately, the people she meets at work have not been the sort who excite her. Engineers tend to be 'serious', she says: they do not do a job that encourages them to be reflective. She went on holiday with an English engineer (that added a touch of exoticism), but he lacked imagination; 'he was almost sad, too academic in his attitude to what is forbidden'. The French are less disciplined, she thinks. But she has had trouble finding a truly undisciplined Frenchman.

There are subjects she cannot talk about with her colleagues. What is the point of having discussions with practical people who always say you cannot change the world? The sort she likes are not engineers, but men who have jobs which leave them free, who have travelled a lot, who can talk about many subjects, who have 'left the mould. But because I have not done this, perhaps they think I am boring.' She longs to be completely different when she goes home after work, to be able 'to surprise others'. Becoming a professional narrows one's world. 'At school I had friends from all classes; now I know only those who have been through higher education; and we all do the same sort of thing. There is no real choice. The world of art and painting is one I don't know at all. I would love to meet people who have taken control of their lives, and who have similar ideas to mine. I would like to have a family, but I don't know that I am going in the right direction. What I enjoy is human contact. We are not made to be alone. So my work does not satisfy me completely.'

There is no going back to watching the computer screen, nor to St Jean de Luz (population 15,000) where time moves at a gentle pace, where she used to go to the beach at lunch time and in the evening, but where there are few interesting jobs. It is true the provinces are improving, the bigger cities offer more opportunities, but it is much harder changing jobs there than in a national capital. Going abroad is impossible, her languages are inadequate. America is not attractive, because the contrast of wealth and poverty is too great. 'So we are condemned to this. But I do not feel that I am in a prison.'

That can have two consequences. Either her life is not a bad one, and she has no desperate need to escape. Or she is free to escape because there are no walls to prevent her, but she first has to decide where to go.

Over a century ago the historian Hippolyte Taine (1828–93) said there were six kinds of tourists. The first travel for the pleasure of moving, absorbed in counting the distance they have covered. The second go with a guide book, from which they never separate themselves: 'They eat trout in the places it recommends and argue with the innkeeper when his price is higher than the one it gives.' The third travel only in groups, or with their families, trying to avoid strange foods, concentrating on saving money. The fourth have only one purpose, to eat. The fifth are hunters, seeking particular objects, rare antiques or plants. And finally there are those who 'look at the mountains from their hotel window . . . enjoy their siesta and read their newspaper lounging in a chair, after which they say they have seen the Pyrenees'. There will doubtless always be tourists wishing to repeat these routines, but there are other possibilities. Tourists may be content to look at places and things, but travel is also, more interestingly, the discovery of people: it is travail, it requires effort, and its reward is a transformation of both the visitor and the host. So I am choosing six examples – out of the thousands whose lives have been changed by travel – to examine the effect that taking a journey and meeting strangers can have.

'Life is a perpetual journey,' said Vincent Le Blanc, born in Marseilles in 1554, who ran away to sea at the age of fourteen and did not return till he was seventy. After visiting every known continent, he at last found a wife in Brazil, but she turned out to be, as he said, 'one of the most terrible women in the world'. The first characteristic of travellers who have been more than tourists has been that they have not found what they expected, or what they were looking for. The ability to realise that one is faced by something new has not been easy to acquire, since most people see what they want to see. Travel became an art when surprises were transformed into advantages.

'In the desert there are many fantasies and goblins seen,' complained Le Blanc, 'that strive to seduce travellers and cause them to perish in hunger and despair.' A modern psychiatrist would say that a journey reveals hidden fears. The Bedouin spoke to Le Blanc not of fears or fantasies but of djinns, and claimed that if one faced them, they disappeared. However, to get rid of one's fears cannot be more than half the purpose of a journey – then comes the discovery of unexpected friends.

All religions have encouraged travel in the belief that it is good for the soul, even though governments have consistently believed the opposite, obstructing it with taxes and red tape. Travel began as pilgrimage. Islam, which was the most systematic in codifying the obligation to travel, required much more than a pilgrimage to Mecca (the *hajj*). Muslims who lived in lands where they were not free to practise their religion were urged to move to where they could (the *hijra*); and some interpreted that to include the duty to move from a poor to a richer country, even if it was not Islamic, with the justification that one could then spread the faith more widely. Visits to local shrines were a third duty (*ziyara*), where the oppressed could feel, at least momentarily, that all were equal before God, women as well as men. And, not least, everyone was urged to travel in search of knowledge, visiting wise men all over the world.

In practice, however, travel has not automatically improved the soul, or at least not in the way expected. There were complaints that the search for knowledge sometimes gave the faithful subversive ideas. Pilgrimages were variously experienced as atonement, rebirth, liberation from materialism or hatred or jealousy, but also as an opportunity to enhance social status, to trade or to smuggle. When the slow, arduous travel of medieval times, which used to transform people physically as well as mentally – the *hajj* to Mecca might take several years – is replaced by a four-week government-sponsored package deal, it is harder for the pilgrim to acquire the sense of belonging to a great international community.

To discover what travel has done to people's souls and minds and habits, it is necessary to go beyond the technical achievements of explorers, the drawing of maps where none existed before, the reporting of unknown facts. For example, how did Ibn Battuta (1304–68), who visited the territory of what are now forty-four different countries, covering a distance of 73,000 miles, improve his life or other people's by travelling? Originally 'swayed by an overmastering impulse within me', this son of a judge went to Mecca at the age of twenty-one simply to complete his education and to qualify as a judge in his turn; but an itch for novelty, or mere bravado, made him resolve never to walk the same road twice. He became a professional traveller when he realised that he was not a born student, that his talent was more for the observation of daily life, for gossip, for being a walking newspaper, reporting amazing sights and customs, making mouths gape in wonder. It was an act of bravery not to try to become learned in the traditional way, out of books, and to rely instead on being able to impress strangers by mere force of personality. However, six years passed before he encountered someone who was anything like his soul-mate, Muhammad Tughluq (1325–51), sultan of India.

This eccentric visionary was deliberately recruiting foreigners to run his empire, being more interested by distant cultures than domestic realities, learning Persian and Arabic, writing poetry, practising calligraphy, studying Greek philosophy, inviting Hindu sages to discuss with him – a sultan whose mind could not stop travelling. Ibn Battuta was appointed an official of his administration in Delhi, and lived in great style, building his own private mosque, only to discover that politics did not suit him: a powerful person needed to pretend that he was more powerful than he was, and Ibn Battuta eventually ruined himself with ostentatious extravagance. There was no escape from the road: he resumed his travels, was robbed of all his possessions, narrowly escaping death at the hands of bandits several times, but continued to be treated with extreme generosity by complete strangers. He returned home to Morocco after an absence of thirty years. Nobody was much interested in him, though he was asked to dictate his memoirs; and he then lived out his days as an obscure judge 'in some town or other', which he might have done if he had come straight home at the age of twenty-two.

There were certain things Ibn Battuta did not want to discover. He married a considerable number of wives in different places, some for only a few weeks, until he moved on, leaving a trail of forgotten descendants behind him; but there were limits to how much novelty he could swallow. In Mali, an African Muslim received him in the presence of female friends: 'Associating with women is agreeable to us,' said his host, 'and a part of good conduct, to which no suspicion attaches. They are not like the women of your country.' Ibn Battuta left the house at once and, though invited back several times, never accepted.

Apart from a brief mission to China, he never stepped outside the Muslim world; even in Hangchow, he stayed with an Egyptian family; and everywhere, as a Sufi, he preferred the hospitality of Sufi fraternities. After no fewer than four pilgrimages to Mecca, his spirituality still failed to blossom. Meeting a Sufi anchorite who lived entirely on fish in the marshlands of Abadan, Ibn Battuta was so deeply moved that 'for a moment I entertained the idea of spending the rest of my life in the service of this sheikh'. However, humility never appealed to him for long. An African king who welcomed him with only three loaves of bread, a piece of fried beef and a gourd of yoghurt aroused his contempt, as 'mean and of feeble intellect': 'I have journeyed,' he replied derisively, 'to the countries of the world and met their kings. I have now been in your country for four months without your giving me a reception gift or anything else. What shall I say of you in the presence of other sultans?'

What travel did for Ibn Battuta was to loosen him from the conventional obsession with a safe, lifelong career. Because travellers were still rare, and therefore valued as entertainment and rewarded with presents and free hospitality, he financed his journey simply by keeping on the move. He discovered that as a wanderer he could lead a more interesting life than he would have done at home. That was all he wanted.

Others, however, less well known, travelled to liberate themselves more profoundly. Shakespeare was right to call travellers rebels, 'out of love with their nativity . . . almost chiding God' for making them the way they were. And his Chinese contemporary Tu Lung described how far such a rejection of ordinary existence could lead. *The Travels of Ming-Liao-Tzu* is the story of a man who was tired of hypocrisy and of being unable to converse freely 'when we have so much we would like to say to each other', who felt 'like a caged monkey' to the extent that 'even when a louse bites our body and our skin itches, we cannot scratch it', tired of 'the desire of possession and the fear of loss'. So, hoping to 'emancipate his heart and liberate his will,' Ming-Liao-Tzu 'sets forth to travel in the Land of Nonchalance'. A hundred coins were all he took with him, and whenever a present caused him to exceed that total, he gave it away to the poor. For to him travelling was flight from normal living, from 'wealth and power and the glories of this world in which people are easily drowned', flight from worry about the morrow. If disaster struck, he would either die or he would not, and if he did not, he would continue his journey. His purpose was to train his mind to be immune to the tragedy of life, to learn to commune with nature, to be able to see all he wanted to in every plant or insect, and to be content to spend a whole day counting the pistils of flowers. They, and not humans, were his soul-mates. In their company he was not lonely. Once he had acquired true peace of mind, he was ready to go home, build himself a hut, and never move again.

Travelling in groups has been a way of making discoveries about one's companions as much as about the countries visited. The first European expedition to Arabia, in 1761–7, commissioned by the king of Denmark, was important not just for its geographical discoveries, but no less for its dramatic revelation of what independence of spirit does to travellers. It was not because they were variously Danish, Swedish and German scientists that the members of this expedition found it so hard to tolerate each other. Its leader, the botanist Peter Forsskal, 'despised all dangers, obstacles and deprivations [but] his shortcomings were his argumentativeness, his stubbornness and his temper'. He had written a thesis on civil liberty, demanding unlimited freedom of expression and arguing that 'the sole danger to human liberty comes from those who by their office, rank or wealth have become all-powerful in their country'. The faculty of Uppsala University forbade him to publish it; he did so regardless, distributing it among the students, and continued to protest until Sweden abolished censorship in 1766. A pioneer of what later came to be called youth culture, he refused to admit that youth could be too arrogant. His condition for joining the expedition was that all its members should be equal, that he be given the title of professor – though he was only twenty-seven – and that on his return he should receive a substantial pension which he could draw in any country he pleased, because 'I will not subject myself to the very restricted freedom of thought and expression one finds in Sweden and probably in Denmark also'. He got everything he asked for and, before leaving, had his portrait painted.

Though science was rewarded with many crates of unknown plants, maps and information of all sorts, these explorers did not discover how to tolerate each other. The anthropologist Von Haven, a cultured man, who enjoyed calling on every French diplomat on his route to make urbane conversation on such subjects as the ideas and private life of Voltaire, equipped himself with 'enough arsenic to annihilate two regiments' in order to protect himself from his colleagues. 'We are living in the greatest danger' was the report they sent home, not because Arabia was hostile, but

because they were threatening to kill each other. The royal organiser replied from Copenhagen, 'Nothing seems to me so easy as to
live in understanding, peace and harmony with all men. All it
demands is that one should cast off all such prejudices as prevent
this and pay heed to the call of reason.' But how could Von Haven
be reasonable when he was additionally obsessed by the sand in
the food, the polluted drinking water and Arabs who he complained would not let him eat in peace, obstructed his research and
refused to let him retrace the steps of Moses? The doctor on the
expedition made friends by prescribing aphrodisiacs to tired old
sheikhs, but his skills were not adequate to prevent all but one of
the scientists succumbing to disease.

The survivor, Carsten Niebuhr, was the only modest member of
the group: he had refused the title of professor and never indulged
himself with tantrums; caring nothing for philosophy or poetry,
his pleasure was to record and measure what he saw, making a
map of every town and compiling five volumes of dispassionate
information. Like Tu Lung's hero, he did not want honours from
the world, asking as his reward only to be appointed clerk of the
council of a village in the most desolate marshlands of Denmark.
Even in that silence 'he suffered greatly from a longing for the
dignified peace of the Orientals'. The French Academy elected him
a corresponding member, but he declined to visit Paris, not wishing to enjoy fame. His conclusion was, 'If only we had been more
on our guard against colds, and also right from the start tried in
general to live more in conformity with Oriental custom, and
if only the various members of the expedition had a little more
confidence in one another, and had not so filled the journey with
frustration by their suspicions and their quarrelling, then perhaps
we might all have arrived back happily in Europe.' It was a conclusion which applied to much more than his own adventure.

Living according to local custom: is that the clue? Most of these
travellers could not speak Arabic. But has knowledge of foreign
languages been a road to a different kind of journey? The answer
can come from Sir Richard Burton (1821–90), one of the most
gifted linguists of all time, who mastered twenty-five tongues, or,
including dialects, forty, with such perfection that he could pass as

a native; two or three months were enough for him to become fluent. Experts found him weak only in two languages – Russian, inexplicably, and German, whose sounds he said irritated him. 'Nothing,' he believed, 'goes to the head of a man so much as to speak to him in his own patois.' He was therefore a disaster as an undergraduate at Oxford, because he liked to learn a language as it was spoken: he told the professors that their pronunciation of Latin was absurd (which it was, having been artificially invented after the Reformation). So he was spared the fate of becoming a professor himself, and was free to use his talent to search for something more elusive than knowledge.

It need not be harmful to discover, at the beginning of one's career, without waiting for disappointment, that one is, as Burton said of himself, 'a waif and stray'. The reason was not that he was (supposedly) descended from one of Louis XIV's bastards, sent away to Ireland; nor that he was the son of rich parents who did not know what to do with themselves, or with him. They brought him up not in England, but in Tours, where they indulged their hypochondria: 'We never thoroughly understood English society,' he wrote, with a certain satisfaction, 'nor did society understand us.' So he was born to be a traveller. 'Voyaging is victory,' he tried to convince himself. However, he was not self-sufficient: he needed an audience to applaud him. 'It is a real advantage to belong to some parish. It is a great thing when you have won a battle, or explored Central Africa, to be welcomed home, by some little corner of the Great World, which takes pride in your exploits because you reflect honour upon itself.' The price to be paid was respect for that little corner, deference to its authority, but Burton was too sceptical to be respectful of anybody, and least of all of his parents. As a boy he became an accomplished liar. Later he said he devoted himself to speaking the truth, but that it only got him into trouble. It was symbolic that the sport of fencing was 'the great solace of my life'.

The Indian army gave him the opportunity to learn half a dozen languages: and at once he began disguising himself as a Persian merchant of cloth and jewellery, which enabled him to enter the closed world of women, even the harem. The way to get to know a people, he claimed, was to know the women. So began a lifelong devotion to sexology, in the course of which he translated the *Kamasutra*, the *Perfumed Garden* and the *Arabian Nights*. However, the brilliance of his scholarship did not teach him much about women. The great disappointment of his Indian career was the revelation that he was an incompetent lover. 'While thousands of Europeans have cohabited for years with, and have had families by, native women, they have never been loved by them, at least I never knew a case.' Sexology was no more the key to 'understanding women' in the Victorian age than in the Swinging Sixties. To listen to what women said, and thought, and did not dare say, was not part of his programme. His view of women showed itself in his marriage to one who despised women, who, since she could not be a man as she would have wished, imagined herself to be part of him and devoted her life to his pleasure, drawing up a code of conduct for herself in which she promised to hide his faults from the world, never to reproach him, never to answer when he found fault with her, to conceal her own ill-health so as not to annoy him, never to ask him for anything, but to 'let him find in his wife what he and many other men fancy is only to be found in a mistress'.

Burton's most famous exploit was to travel to Mecca as a pilgrim, disguised as an Afghan doctor and dervish, darkening his face with walnut juice, growing a large beard, shaving his head: he was unrecognisable. He had himself circumcised, 'taking care that it was in the Muslim rather than the Jewish fashion'. There was no ritual or prayer with which he was not familiar, and quotations from the Koran filled his conversation. He even turned himself into a doctor for the occasion, treating patients. Despite the heat – which he likened to the breath of a volcano, causing several of his companions to drop dead of exhaustion as they marched – and despite the terrifying attacks of bandits – who ripped open the belly of a companion, leaving the vultures and jackals to kill him off – he never betrayed himself. He probably got as near to feeling what it means to be a Muslim pilgrim as any Christian has. But one obstacle remained.

When he finally reached the holiest of the Mecca shrines, he was deeply moved, but he recognised, 'to confess humbling truth', that while the emotion experienced by his fellow pilgrims was 'the high feeling of religious enthusiasm, mine was the ecstasy of gratified pride'. The ruse counted for more than the insight. His elaborate respect for Islam was ultimately superficial. An Arab who learnt of his identity said, 'He hath laughed in our beards.'

That is the significance of Burton's life. Even though he was admired as the greatest interpreter of the East to the West – and he certainly conveyed a lot of accurate information – he did not find a satisfactory way of reconciling his participation in differing civilisations, which is the ultimate goal of a traveller. He said that the Indians secretly thought of themselves as superior to their British rulers, who despised them: the only solution was to fight it out, to rule with an iron fist; liberalism would be mistaken for weakness, because 'Oriental discipline is personal respect based on fear'.

It is perhaps because some women travellers saw journeying as an alternative to marriage that they came nearest to dissolving the frontiers. For them it was doubly an act of defiance, of convention as well as of danger. The Viennese Ida Pfeiffer (1797–1858), for example, seemed to her neighbours to be a staid, practical housewife, 'lacking in outward charms', forced to marry a widower twenty-four years older then herself, who proceeded to lose his fortune; but after painfully bringing up her children, she found a new life as a traveller. 'I had learned,' she wrote, 'to fear my parents rather than love them.' Now she searched for other kinds of relationships. On her own, with minimal savings, she circled the globe twice over, visiting lands where no European had been before her – a diminutive, unthreatening, elderly lady, possessing nothing but a 'talent for awakening and profiting from the sympathy of those she came into contact with'.

After the history of nations and of families, there is another history to be written, of those who were misfits in one or the other, or felt incomplete in them, and who created new affinities far away from their birthplace. Travellers have been a nation of a special kind, without frontiers, and they are becoming the largest nation in the world, as travel becomes no longer a mere distraction but an essential part of a whole person's diet. Today over 400 million people travel between continents each year. The most admirable characters in the history of travel are those who have been most useful to their hosts. A journey is successful when the traveller returns as an ambassador for the country he has visited, just as an actor is most successful when he enters into a character and discovers something of himself in the part he plays.

Travel does not necessarily involve going to distant parts. I now examine the most secret of all destinations, the journey which men and women have made into each other's minds.

Ibn Battuta, *Travels*, ed. H. A. R. Gibb, Cambridge UP, 1958; Rose E. Dunn, *The Adventures of Ibn Battuta*, Croom Helm, 1986; Dale F. Eickelman, *Muslim Travellers: Pilgrims, Migration and the Religious Imagination*, Routledge, 1990; Richard Trench, *Arabian Travellers*, Macmillan, 1986; Thorkild Hansen, *Arabia Felix: The Danish Expedition of 1761–7*, Eng. translation, Collins, 1964; Richard Burton, *Narrative of a Pilgrimage to Meccah and Medinah*, 3rd edn., William Mullan, 1872; Frank McLynn, *Burton: Snow upon the Desert*, John Murray, 1990; Dea Birkett, *Spinsters Abroad: Victorian Lady Explorers*, Blackwell, 1989; Anon., *The Story of Ida Pfeiffer*, Nelson, 1879; D. Murray Smith, *Round the World*, Nelson, 1968; Valerie L. Smith, *Hosts and Guests: Anthropology of Tourism*, Pennsylvania UP, Philadelphia, 1977; Daniel J. Boorstin, *The Discoverers*, Random House, NY, 1983.

18

Why friendship between men and women has been so fragile

Absa says she has no friends. She has never been to school, so it was a great event in her life when her seventeen-year-old daughter (who comes top in her class) passed her examinations. They decided to celebrate by inviting a dozen of her schoolmates to a party. She prepared a vast array of cakes and lemonade. Only two of the guests came.

Absa N'dai, aged thirty-four, with four children, has lived in the same block in Bordeaux for ten years. Never in all this time has she had a conversation with her neighbours. She is Senegalese: there are four French families on the same landing and one Portuguese. She has never heard them talking to each other either. They never said good morning to her. 'It is I who first said good morning, and sometimes they don't answer. Once one of them got stuck in the lift and I called the fire brigade, but all he said was Thank you. Once I smelt gas, and my daughter phoned for help, but for this we got only scowls. My immediate neighbour, whose front door is only one metre away from mine, has two children, who have never spoken to mine.'

But then there is not much talking in her home either. Her husband is a labourer in a biscuit factory, on the minimum wage, threatened by redundancies all round him. He comes home tired and sits in front of the television, and 'each one of us is quiet in our corner. He is sad, and he does not speak even to his children. I do not speak very much with my husband. He is quiet. This makes me very tired. But I have no problem speaking with my children. My son is nice to me: he asks, Mother, Why are you quiet?' She smiles. The children are her pleasure. But finding the money to clothe them and give them what they need for school is like holding up an umbrella in a storm.

Her husband came to France in 1976; she stayed in Senegal with the children till 1983, when she joined him. The first two years she devoted to learning to read and write French, and since then she has been searching for work, in vain. 'I have put myself down for every kind of job. I have offered to be a cleaner; I look at the vacancies every week, though it takes me an hour to get to them by bus. I fill in forms, and they say they will telephone and never do. I have done everything to find a job. I don't know why I can't get one. I paid money to put an advertisement in the paper. I'm all alone. I don't know anybody.'

In Senegal, things were not much better; there was no work there either, though she earned a little as a dressmaker. In France, the government has put her on a training course, which involves work practice in a factory. Her boss says he is very pleased with her, but he has no vacancies and cannot keep her on. She would like to open a little dressmaker's shop: the six of them live in three small rooms; there is no space at home; but even if she could afford a shop, the taxes she would have to pay would immediately ruin her. 'I like working. I hate doing nothing.'

Her elder daughter had a friend at school, who was the daughter of the maths teacher. The friend caused great pleasure to Absa by expressing interest in Senegalese food, and a wonderful dinner was prepared for her; but the friendship petered out when the teacher was transferred to another part of the country, and now there are only occasional telephone calls. Absa's daughter says she is thinking of going back to Senegal.

What Absa misses is African family life. 'In Senegal we all live together. We forget everything. We don't think. We eat together. Here one has to think, all alone. When my children go to school and my husband is at work, I think a lot and cry. When my father died, I cried for a week, I was all alone. In Senegal the family helps, but here no one spoke to me. When my husband is on night shift and I am alone with the children, I think, Why is there nobody to help me, if something were to happen . . .'

The organiser of the training course which is supposed to prepare Absa for work (who admits however that only one-fifth of the trainees will find jobs) is by birth a Kabyll, and a distinguished economist. He says that when he lived in Paris and said good morning to his neighbours, no one answered him either.

Do those who are richer or better educated make friends more easily? Martine Bedin, internationally known architect and designer, is a member of one of the bourgeois families of Bordeaux, but as a child the first question she was always asked by the mothers of her schoolmates was, 'What does your father do?' Respectability is still maintained by infinitely subtle gradations of aloofness. So Martine Bedin has tried instead to make friends by seizing the opportunities of chance encounters: one cannot choose the potential friends one meets by chance, but one has a choice as to whom one can try to keep as friends, by putting effort into friendship. So it is not family connections that she emphasises, but the independence and eccentricity of her inheritance: her Corsican mother, born in Venezuela, was a penniless drama student when she met her father; all Corsicans, she claims, have some blood from visiting sailors, that is why she herself is blue-eyed and blonde; she likes to think she might have some Jewish blood too. Her paternal grandmother was the first woman to have been admitted as a student into the Bordeaux School of Commerce, going on to make a fortune as an industrialist. Martine Bedin was inspired by this tough-minded woman, who urged her to become a politician, or a leader of some sort.

'But I have no authority.' Martine Bedin wanted not to be obeyed but to be liked, and not only at home. She learnt four foreign languages. In London at the age of thirteen, improving her English, she met the film director Visconti's nephew, who was attending the same language school, though living at Brown's Hotel. That friendship opened Italy up to her. She pursued her study of architecture there, attended the lectures of Natalini and was captivated by his charismatic oratory. 'I am fascinated by people who have done things'. So much so that she is brave enough to speak to them. She asked Natalini if she could help him. Working as a waitress to keep herself, she secretly slept in his office; when he found out, he took her into his house. Then Sottsass, the designer of Olivetti's office furniture, seeing her plans for a house at an exhibition, complimented her, saying, 'We must build this house together.' 'Put your signature to it, and it will be possible,' she replied. Young disciples are difficult to resist. Natalini had protested, 'But I am no better than your father.' Sottsass protested, 'But what I do is boring.'

At twenty-three she became one of the founding members of Sottsass's Memphis Design Group. 'You will need courage to join me,' he said. But it was she and her colleagues who gave him courage. They were nearly all of them young foreigners who were able to do in Milan what they could not dream of doing at home. Sottsass had no children, he was like a father to them, except that they had adopted him of their own volition, and 'he was jealous, worried that we would abandon him'. Martine Bedin gave herself up to a 'frenzy of drawing'. The group's radical vision of beauty, its insistence that beauty could be found in everyday objects that no one had ever thought of as worthy of attention, the mixing of expensive and cheap materials, the use of strong colours, the deliberate destruction of accepted taste, won international interest, but after seven years they decided they had run out of ideas, and disbanded.

Is this kind of association of artists and craftsmen working together the most satisfying form of friendship? There is still a feeling in Milan that the designers of the city, working independently, but generously helping one another, have discovered what fraternity really means. For Martine Bedin, however, becoming famous too young created problems. She continues to design everything from furniture and houses, buses and public lavatories to jewellery and lamps, carpets, taps, biscuits, handbags and sun glasses. She has been a consultant to, and received commissions from, the world's most prestigious and luxurious firms. And yet she yearns for a closer intimacy with those who use her products. So she has established La Manufacture Familiale, which makes domestic furniture and sells it without intermediaries to people who come to her own home to see what it looks like.

Does the designer then become the customer's friend? Martine Bedin designs what she likes, not what the customer wants, because the customer does not know and could never explain, just as she cannot explain what she will do until she has done it. A designer, she thinks, is a sort of peaceful terrorist, who creates what nobody expects, what nobody could foresee: that is the whole point of originality. But when she creates her own design, there is always the fear of not being understood, of the customer not seeing how she has given each object its own personality, its own independent dignity. Perhaps, she thinks, the only way of gradually winning that understanding is to enter into people's consciousness without their noticing what one is doing, like a parasite. It is only other experts, who have tried to do the same thing, who can comprehend what an expert has achieved. That is why modernistic art does not sell to a wide public: she prefers to shock or surprise, to remain 'profoundly individualistic'.

Her two husbands, by contrast, have been successful industrial designers whose work has flooded the supermarkets. Her friendships with them have been of two kinds. The first husband was older than she, an Italian, an introvert: 'I am not sure I understood him; he had no need to communicate with others; he taught me to work alone.' After twelve years of contented marriage, one day they each fell in love with someone else, parted amicably and remain friends, for twelve years of one's past should not be wasted. Likewise her second husband's daughter and former wife have, after some hesitation, also become friends, and come to stay. He is Piotr Sierakowski, who gave up his prospering career in the USA to live with her in the rural peace of the Gironde, though he is a lover of cities: they have different styles, different ideas, he is colour-blind whereas colours are the heart of her work, and he has no sympathy at all for the art of Memphis. Emotionally they are together, but intellectually they remain independent. Their child, their beautiful eighteenth-century home, which they have modernised internally themselves, their business partnership in La Manufacture, are domains where they meet. However, when both were asked to give lectures at the art school and she suggested they should put on a joint course, he preferred to perform separately. He never lies: that for him is the great problem for a designer in need of commissions from industrialists, who want you to pretend you share their ideas, who want you to repeat a previous success rather than venture into the unknown, whereas he likes to reflect on the essence of objects, and his ideal is to be a Mozart in wood, to have a style which is full of invention and surprise.

Martine Bedin sometimes thinks of La Manufacture as reviving the Corsican clan or the old artisan family workshop, but at other times says, 'I don't really like working with men. I want to be able to announce that I don't feel like working today, let's go into the country: you can only do this with women; and I have worked above all with women, with whom I have been very close, transforming work into a feminine act, abolishing the distinction between work and life. The men I have worked with have all had a feminine side, they liked being with women and talking about feminine subjects, for creativity is the application of the feminine side of men. When I design my furniture I see it in my mind's eye as part of a domestic scene. Sottsass too explained his furniture with a scene or an anecdote; he had his feminine side, was emotional, cried, got angry. Men and women have different rhythms of thought: I never feel my mind is totally free; I am always preoccupied with something.'

It seems that even educated men and women, quite often, have only partly met.

Why has friendship between men and women been so rare and difficult? The usual answer is sex. But that is to forget that friendship between men has not been easy either. The very first friendship of which there is a historical record, between the Babylonians Gilgamesh and Elkidu, around 2000 BC, immediately ran into trouble, because their temperaments were too different. Elkidu was a 'wild man'. They had to negotiate before they eventually agreed to set off together 'to rid the world of evil'. It has not been sex that has thwarted friendship, so much as fear of people who are different.

In 1936 Americans hailed Dale Carnegie's *How to Make*

Friends and Influence People as the solution to their frustrations. Five million copies were sold over the next two decades. It was almost like another Bible, for though it did not open the doors of heaven, it did teach how to knock on the doors of strangers and not have them slammed in one's face. Carnegie (1888–1955) was a commercial traveller, upset by his short stature, who triumphed over his misery by giving evening lessons in the art of public speaking. In a country of immigrants, he spotted that what held people back was their fear of making fools of themselves when they opened their mouths. His prescription was simple: Smile, never argue, never tell anyone they are wrong, never find fault, be a Nice Guy. Do not be different from others and others will be your friends. In other words, do not be yourself: the great obstacle to friendship was the misfortune that people were different and that they did not do enough to hide it. Carnegie went on to write a book saying the same was true for women – *How to Help Your Husband Get Ahead in His Social and Business Life*: they too must learn to play a part. In accordance with the principle that the easiest cure for a fear is to switch to another fear, he replaced the fear of making a fool of oneself by the fear of being found out.

There was nothing specifically American in Carnegie's remedy. Shakespeare wrote that friendship is 'mostly feigning'. Like Europeans, Americans were half-Christian and half-pagan, unequivocally pagan in their devotion to success and their tacit tolerance of insincerity. Those who accused Carnegie of preaching hypocrisy made no impression on him: if success was what Americans wanted, he was showing them how it could be achieved: their private thoughts did not concern him. For most of history, indeed, friendship has had nothing to do with affection. A friend used to be above all a protector or someone useful to whom one sold one's allegiance in return for favours, for as long as the favours lasted. The ancient Romans, who called almost every one they had dealings with friend, were not ashamed to divide them into three categories, those they welcomed into their homes, those permitted to enter the open courtyard in front of their houses, in groups for a morning salutation, and the lowly clients who waited outside, under the supervision of servants.

It is not ideals that the world is short of, but methods to stop them becoming a sham, a lie. In Italy today, says that country's expert on friendship, Francesco Alberoni, the word 'friendship' still carries a connotation of cheating, of buying privileges by underhand means. Indeed, wherever advancement has depended on knowing the friends of friends, liking them has been a luxury. A specialist in licking boots can say, 'He is my friend and I despise him.' It is only quite recently that the services of government officials, bankers, lawyers, hotel-keepers or insurers have not had to be won as a personal favour and paid for with a part of one's independence. So long as people are hungry for flattery, they will continue to savour Carnegie's sweet and sour recipe. So long as their choice of friends is limited to those who have power, they have no choice.

The link between friendship and fear emerges clearly among the Mayas, or at least among their descendants observed in Guatemala, who had an atmosphere of universal mistrust hanging over them like a permanent mist, perpetuated by 400 years of conquest. The young men tried to fight it by becoming *camarada*. They went about in inseparable pairs, publicly embracing each other, dancing together, even saying they would willingly marry if the other were a woman. These were not sexual relationships, for they helped each other in their amorous advances to the girls; but marriage did not offer all that they wanted in terms of intimacy. A man was made to feel that he needed a *camarada* to ensure that he did not stand emotionally alone. Choosing one was like courtship, and a formal contract sealed the union. 'Everybody has had a *camarada*,' explained an old man, 'and has done the same thing. But it is difficult to keep a *camarada* for life.' For despite their loving devotion to each other, they were unable to avoid the mistrust which permeated their whole society. While each was proud of having a friend, he was also intensely jealous of his partner showing any interest in anyone else: every act not directed to the other's exclusive pleasure seemed to pose a threat. Though they enjoyed confiding in each other, they were unable to get it out of their heads that their friend was a potential enemy, and it was seldom that they had the courage to confide completely. Their idyll usually ended in recrimination: friends turned into enemies. Yearning for an impossible perfection in their friend, demanding wholehearted admiration from him, they preferred to end their friendship rather than share it with another. The moral the old man drew was, 'One cannot and should not trust another person all the way.' So the mist of mutual suspicion could not evaporate.

Neither did it among the Greeks, though they were passionately interested in friendship; but they were even keener on winning admiration, and in addition they wanted justice, so they continued to be haunted by the worry that everybody might not get his fair share of each of these three delights. Aristotle said that he could be friends only with a good man, like himself. That limited his choice very severely. He thought it best to have only a few friends: the possibility that a democracy should be a big friendship between citizens was an idea he raised, only to dismiss it. What should happen, he asked, when two friends are not completely equal? How should they decide if the more virtuous one was receiving more or less than his due reward of admiration and respect? This same problem tormented the Greeks in their homosexual loves, which were not just a matter of physical desire, of being, like Zeus, 'inflamed by a thigh': the adult who won the love of a youth was proud above all of having proved that his merits and experience were worthy of admiration; the more the youth resisted, the more suitors he had, the prouder was the victorious one. So friendship too became confused with pride and clashed with competitiveness. Though Aristotle was undoubtedly one of the most intelligent men who ever lived, though he wrote very perceptive chapters on friendship and though he could talk impressively (after a long pause for reflexion) on every subject under the sun, in old age, indulging his favourite occupation of taking baths in warm oil, he said, 'The more lonely and isolated I am, the more I have come to love myths.' Friendship remained a myth.

The Persian Abu Hayyan Al-Tawhidi (932–1023), whose *Epistles on Friendship* are also an autobiography of exceptional honesty, could not see how to reconcile friendship with his other yearnings. 'I am a man dominated by a desire for security,' he wrote. Friendship did not give it to him, because he believed (following Aristotle, whose opinions blinkered thinking on the subject for two millennia) that friends should be 'one soul in bodies twain', as similar as possible. His ideal pair of friends –

Suleyman the philosopher and Ibn Sayyar the judge – had the same desires, the same passions and the same fears, even their dreams coincided; they told each other everything 'as though he was me', sharing everything, never getting angry with each other. But such a model was unhelpful for Al-Tawhidi's own social life. He was by profession a copyist and calligrapher, keen to better himself; he lamented that to resist ambition one had to be deeply religious and care nothing for the pleasures of the world, 'which is difficult'. He could not make friends with the grand vizier from whom he sought employment and honour: they were temperamentally too different. When he had outlived his son and the men of letters with whom he consorted and corresponded, he felt profoundly alone: 'My energy is failing, my mind is set in its ways, my eloquence is gone, I am growing obsessive and I despair of all human beings.' So he burned his library of books, the possession he cherished above all else, 'though it was like murder ... [but] I did not want to leave the books to people who would have mocked them and would have pointed out my ignorance and my defects.' Friendship had been his 'consolation' but it left him as insecure as when he began. The flaw in it was this ideal of total harmony, which made it impossibly rare and which, when achieved, cut off the united pair from the rest of humanity.

If friends have to be identical, then men and women must despair of being friends. And yet it is reported that in parts of Africa, among the Bangwa in the Cameroons, for example, and among the Nzema in Ghana, men and women enter into close friendships which last a lifetime. When they are together, 'the woman relaxes her usual demeanour of almost theatrical deference before any man and they are permitted to joke, talk frankly and even eat together – a practice normally tabooed between men and women, particularly husband and wife.' The friendships continue after they marry someone else; they intervene to protect each other in their quarrels with their spouses. When Ibn Battuta (discussed in the previous chapter) visited Mali in the fourteenth century, he was surprised to find men and women enjoying each other's company outside the restraints of marriage.

Men have not always been obsessed by sexual intercourse when they have come close to a woman. In the twelfth century, many knights did not expect their devotion to the woman they admired to be sexually consummated. In seventeenth-century England marriage was postponed till the late twenties and yet the illegitimate birthrate was only 3 per cent, a mere fraction of what it is today, despite contraceptives. Samuel Pepys, one of the most highly sexed of men, who kept a diary, in minute detail, of his flirtations, kissed and fondled almost any woman he met, but seldom went further. In his day, it was customary for a man courting a woman to be invited by her to 'bundle' in her bed, which meant that they cuddled and talked and slept but kept their clothes on, though sometimes she might strip to the waist or take her shoes and socks off; but there was a 'clear understanding that innocent endearments should not be exceeded'. The practice was considered 'as little dangerous as a tête-à-tête in a drawing-room', but preferable because it was warmer; bundling was for winter only, often done after church on Sunday, and not confined to courting couples, for a husband might invite a visitor to bundle with his wife or daughters. It was common in England, America, Holland (where it was known as *kweesten*) and apparently Afghanistan. The Frenchman La Rochefoucauld refused to believe it and attributed the populousness of Massachusetts to bundling. The practice survived there until 1827, Cape Cod being the last place to resist the genteel view that sitting on a sofa in a drawing-room was more proper. It is only in the last two centuries that a taboo against touching has been established, and that copulation has replaced it as the symbol of intimacy.

Intimacy has meant different things over the years. It was originally concerned with space and objects – an intimate room, for example, into which one withdrew away from the hubbub of relatives and neighbours – or intimate souvenirs and relics – like a lock of hair – which one cherished as though there was magic in them. Within marriage, intimacy meant domesticity. Friends expressed their intimacy by falling into each other's arms, lying in each other's bosom. In countries where old mores prevail, touching remains the sign of intimacy.

The romantics said this was not enough and invented a second kind of intimacy, which was truly revolutionary because it claimed it was possible for a man and a woman in love to experience that union of two souls which the ancients had admired, but which they had restricted to men. The romantics added that sexual intercourse was the means of achieving that union. Previously a man who claimed to be in love had to prove he was serious by showing that he was wealthy enough to support the woman; he had to be practical and give respectability to passion – which used to be expressed essentially outside marriage – by proclaiming a willingness to marry. She was supposed to fall in love with him before engaging in sex. The great innovation was to say that sex would keep them happy ever after. That way there was no need to worry if they had nothing to say to each other: the union of souls was declared to be so ethereal as to be beyond communication; in the process, the two ideally lost their sense of separate identity. There was no need to love one's partner as an individual, because what one loved most of all was love, being in love and being loved.

All this was a transitional phase in the history of the relations between men and women, produced by the difficulty they had in communicating. It made it unnecessary for them to get to know each other; it made it possible for an unbalanced relationship to be experienced as companionable and selfless; above all it abolished all preconditions to love, by making it defiantly inexplicable, by idealising the loved one, by enabling anybody to fall in love with anyone and by abetting the pretence that the love would never end. This was one of humanity's most wonderful inventions. Intimacy became like a pair of wings and there was no need to learn to fly. But those who succeeded in incorporating this ideal into the day-to-day stresses of marriage were rare indeed. There was a long debate in the eighteenth century as to whether friendship or passionate love held men and women together more firmly. The matter was settled by the acceptance of this idea that sex was the best guarantor of harmony.

Since then, however, a third kind of intimacy has been imagined, an intimacy of minds, for people who read and think and observe both others and themselves, and for whom life is an exploration. Instead of constantly asking each other, 'Are you still besotted with me?' the question has become, 'Do you still interest me, stimulate me, help me, comfort me and care for me as I change and grow, and do I still do the same for you?' This intimacy is a partnership in the search for truth, enabling each to see the world twice over, through the other's eyes as well as one's own. Penetrating into each other's mind involves neither submission nor domination: the partners try to listen to each other, while each remains a separate person, conscious that intimacy can be a cause of conflicts, or become too close and stifling, or too defensive. It cannot be a totally safe retreat from the hostile world, and the implication is that two people are unable to satisfy each other's needs in their entirety. But their differences make it possible for them to help each other to explore, together and separately, what they could not attempt on their own.

How to combine all three forms of intimacy and get the best of all possible worlds? For at least two centuries, and probably much longer among the poor, there have been attempts to combine marriage and friendship in one. In 1782, at the same time as an English judge held that it was lawful for a husband to beat his wife provided the stick was no thicker than his thumb, an Englishwoman replied, 'I believe it . . . absolutely necessary to conjugal happiness that the husband have such an opinion of his wife's understanding, principles and integrity of heart as would induce him to exalt her to the rank of his first and dearest friend.' There were enormous prejudices in the way: Michel de Montaigne, whose life was dominated by an Aristotelian attachment to a like-minded man, thought that friendship with a wife might indeed be the most perfect arrangement, but unfortunately women were 'not ordinarily capable' of it: 'their soul does not appear firm enough to support the strain of so hard and durable a knot'. Nevertheless the presidents of the USA began appearing on the hustings hand in hand with their wives, and in 1980 an American enquiry found 60 per cent of married men and 50 per cent of married women describing their spouse as a very close friend.

However, companionate marriage proved to be a fragile construction, collapsing ever more frequently into divorce. The reason was that friendship propped it up only from the inside; there was no provision for friendships with the opposite sex outside the marriage, which was therefore liable to crumble at the merest puff of jealousy. Outside friendships remained suspect, and many people declared them to be impossible. The accepted opinion was that women and men had fundamentally incompatible attitudes towards friendship: men understood it to mean doing things together, keeping their innermost thoughts (if they had any) to themselves, while for women it implied exchanging intimacy and emotion, talking frankly about their real preoccupations. But recent research has not confirmed that this is an unalterable fact. Women historians, while revealing the richness and intensity of female friendship, have also warned against idealising it, showing that when it has taken the form of an exchange of grumbles, it has often encouraged 'psychological hypochondria' and 'dependence on therapy'. Sociologists have calculated that only half of women's friendships can be regarded as genuinely supportive: the rest, they say, are seen as a drain on energy, tolerated only from fear of having no friends. One survey found that many of its married women respondents would have ideally preferred to confide in their husbands, but turned to women friends because they were not listened to. Janice Raymond, a former nun, one of the writers who has celebrated friendship among women with particular brilliance and learning, insists nonetheless that it needs drastic reform, to become more 'thoughtful'. Women's friendship has emerged less as an intuitive passion than as a work of art delicately and laboriously created, and therefore essentially flexible.

A proportion of men continue to repeat that they do not want the friendship of women, claiming that they get their friendship from their cronies, and that women are for children and sex. But the research of Karen V. Hansen, for example, into the correspondence of ordinary working men in nineteenth-century America, has shown that the stereotype of male friendship is not as unchanging and unchangeable as used to be thought. Their private letters reveal them as having close and intimate friendships among themselves which were not unlike those of middle-class women, nor did the suppression of emotion seem to them to be essential to demonstrate manliness. Visiting, exchanging services, nursing the sick with tender care and organising hospitality were all part of their routine, and often they had warm relations with their sisters too. There are not enough of these letters to make it possible to say that this was typical, but they demonstrate what was possible, and how fashion can alter behaviour as well as clothes. When men had to fall romantically in love in order to get married, the almost-romantic relations between men themselves declined, and it was then that they flaunted their toughness more aggressively to compensate for what they feared might be taken for weakness. Recent investigations into where men get their emotional nourishment, whom they frequent and what they talk about, have shown that a not negligible minority do have intimate, self-revealing discussions among themselves, particularly while they are unmarried. It is marriage which brings the shutters down. Then it is largely within the privacy of marriage that they are undressed of their hypocrisy. As history's manly heroes are transformed by modern biographers into pathetically lonely and flimsy waifs, it becomes uncertain whether a marriage, even combined with a friendship, is enough to prop men up. The companionate marriage has been only a partial solution.

What it has left unsolved is the riddle of friendship between the sexes outside marriage. That is the next item on humanity's agenda. When novelty was feared as a kind of sacrilege, the ideal friend was a clone of oneself, but once boredom becomes the great menace, and originality the only protection against it, friendship has to be a partnership which kills boredom. For those who separately do not know what they want, friendship has the potential to be the means through which, as a pair, they can find a purpose. When friends have differing ideas, when they are of different sex, when they enjoy their differences, when they become curious about each other's uniqueness, curiosity can become the moving force in their relationship; beginning with curiosity about each other, it can grow into curiosity about what lies outside them, leading to untried experiences which are too frightening to confront alone. Curiosity is the energy out of which a purpose grows, when its satisfaction is not a purpose in itself.

It implies a vision of life in which the past, the present and the future are like a giant snowball, a ceaseless accumulation of new experiences and ideals, or like the universe itself, ever expanding. Friendship between the sexes goes beyond the romantics' absorption in the ephemeral moment, forgetting all else, dreaming only of passion. If passion is the only reality, life can have no other meaning, and everyone knows that passion fades. The growing number of choosy individuals who cannot find partners to suit them exactly are searching for something more than the mutual admiration of a closed couple; trying to fulfil seemingly irreconcilable ambitions, they need relationships which will not set hard like rigor mortis.

However, friendship between the sexes requires a willingness not to think in stereotypes, not to insist that one is either a lover, or a friend, or family, and that there is no ebb and flow between such categories. The Marquise de Lambert (1647–1733), whose husband was governor of the duchy of Luxembourg and who tried to work out from her own experience how, side by side with love, mixed friendships could be made to flower – she was convinced that there was always a 'liveliness' in them which was absent from meetings of the same sex – concluded that the heart needed to be educated, just as the brain was, and that affection should be studied as a form of art. A loving friendship (*amitié amoureuse*) is indeed a new form of art, as interesting as a companionate marriage.

Soon after she wrote, a few men and women did begin to seek each other's company deliberately for friendship's sake. In 1765 the first mixed club, Almack's, was founded in London, the men members being elected by the women, and the women members by the men. And then sport and hobby clubs, though they often kept the sexes apart, occasionally did bring them together. The possibility of a new era in human relations appears every time a Berlin Wall is knocked down.

Probably the most successful experiment in mixed friendship, which yielded the most interesting practical results, was that of the Society of Friends (alias the Quakers). It was based on the principle that individuals must make up their own minds about how they lived: it had no doctrine, no book of rules, no priests, and in its marriage ceremony there was no promise of obedience. Run on democratic lines, its members said what they thought at meetings in which no decision was taken until everyone was agreed. They ignored rank and status, calling everyone Thou. Their way of dealing with their persecutors was to meet them personally and talk with them face to face, which surprisingly did sometimes

work even with fierce opponents and even though they were challenging the very foundations of society. The explanation of their
success was a friendship between their founder, George Fox
(1624–91), a shoemaker's apprentice, and Margaret Fell, the wife
of a judge. Equality between the sexes was a basic belief,
reinforced by equal education, so that the society produced some
very remarkable women. One of them, Mary Fisher, travelled
1,500 miles on foot to urge the Sultan of Turkey to change his
ways, and he received her: when she asked him whether he understood her message, he replied, 'Every word, and it was the truth.'
But of course nothing changed in the Ottoman Empire.

It may seem that it was foolish to hope that anything would
change as a result of a few people practising friendship, which has
generally been assumed to be a private matter, of minor consequence in public affairs. The Society of Friends, after over three
centuries of existence, has indeed fewer than a quarter of a million
members – spread thinly over the globe, though strongest in Boli-
via, the USA, Kenya and Britain – but it has had more influence on
how human beings treat each other than any government ever had,
of however powerful an empire. It achieved nothing by force or
decree; it did not move mountains, but it did show how exemplary
deeds could at least cause them to crumble a little, gradually.
Quakers set up the first anti-slavery society, which led to the first
law challenging slavery: any slave who set foot in England became
a free person. They were the first to organise a boycott, in the
eighteenth century, against products originating from slave-
owning countries. They were the first people in the world to plead
for the abolition of the death penalty. In the eighteenth century,
John Belley proposed a free national health service, as well as a
study of Indian and American medicine to complement that of
Europe. Elizabeth Fry (1780–1845) was one of the earliest insti-
gators of prison reform: the Friends did not argue that humans
were basically good, or bad, but that one should try to draw out
what good there was in them, irrespective of the crimes they had
committed. They invented the idea of offering humanitarian help
to civilians devastated by war: in 1870–1 they brought food, clothing and medicine to both sides in the Franco-Prussian War. They
went to jail in 1914 to establish the rights of conscientious objec-

tors. Four of the five leaders of the feminist movement in nine-teenth-century America were Friends, a third of the pioneers of prison reform, 40 per cent of the abolitionists. They wrote the Equal Rights Amendment. Amnesty International is their child.

However, they have never tried to keep control of the humani-tarian institutions they have given birth to, partly because they discovered their inability to play at politics at a very early stage. In Pennsylvania, founded as a Quaker colony, outstanding for its unusual, peaceable attitude towards the Indians and for its excep-tionally democratic government, it became clear that friendship and the giving of orders were incompatible, and that friendship was not a system for large groups of people. It was not just because they refused to pay taxes which might be used for war that they lost control of Pennsylvania. Quakerism also faded out of business, where it was originally very successful, for it offered complete integrity, but it implied that the quality of relationships with employees and customers mattered more than profit, and it was incompatible with expansion at all costs. Most Quakers have now moved to the service and caring professions.

The experience of the Quakers, whose tolerance of internal dis-agreement is outstanding, suggests that there is no need for friends to think alike, if friendship is seen as an exploration, rather than a search for security, and if each partner is recognised as having an equal dignity. While making up their minds for themselves, they made listening to the opinions of their friends an essential part of their method, and it was particularly important that these friends were of both sexes.

The history of friendships within other groups – nations, towns, families – has still to be written, and when it is, the obstacles and the prospects may be better understood. Poland, for example, obviously has a rich history of friendship. It used to be famous for its unique constitution, which made it impossible to pass a law unless every single member of the ruling nobility agreed: this was the most perfect political expression of friendship ever devised, and it brought chaos; but there is something precious in such a heritage, which takes respect for the individual to its ultimate conclusion. Today's political parties in Poland are parties of friends, perhaps because they are young; again chaos follows, but the atmosphere seems to be less poisoned than it would otherwise be. The *srodowisko*, or social circle, uniting family, friends, colleagues and acquaintances, usually decays into an exchange of influence and services, a way of circumventing state regulation, but there is a noble side to it. Boys and girls play together, but it does not automatically follow that adult relationships are easy. Friendship has not yet borne the fruits which might be expected of it; it has been like a tree whose buds are repeatedly destroyed by the frosts. Tyranny and poverty have sometimes been destructive of it, but have sometimes stimulated it. There can clearly be no five-year plan for creating friendship; it is not an ideology, which is why it suits an age which is tired of off-the-peg solutions. There is no way of foretelling the undesirable results which it may induce, but the unexpected is now something, at last, that people expect.

In the 1970s, Igor Kon, Russia's expert on friendship, reported that Russian girls in their late teens had twice as many mixed friendships as boys did. However cut off they might have been in other ways, the Russians have tried to establish equality between the sexes, and these girls were expressing aspirations which followed from the partial failure and partial success of that endeavour. Those aspirations are being echoed, more or less, in all other countries on the same path, when young women talk about establishing 'relationships' rather than focusing on marriage, not because they reject marriage, but because intimacy and exchange have replaced the old conventional obsessions. The most recent American study of mixed friendship reveals that many men still have enormous difficulties with the idea. The lack of models to imitate holds them back. But everything that is happening between the sexes turns around the issues that friendship is concerned with. After two phases, when the sexes affirmed their equality and then their differences, the next phase can only be a new synthesis. Mixed friendship now stands in the avant-garde not just of private life, but of public life too.

The French were recently asked what they valued most, friendship, or family and children, or freedom and independence, or justice, or loyalty, or work, or the sense of duty, or love, or honour, or success and climbing the social ladder, or money, or sacrifice and devotion, or marriage, or sex, or patriotism, or religion. The relative popularity of these ideals varies a little from poll to poll, but not greatly: friendship always comes very high, and in one poll it was top with 96 per cent, the others following in the order I have given, religion coming last with 51 per cent, a majority nonetheless. Only the French ask themselves such questions in such an abstract manner. But it may be that they are once again speaking for humanity as they did in 1789, or at least for that section of humanity which likes to know where it is going.

One of the main obstacles in its way has always been the conviction that there are forces too big to overcome, that fate cannot be defied; but, as will be seen in the next chapter, such prohibitions have sometimes been circumvented.

Roy Porter and Sylvana Tomaselli, *The Dialectics of Friendship*, Routledge, 1989; Robert R. Bell, *Worlds of Friendship*, Sage, Beverly Hills, 1981; Francesco Alberoni, *L'Amitié*, translated from the Italian, Ramsay, 1984; Janice Raymond, *A Passion for Friends: Towards a Philosophy of Female Affection*, The Women's Press, 1986; Yan de Kerorguen, *Le Plaisir chaste*, Autrement, 1984; Pat O'Connor, *Friendships between Women*, Harvester, 1992; Peter M. Nardi, *Men's Friendships*, Sage, Newbury Park, 1992; David D. Gilmour, *Manhood in the Making: Cultural Concepts of Masculinity*, Yale UP, 1992; Igor S. Kon, 'Friendship and Adolescence' (in the USSR), *Journal of Marriage and the Family*, Feb. 1978, 143–55; Christine Castelain-Meunier, *L'Amour en moins: L'Apprentissage sentimental*, Olivier Orban, 1991; Ezra Vogel, 'From Friendship to Comradeship: The Change in Personal Relations in Communist China', *China Quarterly*, Jan.-March 1965, pp. 46–60; Maryon Tysoe, *Love is Not Enough*, Fontana, 1992; Ferdinand Mount, *The Subversive Family*, Unwin, 1982; Steve Duck, *Friends for Life: The Psychology of Close Relationships*, Harvester, 1983; Anthony Giddens, *The Transformation of Intimacy*, Polity Press, Cambridge, 1992; Laurens J. Mills, *One Soul in Bodies Twain: Friendship in Tudor Literature and Stuart Drama*, Principia Press, Bloomington, Indiana, 1937; David W. Plath, *Long Engagements: Maturity in Modern Japan*, Stanford, 1980; Robert R. Edwards and S. S. Spector, *The Olde Daunce: Love, Friendship, Sex and Marriage in the Medieval World*, State University of NY Press, 1991;

Ruben E. Reina, *The Law of the Saints: A Pokoman Pueblo and its Community Culture*, Bobbs Merrill, Indianapolis, 1966; M. E. Mullett, 'Byzantium, A Friendly Society?', *Past and Present*, Feb. 1988, pp. 3–24; Richard P. Saller, *Personal Patronage under the Roman Empire*, Cambridge UP, 1982; K. J. Dover, *Greek Homosexuality*, Duckworth, 1978; Robert Garland, *The Greek Way of Life*, Duckworth, 1990; Aristotle, *Ethics*, Penguin translation, 1953 (chapters 8 and 9); Gabriel Herman, *Ritualised Friendship and the Greek City*, Cambridge UP, 1987; Janine R. Wedel, *The Unplanned Society: Poland during and after Communism*, Columbia UP, 1992; Robert Brian, *Friends and Lovers*, Granada, 1977 (especially on Africa); Marc Bergé, *Abu Hayyan Al-Tawhidi*, Institut Français de Damas, 1979; L. Stone, *The Family, Sex and Marriage in England, 1500–1800*, Weidenfeld, 1977; Maurice Aymard, chapter on 'Friends and Neighbours', in P. Aries and G. Duby (eds.), *History of Private Life*, Harvard UP, 1989, vol. 3, pp. 447–92; Robert L. Selman and L. H. Schultz, *Making a Friend in Youth: Developmental Theory and Pair Therapy*, Chicago UP, 1990; Wayne Koestenbaum, *Double Talk: The Erotics of Male Literary Collaboration*, Routledge, 1989; Dana Vannoy-Hiller and W. W. Philliber, *Equal Partners: Successful Women in Marriage*, Sage, Newbury Park, 1989; Marie José Fasiotto, *Madame de Lambert, ou le féminisme morale*, Peter Lang, NY, 1984; Madame de Lambert, *Oeuvres*, ed. R. Granderoute, Champion, 1990; Niklas Luhmann, *Leibe als Passion*, Eng. translation *Love as Passion*, Polity Press, 1986; P. G. Zukow, *Sibling Interaction across Cultures*, Springer, NY, 1989; Judy Dunn, *Siblings*, Grant Macintyre, 1982; George Levinger and H. L. Raush, *Close Relationships: Perspectives on the Meaning of Intimacy*, University of Massachusetts Press, Amherst, 1977; John Nicholson, *Men on Sex*, Vermillion, 1992; Margaret Hope Bacon, *Mothers of Feminism: The Story of Quaker Women in America*, Harper, 1986; Harold Loukes, *The Discovery of Quakerism*, Harrap, 1960.

19

How even astrologers
resist their destiny

To have a purpose in life means deciding, but nothing is more elusive than a firm and fully satisfying decision. Even the brilliant managers of the Citroën car company could not find a way of deciding on which day to launch their new model: so they asked an astrologer. Among so many applicants for jobs, whom should the Michelin tyre company select? They asked an astrologer. It is easy to assume, when suffering from indecision, that one has something wrong with one, so there is a Parisian astrologer, formerly a philosophy teacher, who treats her clients as 'patients', and says their disease is that they have not grown up; they must 'stop behaving like children'. On the other hand a French Protestant priest resigned to become an astrologer, deciding that only reincarnation could explain his having a seriously handicapped son, and now he is back practising both professions simultaneously.

Ghislaine Bourgogne shows how a woman joins the ranks of these shadowy figures lurking behind the world's important decisions, the public reasons for which conceal quite different ones. Astrology is supposedly about destiny. But her whole life has been a protest against that, and a series of deliberate personal choices. Her main qualification for advising people on their future is that she has been unhappy.

She was brought up wearing the mask of bourgeois respectability, but did little work at school and still less at university, where she studied medicine. Her parents divorced, her father's remarriage was a failure, she moved out to her grandparents, and at sixteen began living alone, 'profoundly lonely, asking myself questions about the future'. Then someone gave her a book on astrology, which seemed to provide some answers, not quite sufficiently personal answers, but at least more pertinent ones, it seemed to her, than those of her professors of medicine. Skipping lectures, she devoted eight hours a day to studying astrology. But she disliked being defined as a 'type'; moreover, her horoscope revealed that she was intelligent and sensitive and artistic and intuitive, but had trouble in her emotional relationships: all too true. It annoyed her, because it offered no hope; above all it frightened her.

To reach a decision on her future, she consulted a professional astrologer, who reinforced all her fears: her life would be hard, with 'very disagreeable events', though it was not clear exactly of what sort; everything would be in ruins for her by the age of twenty-four. She was told she was wasting her good luck, missing her vocation to be a doctor. Her boyfriend, with whom she was getting on badly, was the only one for her, and all her future loves would be failures. 'Don't think I'm trying to stab you,' said the astrologer, 'but that is what I see. And it is your own fault: when you chose to live as you do, you knew what you were doing.'

Ghislaine was flat on her back for forty-eight hours after that. 'I have drawn an unlucky number, I said to myself. I can never be happy.' Now she thinks the astrologer was 'inhumane'; and besides 'it is impossible to predict a whole life for thirty years ahead'. She determined to study astrology seriously to prove the astrologer wrong, and she abandoned medicine for ever.

The purpose of astrology, in her case at any rate, was not to know her fate. It is true that from time to time she said to herself, 'If life is going to be like this, I accept it.' But she did not accept it. 'Our lives are made by free will as well as being determined. I discovered that I was the only person who could help me. I had to wait for the opportunities and then swim with the tide.' She claims that astrology enabled her to discover herself, but she wanted more than that. In the distance, she saw herself becoming serene. But so far serenity has visited only in rare moments. Every part of her life has been difficult. Every time I have spoken to her, she has told me that she has been passing through 'difficult moments'.

Her diagnosis is that she 'cannot choose between two different types of men'. Her husband wanted her to be interested exclusively in him. 'I could not. I needed contact with others. He said I didn't love him, forbade me to meet other men. He felt threatened, as though I was trying to escape from him. I thought about it, and realised I did not wish to commit myself, and that was why I chose him.'

Being locked into a destiny is as bad as being locked into a stereotype. One side of her is indeed pessimistic. 'We have the illusion that we meet other people, but we see in them what we are looking for. Even when two people do manage to have a real meeting, there is a space between them; they cannot open themselves to each other; we meet only ourselves; we cannot eliminate loneliness.' Nevertheless she longs to find someone whom she can recognise as a coherent person, not in order to 'fuse' with him: she needs someone who is different from her, so that they can each 'find their own truth'. Independence is necessary too. The inevitable is accepted only if it leads to greater freedom beyond. 'It is just as well that attraction is chemical,' she says, 'and that one meets those whom one is destined to meet: I am happy with this limitation on my liberty, because it means I do not have to exhaust myself consuming men.' Then she makes her choice.

After five years of marriage, she is divorced, living with a child who is sometimes turbulent and impulsive but who has taught her a great many things, and from whom she gets much happiness. But she adds: 'The love of a man is essential.' Practising astrology fulfils her need to express herself; she does it primarily as a therapy for herself, as she also sings and dances; if others benefit, all the better. And indeed the stream of words that pours out of her has the feel of menthol, cool and soothing. Advertising has not been necessary; clients from every walk of life, managers, doctors, artists, foremen, secretaries, come 'naturally'. Being busy convinces her she made the right choice in sacrificing medicine and in taking the great risk of becoming a marginal with no social status. It is with some pride in her courage that she says that happiness cannot be found where she once thought, in doing what other people do: it has to be approached by self-reliance; and her clients know it too. What they want above all is 'to meet themselves, because only then can they meet others', in other words, they want to become conscious of their abilities. Her own gift is to have 'intuitions' about what has happened to them, without being told. She simply passes on to them what she 'sees', and they must stop her if they cannot recognise themselves. There is no need for them to say what they think about themselves: 'I don't want to hear that.' She 'listens' without them needing to speak. This listening is the meat in the sandwich of her astrology. She knows a great deal about the stars, but she does not rely on them to give her what she is searching for, which is a very individual choice. Her conclusion is that 'there are as many truths as there are humans'. And her conversation is full of maxims which have nothing to do with the planets: 'Learn to live in the present; though that is very difficult, because the present is conditioned by the past; listen to yourself; communicate; listen to others.'

Is it relevant that she cannot take her own advice? That she remains, by her own admission, timid and vulnerable and lonely? The characteristic of successful astrologers is that they do not judge their clients, and they are not judged as private individuals. They are valued, paradoxically, for helping people, overwhelmed by uncertainty, to make choices.

Deciding to turn over a new leaf in one's life has always presented two difficulties: how to throw away old habits, and how to beat down the feeling that one was born lucky or unlucky and that there is no way of changing it. 'The fates rule the world,' said a North African citizen of Rome in the year AD 10. 'Our end depends on our beginning: from it flow wealth and power and poverty; from it all are given their skills and characters. . . . No one can renounce what he is given nor possess what he is not given, nor can he grasp by his prayers the fortunes denied him. . . . Each must bear his lot.' Today, the words used to say this may have changed, but the idea remains.

To understand what can be done with old ideas which linger on so obstinately, it is helpful to look more closely at one of the oldest of all ideas: astrology has survived despite all its mistaken predictions, and even though it has been condemned repeatedly by religion, science and governments. It suggests that new ideas by themselves are not enough to change behaviour, because old ideas cannot simply be swept away. Astrology shows how the old and the new have, in the past, combined as well as fought.

'It is possible to write the history of astrology now that it is finally dead,' said a Sorbonne professor in 1899. But in 1975 a group of 192 eminent scientists, including nineteen Nobel prize-winners, led by a Harvard professor, published a manifesto declaring that they were 'concerned about the increased acceptance of astrology in many parts of the world. . . . Acceptance of astrology pervades modern society. This can only contribute to the growth of irrationalism and obscurantism.' And theirs was not a war simply between scientists and the rest. One of the signatories, a professor of astronomy at UCLA, complained that one-third of the students attending his lectures professed a belief in astrology, and so did his wife. Paul Feyerabend, professor at Berkeley, refused to sign, saying, 'Science is one of the many modes of thinking developed by man, but not necessarily the best.'

If all the inhabitants of the world were questioned today, there might well be a majority of believers in astrology, and that would probably still be the case even if they were all given scholarships to study at Harvard. In France, where lucidity is a national virtue, three-quarters of the population say they are not superstitious, but 45 per cent touch wood, and at least one-third say they believe in astrology. (The same proportion of Britons admit to being believers.) Among French people, sometimes the figure goes up to two-thirds, depending on whom they are talking to; certainly 90 per cent know their zodiac sign. What baffles the scientists is that the believers are not illiterate old peasants. The majority of young people (55 per cent) say they believe in the paranormal, as do many graduates in the liberal arts and 69 per cent of ecologists. France now has twice as many professional astrologers and fortune tellers as priests. And it is the same in the USA. . . .

The astrology which the Western world believes in today is

not the same as that which excited the Babylonians, who were not interested in the fate of individuals and did not invent the horoscope, which first appeared only in 410 BC. They did not believe that the gods were bothered with anything so trivial as the details of daily life: ordinary people had no destiny, because nobody cared about them. What mainly concerned the first astrologers was the outcome of wars and harvests, and the secret rules of prediction, which only they could understand. As the world appeared more and more complex, so the rules became more intricate. The Mayas went further than the Babylonians, discovering cycles of 374,440 years to explain events, with unbelievably abstruse mathematical calculations. But that was not of much use in solving the private problems of the slave in the street. The inventors of astrology had a horror of uncertainty, but also no sense of ever attaining certainty; so every form of divination seemed worth trying at the same time – oracles, however imprecise, haruspication, however perplexing, examining chicken livers for no less than 6,000 warning signs. From the beginning, astrology was combined with other forms of reassurance, not necessarily compatible.

The Greeks, being interested in the individual, turned to astrology when their faith in the Olympic gods began to wane, but they did not just switch from one system to another: rather, they mixed new ideas with old. Thus they began by adding small doses of astrology to the best medicine of their time. The Chaldean refugee who introduced them to astrology, Berosus, settled on the island of Cos, which was the home of the celebrated physician Hippocrates, and that was how medicine took to using astrology as a tool of diagnosis, shaping it into a clinical art, not offering rigid predictions, but interpreting possibilities in each individual case.

It was Ptolemy of Alexandria (fl. AD 127–51), the most successful writer of textbooks of all time, who made astrology into an international creed. All that was known on mathematics, astronomy, geography, history, music, optics, was laid out in Ptolemy's works, which were the world's do-it-yourself manual of information for 1,400 years, and astrology was included as a branch of science. He tied everything up so neatly that it became almost impossible to think except in the context of his picture of the world, or to embark on any technical exploit, or to sail the seas, or to find a place on the map, without his help, all the more so because he provided the most comprehensive description of the scientific instruments that existed in the ancient world. Of course, he often did not know what he was talking about, pillaging his erudition from other books, drawing his famous world map without ever having travelled, relying on hearsay; but he was trusted because he made everything sound reasonable. Astrology was not meant to be irrational. Never had fantasy been so elaborately interwoven with fact as in Ptolemy, with such ingenious mathematical calculations. Never has a man's work contained so many plausible mistakes – such as underestimating the distance between Europe and Asia by 50 degrees of longitude, which encouraged Christopher Columbus to sail for what proved to be America.

Astrology nevertheless was presented as a form of technology, which made the facts of life awe-inspiring, hauntingly mysterious and practically useful. It could appeal to those who had no astronomical knowledge because it grafted itself on to the traditions of the oracle and reinforced fascination with the exotic. The future was effectively put on the map as a foreign country, needing foreign eyes to explain it. So whereas in the Western world astrology was perceived as an Eastern discovery, in the East it spread as a Western science. Ptolemy's system, written in Greek in the second century AD, reached India around AD 500; there it was adopted, after combining with local customs, as the backbone of a faith in the stars which remains to this day. He was translated from the Sanskrit into Arabic in the eighth century and became part of Islamic culture. Acclimatisation always involved intermediaries with foreign curiosities: in Islam's case, it was achieved by Abu Mashoor (805–85), an Afghanistani settled in Baghdad, who came from 'the mother of cities', Balku or Bactra (once the rival of Babylon and Nineveh, and the capital of the Zoroastrian religion), a city peopled also by Buddhists, Hindus, Jews, Nestorian and Manichaean Christians, well used to mixing Greek, Hebrew, Indian and Persian ideas. Only when strengthened by the prestige of being part of Arabic science, and expanded to offer advice on a wider range of worries, was astrology adopted by Europe, in the twelfth century. Ptolemy was translated into Latin from the Arabic

by travellers to the East who came from Bath, Seville and Car-
inthia.

Wherever it went, astrology remained foreign. Had it been an
official creed, people would have searched for a more exotic alternative. When they could not make up their minds, they turned to it
as something completely outside their normal lives, like a court of
appeal with the power to snatch them from the bewilderments that
troubled their daily drudgery. Astrologers were modernisers of
paganism, adapting it for the use of city dwellers who were overwhelmed by having more and more options, but who, being
pagan, preferred not to commit themselves to a single god or a
single creed. In making use of astrology, there was no requirement
that other beliefs should be abandoned. Ptolemy, for example,
married astrology to Stoicism, a favourite philosophy of the dis-
abused men of his time, whose ambition was to avoid anxiety and
who valued being given warning of forthcoming disasters; they
believed in living according to nature, and he told them how
nature worked. Nothing is known for sure about Ptolemy as a
person: he was just a textbook, and his anonymity inspired trust.

Originally, Christianity and Islam rejected planetary influence as
a notion offensive to God's omnipotence, but both eventually
found room for it as a part of astronomy, subordinated to God's
will. In 1348 the medical faculty of Paris University, bewildered by
the Black Death, fortified itself with a dose of astrology and
explained the calamity as due to a conjunction in the influence of
Mars, Saturn and Jupiter. Cultured Europeans fell in love with
astrology when it was clothed in the fashionable robes of classical
learning; from the thirteenth to the sixteenth centuries almost
every prince, and even popes, used the science of the stars as a
secret weapon, like a new sort of artillery.

Eventually, however, Ptolemy's credibility collapsed because he
had put the earth in the centre of the universe. His map had
claimed that it was impossible to sail round Africa, and when that
was achieved, the map, and everything else he wrote, were thrown
away. Three centuries of lonely ridicule followed, and astrologers
almost vanished. It looked as though old ideas could be consigned
once and for all to the dustbin. But no, they do not vanish, and
when there is a crisis, and when people lose hope, or when they

feel that the world is changing too fast and not giving them what they want, when they do not know where to turn, they discover that the old ideas were only packed away in their bottom drawer. They fetch them out, and try them on again.

Modern astrology was reborn during the Industrial Revolution. England was the pioneer in both domains: Moore's *Almanach* sold 393,750 copies in 1803, 560,000 in 1839; *The Straggling Astrologer*, the world's first weekly devoted entirely to the subject, appeared in 1824. France, disillusioned with its Enlightenment, produced 170 authors on astrology between 1890 and 1941. Germany became at the same time the most powerful industrialised nation and the one most devoted to astrology, the rise of Hitler occurring at the very peak of a movement of interest in the paranormal. Ernst Röhm, chief of Hitler's private army, the SA, asked for a reading of his horoscope saying, 'Then I might learn what sort of person I am: frankly I don't know.' Who else could tell him? The Führer himself was a sceptic, and indeed ordered the liquidation of astrologers; but Hess, Goebbels and many other Nazis were interested, and many opponents of Nazism too.

Today, of course, the USA invests lavishly in astrology, and no one has called it un-American. When President Reagan became a regular client of the Hollywood astrologer Joan Quigley, and when her horoscope of Gorbachev decided that the Cold War should end, he was retreading the medieval footsteps of the Holy Roman Emperor Frederick II, King of Sicily and Jerusalem (1194–1250), who, having married the daughter of King John of England, 'refused to know her carnally until the fitting hour should be told him by astrologers'. The emperor, in a world dominated by religion, could not make up his mind on the basis of the contradictory advice of his priests, so he turned to what he believed were the scientists of his day, but without ceasing to be religious. The president, for his part, was finding that science had become too complicated to make decisions for him, and besides it did not completely satisfy the emotions; but he was reassured by claims that certain scientific discoveries were interpreted by a few scientists as supporting astrology's claims, and he remained a pious believer in Christianity at the same time. In this rather confusing way, an old idea, astrology, made possible the acceptance of a new idea, that a leopard could change his spots and that Russia was not necessarily the eternal enemy.

Heads of state used to try to keep the findings of astrologers secret, persecuting those not in their own employ because they feared prophecies of assassination and doom. But since 1930 mass circulation newspapers have become the astrologers of democracy, after the *Sunday Express* published a horoscope on the birth of Princess Margaret which aroused enormous interest. Now the vast majority of newspaper readers, battered by ceaseless information about how much the world is changing, look at their horoscope to see what remains constant, at least in their own personal lives. The Nobel prize-winners were too simplistic in calling this a return of medieval superstition and irrationality. As the jumble of facts fed into the brain has increased, people have become more contradictory. They have therefore turned to 'intuition' to resolve insoluble conflicts. Astrology shows them simultaneously seeking certainties and battling against certainties which they dislike, searching in every direction for something to look forward to. Astrology has become the black market of hope.

One conclusion that can be drawn from its history is that reformers intent on changing humanity's habits have made the same mistake again and again, forgetting that new habits of mind cannot be put on like a clean shirt. Innumerable historical precedents should have warned them that the Communists of Russia would behave in at least some ways like the Tsars they imagined they had got rid of. Relics of paganism have always survived even in strict religions determined to extirpate them. Ancient male prejudices, judging by the history of prejudices, were more likely to go into hiding than suddenly cease to exist.

There are plants which cannot be propagated easily by seed. For growing new habits in the place of old ones, it is more appropriate to turn to the art of horticultural grafting. The old and the new usually rub against each other, causing mutual pain, but grafting is defined as 'the healing in common of wounds'. The discovery of common suffering has always been the best basis for reconciliation and for peaceful coexistence between the irrational and the rational. New habits, besides, need time to establish themselves. So now I turn to the problem of not having enough time.

S. J. Tester, *A History of Western Astrology*, Boydell Press, Woodbridge, 1987; T. O. Wendel, *The Mediaeval Attitude to Astrology*, Yale UP, 1920; Christopher McIntosh, *The Astrologers and Their Creed*, Hutchinson, 1969; Christopher McIntosh, *Eliphas Levi and the French Occult Revival*, Rider, 1972; Alexander Volguine, *Astrology of the Mayas and Aztecs*, translated 1969, originally published in French 1946; Ellic Howe, *Astrology and the Third Reich*, The Aquarian Press, Wellingborough, 1984; Joseph Head, *Reincarnation: An East–West Anthology*, Julian Press, NY, 1961; Michael Loewe and Carmen Black, *Divination and Oracles*, Allen and Unwin, 1981; Jerrold C. Frakes, *The Fate of Fortune in the Early Middle Ages*, Brill Leiden, 1978; Hellmut Wilhelm, *Change: Eight Lectures on the I Ching*, Routledge, 1960; *The I Ching, or Book of Changes*, ed. Cary F. Baynes, with a preface by C. G. Jung, Routledge, 1951; Alan H. Gardiner and Kurt Sethe, *Egyptian Letters to the Dead*, Egyptian Exploration Society, 1928; Michel Gauquelin, *Dreams and Illusions of Astrology*, 1979 (Hachette, 1969); Vicomte Charles de Herbais de Thun, *Encyclopédie du mouvement astrologique de langue française*, Éditions de la Revue Demain, Brussels, 1944; Judith Devlin, *The Superstitious Mind*, Yale UP, 1987; Gerd Gigerenzer, *The Empire of Chance*, Cambridge UP, 1989.

20

Why people have not been able
to find the time to lead
several lives

'My life will start only when I stop this masquerade.' The world
believes that she is a surgeon. She has indeed passed all the
examinations, won many competitions and does her work with
skill. But she thinks she is a coward because she continues to be a
surgeon.

To be one, you have to play the part. Your colleagues are merciless. You must not show your weaknesses; if you make a mistake,
expect no pity. Your patients know that just one slip of the hand,
one moment of distraction, and they could be paralysed, or die:
they expect you to behave like an infallible god. And so she is a
different person in the operating theatre from what she is in private. To her patients, she is always cool, practical, reliable,
reassuring, interested in each individual's problems, never neglecting those for whom she can do nothing more: 'I have never skip-
ped a dying patient on my rounds, as some do.'

In private, however, she sees herself as nervous, unfulfilled, hesitant, the exact opposite of her cool public self. It is not just that she
is appalled by her timidity in the trivialities of her non-professional
life. 'I hate going into shops, speaking to shopkeepers; a salesgirl
has only to say she hasn't got what I want, in a slightly disagreeable tone, and I feel awful. I say to myself: But I'm a doctor, I

can't react like this. It's as though I'm not really grown up.' The uncertainty of life leaves her bewildered, as when she hesitates endlessly buying a pair of shoes and then regrets her choice. Her husband says it is just that she sees ambiguity everywhere. In fact she is very stylishly dressed.

Her public role is a masquerade, because she feels that she is not really playing the game, being only a spectator, not a real part of it, as though watching the world of medicine in a film. Her colleagues see her as respectful of seniority – they are very conscious of hierarchy – joking sometimes, but keeping her distance. She says her relationships with them are non-relationships. Her husband says perhaps she uses her work to palliate her inner anxieties.

But that is how the world is organised. Private life keeps its secrets even when fragments are blown on to the stage. There is no one who is not a mystery.

When she was a girl, she had no friends, pretending to play with other girls, so that the teacher would not know it; then she locked herself up in her room, from the age of eight, to write poetry. She studied medicine for no urgent reason, with no particular sense of vocation; but once into it, she could not stop, could not be content with 'being mediocre', refused to become a general practitioner, did not want to disappoint her parents or her professors. Knowledge is always a spur to yet more knowledge, and the excitement of discovery. Now her speciality allows her to apply ever newer techniques, and she enjoys the manual side of surgery, its 'handyman side': operating is 'like squeezing pastry into a mould'.

But ultimately surgery is technical. Emotions have no part in it. And yet it is 'the emotions which prevent us from dying of boredom; we are nothing without them; life would have no interest but for them'. What torments her about them is that she cannot decide whether human beings are mere bits of emotion joined together, puppets of external forces, how far they can modify their destiny and whether feelings are innate. So in her spare time she writes short stories, which gives her the power to decide the fate of the characters she creates, who are mostly struggling to escape from their fate. What she wants above all is to be in control of her own life, of all those anxieties that blow her about. Written with great power and feeling, her stories show a free imagination, invariably ending in a surprise, the fascination of the unexpected. Sometimes, she writes in her office, and feels guilty that she is masquerading as a doctor, with her nameplate and her degrees on the door: what would people think of her if they knew her mind was elsewhere? Her next book will be about an impostor. Though publishers frequently return novels by unknown authors unread, one has expressed interest. Until she can devote herself to literature, she will feel suffocated. She has illustrated her writing herself, because she is also a painter with a distinctive style; her portraits often show people with a part of their bodies outside the canvas, people struggling to escape out of the frame. And she has set her poems to music, being a composer too. When she was a medical student, she worked as a jazz pianist in a bar – her mother never knew – because giving others pleasure adds to the joy, and she likes to have her artistic talent recognised. Only in her art does the mask of coldness dissolve, that appearance of perfect self-control which the outside world sees and which even her boss is taken in by, saying, 'You never get nervous.' The way work is organised, it rarely offers scope for people to flower. So she wants not promotion, which has rather been thrust upon her, nor more responsibilities, but days with forty-eight hours. Her sense of emptiness is worse than thirst or hunger; it can be filled only by more time than the world contains. Emotion is insatiable.

Promising herself that she would become a mother only after she was personally fulfilled, she postponed having children. For most mothers, their child is their masterpiece, but she determined, when she finally did have hers, that they would not stop her writing, that they would not be the final purpose of her life. 'I have had very intense joys with the children. . . . I find poetry, humour and imagination with them, and I love them, but I am not a mother who will spoil them. I regret I am not a man who could give less time to the children, though I like being with them.' And of course she is worried by the power to alter another being's fate: if only she could confine herself to giving the children confidence in themselves. . . .

Her husband wonders whether women's liberation is not killing them, by opening up all these choices. But of course men are also wearing themselves out struggling to keep up with all that is expected of them, with all that they want to do. There is no chance of him taking a year off from his work, any more than she could. Why? 'It's impossible.' Ultimately, it is fear of losing one's place, disturbing habits, offending the boss by creating problems: she did not even dare to take maternity leave. In order to change one habit, a large number of others need to be changed too.

My surgeon is anonymous. She asked to see me, but said more to me than she should have done in a world where patients and employers do not want their doctors to have emotions. I should have liked to say who she is, because I hope that one day she will be famous as an author and an artist. When that happens, I shall be free to name her, and those who have benefited from her professional skill will be proud that they were treated by one who is much more than professional.

How can people believe themselves to be free if they have no space in their lives where nothing is expected of them, and if they are always late or in a hurry? The Lilliputians observed two centuries ago that Gulliver's God was his watch, and that 'he seldom did anything without consulting it and said it pointed out the time for every action of his life'. Montesquieu philosophised that the English were impolite because 'they are busy people, who do not even have the time to raise their hats when they meet'. Time was short even in this supposedly relaxed age. So what chance is there of escaping its pressures today?

The author of *Why Superwomen are Fed Up*, Michelle Fitoussi (a star columnist for *Elle* magazine), concludes on their behalf: 'What we lack more than anything else is time.' But she can see no solution. Whereas male tycoons have been willing to sacrifice their families to their careers, superwomen, she says, are determined to be admired in every one of their roles, refusing to choose between being helpful, pretty, intelligent, amusing, hardworking, tough in business, and ever so refined. 'The prison bars have changed. In the past, chained to men . . . we fought for just causes. Today, we tie ourselves up. . . . The desire to dazzle has become the hard drug we are addicted to.' She can see no hope for her own generation, nor for that of her daughters: perhaps her granddaughters will have an idea. . . .

Before clocks were invented, frustration had a different shape. Time then was not made of little pieces, of hours and minutes, needing to be saved and accounted for, but was like a huge cloud enveloping the earth, and humanity was waiting for it to clear. The past was a part of the present; individuals lived surrounded, in their imagination, by their ancestors and their mythical heroes, who seemed as alive as themselves; they often did not know exactly how old they were, being more preoccupied with death than with time, which was only a music announcing another life that would last for ever. Every civilisation has made a different prediction as to how long it would take for eternity to arrive. The Hindus spared themselves immediate anxiety because they thought it would be 300 million years; the Chinese insisted that time went round in circles (in cycles of 129,000 years, said Shao Yung), so that nothing ever really changed; and the Zoroastrians said that God took 3,000 years just to create the world.

But then the Jews invented a new idea of time, which has been adopted by all modern societies: they separated the past clearly from the present. Having made a contract with God, they looked forward to its implementation in the future, not in heaven, but in this world. They were the first to imagine a time when justice would be established, when the deserts would become fertile and when there would be an abundance of food and drink for everyone. This vision was their answer to persecution, and the beginning of a new tradition of dreaming about the future, stretching from the Book of Daniel to medieval heresies, socialist utopias, industrial revolutions and science fiction. The early Christians followed the Jews in promising a better future, until they had an established and powerful church, with a stake in the world as it was: but in AD 431 they condemned belief in a millennium – better times in this life – as superstition. Instead they insisted that this world had been made to last only 6,000 years, and would certainly end very soon.

So most of the humans who have lived have not been much bothered by the passing of time. The modern idea of time is peculiar because it includes a new sense that once something has happened, it is gone for ever, that time means change, and so insecurity. Humans welcomed the clock's regular tick, its unalterable habits, its tyranny, because it consoled them for this new insecurity. It was a tyranny which began as a liberation, as so many other tyrannies have. Medieval monasteries were the first to assign a fixed duty to every minute of the day and night, so as to free people from the pain of not knowing what to do with themselves, and from the temptations of idleness. But some thought the price of security was too high: Rabelais protested, 'Never will I subject myself to the hours; the hours are made for man and not man for the hours.' He used the word 'hours' because they were a division of time that people were only just becoming conscious of. 'I treat mine like stirrups, which I shorten or lengthen as I please.' He was announcing a quarrel between the easygoing and the orderly which was to last for several centuries, until the hours won. But now that victory is being challenged.

In 1481 some citizens of Lyon petitioned for the erection of a town clock, in the hope that it would enable them 'to lead more orderly lives', and so be 'happy and contented'. City merchants and industrialists became the main advocates of exact timekeeping. After building cathedrals, they put up clock towers, with the same aim, of showing that there was order in the world. The clock tower of Strasbourg, begun in 1527, took twenty-seven years to build, and perhaps it did increase the orderliness of the inhabitants. But only in the 1770s did the word 'punctuality', meaning exactness to the minute, come into use. Enormous effort went into persuading factory workers to regard obedience to time as a virtue. In the early part of the Industrial Revolution, a Scottish industrialist wrote of 'the utmost distaste on the part of the men to any regular hours or regular habits', and incomprehension that 'they could not go in and out as they pleased, and have what holidays they pleased'.

But regularity has never been found to be a complete answer to the management of time. Nor has saving time, which Japanese emperors first urged on their subjects in decrees issued in the seventeenth century, for even the efficient continued to be harassed by too many demands on their schedules. Fighting time, to remain forever young, did not result in victory. Nor did killing time, for there are more bored people than ever, and time always ends by killing you. Bureaucracies now exist to enforce Parkinson's law that there will never be enough time and that work will always expand to fill the time available. The ancient Chinese idea that frequent sexual intercourse helps one to live for a much longer time was still being propagated by prostitutes in eighteenth-century London, who used to accost customers with the words 'Sir, may I wind your watch for you?', but neither by filling time with pleasure, nor by living to a great age, nor by working shorter hours, have modern people established a perfect relationship with time.

The cow devotes twenty-two hours a day to eating. Is that the natural way to spend time? The housefly spends only one-third of its time eating, and 40 per cent resting, keeping 12 per cent for walking or flying about idly and 14 per cent for grooming itself. That is the female fly. The male rests less and eats faster, so it has 24 per cent of its time available for its promenades and a full 20 per cent for grooming itself. That adds up to 44 per cent of leisure. After many centuries of struggle, Westerners have not done better. They have not been able to cut down on their sleep, which still absorbs 40 per cent of their total existence. They have indeed succeeded in reducing time spent working to about 10 per cent – around 60,000 hours in all, or seven years, which is half of the amount of work done in 1945 – but they now have to add about 12 per cent for the education they need to qualify them for work, and a large part of their travelling (8 per cent, six years in all) is needed to get to and from work. That leaves less than 30 per cent free for leisure, for walking and flying and doing nothing at all.

William Grossin, who spent many years studying how French people spent their time, found that two-thirds of them suffered tension in their relations with time, and that the well-educated and the rich were the most dissatisfied of all. The wider the choices before them, and the more numerous their desires, the less time they have to give to each one. Leisure has become organised, and so full of opportunities too tempting to miss that it does not necessarily offer freedom. The wish to live as intensely as possible has subjected humans to the same dilemma as the waterflea, which lives 108 days at 8 degrees Centigrade, but only twenty-six days at

28 degrees, when its heartbeat is almost four times faster, though in either case its heart beats 15 million times in all. Technology has been a rapid heartbeat, compressing housework, travel, entertainment, squeezing more and more into the allotted span. Nobody expected that it would create the feeling that life moves too fast.

Another sociologist, Stoetzel, claimed that a much larger proportion of French people were relaxed in their attitude to time, 36 per cent, but he included many who did not work. Grossin, who studied only people at work, could find a mere 7 per cent fully relaxed: he concluded that the strain, strictness and monotony of work have a decisive effect on the way people spend the rest of their time: some become passive, losing the capacity to remain in control of their leisure, while others are stimulated to rebel against constraints forcibly accepted during working hours. The rebels are first of all the young, who have not been subdued by the attempts to regiment them at school. They show their resistance by the extent to which they welcome unexpected events and opportunities. In the French population as a whole, 42 per cent welcome unexpected events and 38 per cent do not: the rest don't know. That kind of division is hardly surprising. But 68 per cent of 20- to 25-year-olds like the unexpected. And the second category of people who do are women: one in two, as against only one in three men. The alliance of women with the young, in favour of improvisation and against the laying of fixed plans from which it is impossible to escape, is indeed the new explosive mixture. But it will not explode until they can combine a sense of purpose with the desire to keep their options open.

Moreover, just because they are excited by change does not mean they will change. As they get older, many become more regular in their habits: half of French bachelors and couples without children do not have meals at regular hours, but the more children they have, the more regular they become. The significance of regular habits, Grossin claims, is that they go with the acceptance of constraints, a greater realism, fewer expectations, less hope of a radically better future. When people become skilful at doing what is expected of them, they cease to see such demands as tyrannical; indeed they find a certain kind of freedom in what their detractors call complacency. Forty-five per cent of the French, according to Grossin's calculations, lead orderly lives and 22 per cent are fairly orderly. With such a majority, it is not surprising that new solutions to the torment of being too busy are not seriously thought about. Two-thirds cannot envisage working less than thirty hours a week. They have no desire to emulate the fly.

Nevertheless, there is a new sensitivity to the texture of time, to what makes it flow smoothly, agreeably, sensuously. People do dream of enjoying their work by doing it at a rhythm which suits them and varying their rhythm for different occupations. This notion of personal rhythm was what the Industrial Revolution attacked and tried to destroy. How that happened can be seen in the forgotten history of the weekend. The English word has been adopted by almost every language, but it represents a poisoned gift from the English to humanity.

In the artisan workshops of eighteenth-century Birmingham, wrote a visitor, 'the industry of the people was considered extra-ordinary; their peculiarity of life remarkable. They lived like the inhabitants of Spain, or after the custom of the Orientals. Three or four o'clock in the morning found them at work. At noon they rested; many enjoyed their siesta; others spent their time in the workshops eating and drinking, these places being often turned into taprooms and the apprentices into potboys; others again enjoyed themselves at marbles or in the skittle alley. Three or four hours were thus devoted to "play" and then came work again till eight or nine and sometimes ten, the whole year through.' They worked hardest on Friday, handed their work in on Saturday, but took off Monday as well as Sunday, and often Tuesday as well, and sometimes even Wednesday. This was partly to recover from their drinking, partly to amuse themselves with dog and cock fights and pugilism, but even more because 'the men are regulated by the expense of their families and their necessities; it is very well known that they will not go further than necessity prompts them, many of them.' In other words, they followed the same principles as have guided Indian peasants, who decide what standard of living suits them, and work just as much as they need to maintain it.

The Saturday holiday was invented by manufacturers to elimin-ate this irregularity from their factories. They dethroned Saint

Monday by a trick, offering a reduction of three hours on Saturday afternoon in return for work the whole of Monday. The workers accepted from fear of unemployment, but even more because new 'rational amusements' were offered to them at the same time to wean them away from their 'evil' ones. When the first railway excursions were organised from Birmingham in 1841, they were on Mondays, and club feasts were held on Monday nights. But gradually Saturday was made into a shopping day; more and more money became necessary for the good life; building societies encouraged saving to buy a house through harder work. A long war between the supporters and the enemies of Saint Monday ended with a divorce between work and leisure.

But now people are demanding that their work should be above all interesting. A new attitude to time is required. Saint Monday may be resurrected, though completely transformed, not as the weekly hangover, but as a long, very long holiday people will give to themselves several times in their lives. The present division of a lifetime seems fixed for ever only because it has been forgotten how recent retirement and the weekend are. The English inventors of the weekend (the word dates from the end of the nineteenth century and was imported into French in 1906) have forgotten also that the number of their holidays fell sharply in the sixteenth and seventeenth centuries, that their Bank Holiday Mondays are token survivals of the reign of Saint Monday. But the world, which copied 'the English week' because England was once the richest industrial nation, may remember that it is only a fashion, and fashion can change.

The weekend is only one-half of the sabbath. God also instructed the Jews to take a sabbatical holiday every seven years, in which they should stop tilling the land, cancel debts and release their slaves. The sabbatical year may become the human right demanded by the twenty-first century. Since 1971 the French have been legally entitled to take sabbaticals, to improve their skills, or acquire new ones, or simply to broaden their minds; but in practice few have done so. Now that the expectation of life has been doubled, life cannot be viewed as offering just one chance, in one profession. Experience of more than one discipline has become the key to success. Since knowledge has to be constantly renewed, and since individuals are increasingly unhappy about wasting talents which they cannot use in their jobs, the sabbatical year might have a future, offering an opportunity to change direction, or simply to do what busy people do not have the time to do, namely think, or take a long promenade.

Of course, it cannot be viable unless it is made compulsory by law, as the forty-hour week was, because otherwise the fear of losing to the competition remains too great. If everyone took a sabbatical, as they now take a weekend, there would be no suggestion that it implied disloyalty, or lack of commitment; and there would be no possibility of objection to maternity or paternity leave. The sabbatical could not be financed except by reconsidering the whole idea of retirement, which does not need to come all in one lump at the end of a career. By British law, it is already possible to take a slice of one's retirement pension in advance, postponing one's final retirement by a year or two to make it up. At present one has to wait till the age of fifty, but perhaps the forty-year-olds will discover the attraction in their turn, and gradually one, two, three or more sabbaticals in a lifetime may become the norm, ever earlier. The insurance industry is only at the beginning of its inventiveness. The sabbatical year is much more than a way of coping with the fact that there are not enough full-time jobs for everybody.

Only when the idea of taking a year off from time to time appears natural will people be able to insert family life into a curriculum vitae without loss of status, without having to justify themselves; taking off seven years could be natural too. To follow a time-diet means not just organising the day or the week, but thinking in at least seven-year laps. Habits are comfortable, but when they fossilise, the humanity is gradually drained out of human beings.

William Grossin, *Le Temps de la vie quotidienne*, Mouton, 1974; William Grossin, *Des résignés aux gagnants: 40 cahiers de doléances sur le temps*, Nancy, 1981; Jeremy Rifkin, *Time Wars*, Henry Holt, NY, 1987; E. T. Hall, *The Silent Language*, Doubleday, NY, 1959; Douglas A.

Reid, 'The Decline of St Monday 1766–1876', *Past and Present*, May 1976, pp. 76–101; Michael A. Meyer, *Ideas of Jewish History*, Wayne State UP, Detroit, 1987; Frieda J. Forman, *Taking Over Time: Feminist Perspectives on Time*, Pergamon, Oxford, 1989; V. G. Dethier, *The Hungry Fly*, Harvard UP, 1976; S. G. F. Brandon, *Man and his Destiny in the Great Religions*, 1962; S. G. F. Brandon, *History, Time and Deity*, Manchester UP, 1965; Stephen Toulmin and June Goodfield, *The Discovery of Time*, Hutchinson, 1965; Keith Thomas, *Age and Authority in Early Modern England*, British Academy, 1976; Richard Glasser, *Time in French Life and Thought*, Manchester UP, 1962; J. T. Fraser, *The Voices of Time*, Allen Lane, 1968; Patricia Hewitt, *About Time*, Institute of Public Policy Research, 1993; John P. Robinson and V. G. Andreyenkov, *The Rhythm of Everyday Life: How Soviet and American Citizens Use Time*, Westview, 1988; European Foundation for the Improvement of Living and Working Conditions, *The Changing Use of Time*, Dublin, 1991; Michel Jouvet, *Le Sommeil et le rêve*, Odile Jacob, 1992.

21

Why fathers and their children are changing their minds about what they want from each other

My first story begins in China. He was a communist, one of the first to join Mao's army, while she was the daughter of a rich landowner who had had all his property confiscated and been made to work in the fields as a peasant. She was beautiful, dreaming of becoming an actress; she married against her will, to please her father who wanted to make peace with the Reds. Love did not come into it and soon she had her own men friends, for wives no longer had to obey their husbands. Their daughter Wei-ling was brought up to the sound of constant disputes, which became unbearable, and at twelve she opted to go to boarding school. However, this has in no way diminished her affection or respect for her parents. 'I do not believe,' she says, 'that there are eternal loves between men and women, but I do think there are between children and parents. My parents fed me, took me to hospital when I was ill, educated me, invested a lot in me. The first wage I earned I sent in its entirety to them. Our parents leave us a great deal of independence, but we keep our respect for the aged. Nothing is stronger than the link between parents and children.'

Wei-ling did not grow up into a woman who does what she is told. Her explanation is that, because of her parents' quarrels, she was left to look after herself. It may be true all over the world that when parents have fought, they have done as much to set their children free as any of the great liberators who fought against tyrannical governments, but freedom has never told anyone what to do with their lives. Wei-ling studied foreign languages, imbibed French ideas, saw French films and met French people in search of a better world than the humdrum one she knew. Meanwhile, she was being watched, so when the time came for her to be given a job, it was decided, on the recommendation of her school's 'political counsellor', that she should be sent as a teacher to the far north, the equivalent of Siberian exile. How to escape? She passed the entrance examination into a prestigious post-graduate college. But there she still did not find freedom. A minister welcomed the new students with the words, 'You are soldiers without a uniform', meaning total obedience was expected; all contact with foreigners was forbidden. Wei-ling had many foreign friends and had studied French precisely because to her foreigners were the breath of life. Her new comrades, interested only in advancing their careers, were uncongenial. She refused to join the Communist Party, though it would have given her the key to success, but taking an oath, holding up a hand, that one was willing to sacrifice one's life for the party, was too much for her. They said she was an 'egoist', an 'individualist'. Her brother had none of her problems or her qualms: 'He does not feel uncomfortable in a society of extreme equality, as I did, where everybody is expected to do the same thing, and if you do more than others, people are jealous, and refuse you any reward, which has always to be shared; where you have to make compromises with your superiors, who are often stupid, party members or people with connections, who agree to obey and are chosen to be leaders because they will do what is expected of them. The communists know how to judge people's characters and to give jobs only to conformists.'

For Wei-ling, being born in China makes her a member of another sort of family, to whom she wants to be as loyal and grateful – whatever disagreements she may have with it – as she is to her parents. The disloyal, who do 'scandalous things', are never allowed to forget it: 'you carry the opprobrium on your back for the rest of your life. I am brave: I know when to advance, but also when to retreat. The anti-conformists must fight, but it is a mistake to break off relations with one's country.' Wei-ling succeeded in finding a job where there was a tolerant atmosphere, with only one afternoon of political discussion a week. She made contact with poets and film makers, and visited France. After the 'events' of Tienanmen Square, she decided to stay there. But she has kept good relations with her Chinese professors; she has paid her own air fares from China so that there should be no cause for complaint against her; she wants to go back regularly to see her family.

Now she is going to marry a Frenchman, but it is not mad love that moves her. She is building a new kind of family: 'there are limits to the relations that the French and the Chinese can have. If I behaved with my fiancé in a Chinese manner, we would have separated. He is totally French. It is up to me to make the effort, to behave like a Frenchwoman, and sometimes that annoys me. I try to explain, but little incidents accumulate. He eats cheese: I buy it for him. But when he goes shopping he forgets that I am Chinese and need soya and rice. His father works abroad, his mother is alone 200 miles away: if I were him I would telephone her once a week, invite her for holidays, buy her presents, but he rings only once a month. So he does not understand that when I get a letter from my mother I become very happy: I ring him at his office to ask what present I should send her, and he replies, "But I am working, I have clients sitting in front of me." ' She wants her relationship to contain more politeness towards relatives and friends.

What will hold her old and new families together is a shared ideal of justice. That is the quality she respects in her fiancé. The Chinese, she claims, are more opportunistic, more willing to compromise, to believe that one's behaviour must depend on the circumstances in which one finds oneself; that is how fervent communists have been able to set up capitalist businesses: 'The Chinese are adaptable, in a way that Westerners cannot be in the East. It is all due to their not having one god, but a pig in one village and a horse in another; their gods represent equilibrium, compromise, conformism. The French do of course have pragmatists who care only for money, but they also have people with deep faith, and that touches me.'

Wei-ling is preparing to become a Protestant, because her fiancé is one. She has no religion herself, but 'religion is the basis of civilisation, and if I want to understand civilisation I must understand religion'. For her, the attraction of Protestantism, as opposed to Catholicism, is that it holds that there are many ways of approaching God and that the love of God resembles the love between children and parents. 'My experience has given me a sort of faith.'

In China, she says, she was never 'totally Chinese', developing, learning, changing her behaviour. There are Chinese people who think she has betrayed them. But 'it makes me uncomfortable to be asked whether I am Chinese or French: I want to be myself'. That is of course an increasingly common aspiration. Wei-ling belongs to the immense new nation of people who do not belong wholly to anybody.

By the standards of some people in China, she says, her fiancé is an unsuitable husband, because he is an 'egoist'. The old fashioned Chinese criterion for a good husband is that he should be gentle with his wife, have a good social and financial position, and be honest. But when she lists her fiancé's good qualities, she puts first that he is intelligent.

Sandrine spent her childhood wanting to run away from home for different reasons. Her story may seem quite unlike Wei-ling's, but their ambitions have much in common. Her mother died when she was six, and her father married a woman who, she says, hated her: 'I was always in the way.' That made her think about how she wanted to live. She sees her father only three or four times a year, and she does not find him an easy or congenial person. He is not a bad man, she says, it is just that he lacks courage. 'He does not want anything to disturb his life, to the point that he shuts out life; but of course he does not see it that way; he says he likes to see me, but he does not show his affection by concrete acts, he won't take any risks to help me.' Sandrine wants affection to be clearly demonstrated. Sometimes she thinks she should have no more to do with him at all.

So she too has decided to build up another kind of family. She has 'adopted' an older woman as a mother. Her independence of mind is not challenged, but she does not attempt emotional independence. She divorced her husband when he became a barrier to her being in charge of her life. She tried living alone for four years, but she missed the pleasure of sharing, of doing things together, of having someone listen to her. Now she lives with a man whom she likes to be with, a foreign journalist who often goes abroad. 'The telephone is my anchor.' He is her moral support, a source of encouragement. 'He comes first. I need affection around me. I cannot sacrifice everything to my work. I am not attached to money or to power.'

However, Sandrine often spends less time with him than with her business partner, of whom she says, 'We have the same ethical attitudes, the same convictions about how to negotiate, what we will not do, the same determination not to get what we want by lying. I know she can deal with any problem if I am absent. I like to share and to lean on people in whom I have confidence.' Her colleague is in effect an adopted sister.

Sandrine used to work helping the French government to save French culture from being buried alive in the cemetery where the Assyrians and the Mayas and other dead civilisations rest in peace, forgotten; her job was to persuade you to watch a French film from time to time, incomprehensible though it might be, to rescue you from being an addict of Hollywood. But the woman she worked for dominated her like an anthoritarian parent: 'I was her slave. She wanted to be Pygmalion. She even told me how to dress, questioned me about my private life, took sides in my disagreements with my husband. She wanted to change me: she once said, "You are nothing without me." One day I told her I could not stand her any longer, and I left. I escaped.'

Sandrine has escaped several times in her life. The only kind of security she has, the only one she trusts, is her ability to escape, when the atmosphere around her becomes suffocating and incompatible with what she regards as most essential about her. Her favourite book is Laborit's *In Praise of Flight*, which I discuss in chapter 13. Perhaps the world of television and the cinema, which helps people to escape from reality at the touch of a button, is made for the spiritual descendants of those nomads who throughout history have fled from the country to the city, from the old world to the new, always to an unpredictable destination.

It was only when Sandrine reached her thirtieth year that she realised what her instinct for escape meant. She was offered the choice of going back to the safe and comfortable job she held before marriage, or accepting a risky invitation to venture into something quite new. 'How can you hesitate,' said her old boss, 'between a job that will last till retirement, giving you a guaranteed pension, and the insecurity of television?' 'He made my destiny,' she now says. 'The idea of never changing, of doing the same thing all my life, of being like all those people I knew who had chosen their path and had become fossilised, horrified me.'

Once she had mastered her first job in television, she wanted one that would challenge her more. She moved to a newly established company which promised to give her the opportunity to take part in the making of films. That proved a disappointment: her new boss did not use her experience or her talents enough; he did not know how to delegate, he took it for granted that she would work at weekends, whenever she was needed, which she did willingly because she was part of an excited small team, but he could not accept that she might take a day off when it suited her. 'He used me as though I was a young girl, asking me to find his glasses when he lost them, to get a taxi when he was late for an appointment'. So she escaped again. She gave up being a wage-earner. Divorcing one's job, divorcing one's husband, divorcing one's parents, are part of the same search for a new set of priorities.

She has set up her own company, to use her talents no longer in an organisation that demands sacrifices from her private life, but where, at last, she can create the precise mixture of dedication and relaxation that meets her own needs. The basic principle of this company is that work is not the sole source of joy. She (like those who work with her) wants to have time to read, to travel and to cultivate her affections. Affections come first for Sandrine.

It is time that parents told children the facts of life, not just the facts that children usually discover for themselves, but the facts about what has happened in the past as a result of people living together in families. Judging by past experience, the family exists to produce the unexpected. Getting it to do exactly what was demanded of it has never been possible. Of course, family means the assured embrace, the safe haven and memories that even death cannot destroy; but it has also been a laboratory engaged in adventurous experiments. In particular, it provides training in the art of coping with the uncertainties of life. Uncertainty is a precondition of freedom, but without it everything would be inevitable and there would be nothing to dream about. So it is important to know about the shifting sands in which the family's roots are buried.

When the Swedish playwright Strindberg (1849–1912) denounced the family as a retirement home for women wanting an easy life, a prison for men and hell for children, he was only expressing despair at its not giving people what they expected, despair at 'suffering from the pain of not being able to be the person I wanted to be', even after three marriages. But people's expectations have been too simple, because they have forgotten their own past.

The first fact of life is that fathers have never felt totally in control. In the beginning, in the relations of fathers and their children, there was fear. Dead fathers were feared even more than living ones, for absent ancestors were believed to manage most of what happened in the world. Fathers were originally awesome because they would one day become ancestors, spirits who had to be placated. The Chinese were responsible for the world's first great revolution of the emotions, when they transferred supremacy to living fathers. Chinese fathers fought a battle against fear of the dead, and won. In its place they set up the religion of filial piety, which was simpler and which everybody could practise effectively, without need for priests, without worrying about what exactly the dead were demanding. A good proportion of fathers all over the world have been more or less pale copies of Chinese fathers, though Christian, Muslim or Jewish ones could never be as powerful, because they had a God who was all-powerful and a divine book which demanded total submission.

Fathers yearned to be treated as gods, but children seldom behaved exactly as they were told to. So ways had to be invented to diminish the irritation they caused by disappointing their fathers. The Chinese found not a cure but a palliative. Their solution was to preserve appearances, by swathing filial piety in etiquette. Confucius knew that blind obedience was impossible. 'A father has a critical son,' he wrote, and so he advised the son who had an unreasonable father to redouble his expressions of reverence while continuing to remonstrate; if compromise was unattainable, the son must avoid damaging the father's reputation for wisdom, and should run away rather than publicly dishonour him. Respect, or the pretence of it, was turned into a ceremony. A foreign visitor could therefore say, on the one hand, that in China disobedience to parents was the worst crime and that every fault could be traced to lack of filial piety, and, on the other hand, that 'Chinese children have no proper discipline, no idea of prompt obedience'.

Since paternal authority was always more or less in difficulties, fathers tried changing the basis of their hold on their children from fear to gratitude, and for a long time gratitude has been partially effective, like glue which does not always stick, in preventing families from falling apart. But when, in the West, even God did not receive gratitude, fathers did not fare better. The collapse of gratitude was hastened by cynicism, envy and wit: some condemned it as being inspired only by a secret hope of greater favours, while others argued that humans so hate being inferior that their gratitude is a form of revenge, and that they repay the benefits they receive not from pleasure, but because they find obligation painful. Bernard Shaw asked, 'Do you like gratitude? I don't. If pity is akin to love, gratitude is akin to the other thing.'

The first fact of life is that obedience has never been guaranteed, and gratitude has always been unpredictable. There is nothing malfunctioning in families when fathers do not get what they dream of: rather it is surprising when they do.

Secondly, there have always been outsiders trying to interfere between parents and children, which has made their relations additionally uncertain. Christianity, for example, while urging children to honour their parents, also invited them to worship God as their father and to abandon the bad example of their natural fathers. The Church opposed the idea of the omnipotent human father, which the Romans had enshrined in their jurisprudence and which long survived in many countries. Spiritual fathers competed for influence against natural fathers. From the eighth century, confession, which used to be a public ceremony, was gradually turned into a private moral inquisition (Irish monks were the first to show the way): priests began giving guidance on the most intimate subjects, insisting that children should be named after saints, not after their parents, offering an alternative model for imitation. Godparents were given responsibilities to repair the deficiencies of parents, making children members of the parish as much as of the family. Marriages between cousins were forbidden, to thwart the ambition of fathers trying to turn their families into clans whose religion was self-worship.

In the New World, missionaries saw children as 'ministers for the destruction of idolatry'. They systematically turned them against their heathen parents, winning them over with presents, and then humiliating the fathers in front of the children. Among their methods for converting was 'emasculation', grabbing men by their testicles until they collapsed in pain. They made wives laugh at husbands by altering the traditional division of labour, forcing women to weave, which used to be a man's job, and men to build houses, which women used to do. The Church offered itself as the universal mother, and as the protector of mothers.

In Europe, from the sixteenth century onwards, fathers hit back. The kings who were establishing their own despotisms against the Church supported them, but the more fathers tried to control their children, the more the opponents of despotism protested on behalf of children. The Declaration of the Rights of Man abolished the duty of filial obedience. 'When the Republic cut off Louis XVI's head,' wrote Balzac, 'it cut off the head of all fathers.' Though Napoleon tried to rebuild the rights of fathers when he restored despotism, the children's champions retaliated. The battle continued for a century and a half. The result was unexpected: though fathers lost most of their rights one by one, it was not the children who won; power over them was largely transferred to teachers, doctors, the courts and social workers. But then there was another surprise: a resistance movement by children, who demanded that their opinions should be listened to, limited the influence of experts, making the outcome of their intervention never quite predictable.

Then fathers tried to replace their lost authority by winning the affection of their children. Another uncertainty was now introduced, for affection, like the wind, blows only when it pleases, and no ceremonial pretence can ever be a substitute for it. When fathers and children become conscious that they have different opinions and different personalities, the winds of affection turn into storms. Nevertheless, the attempt by parents to become the friends of their children, instead of tyrants, has been one of the great human adventures, in which the poor were often more successful than the well-to-do, who had other priorities and used their children to satisfy their own ambitions. Though there has been virtually no civilisation in which parents have not enjoyed the company of infants, playing with them as if they themselves were children, severity usually triumphed when the child became capable of earning a wage. However, even the Victorians, now notorious for being, at least on the surface, cold and distant fathers, included many who were the very opposite, backslapping, hearty, frivolous: there have always been families which broke the rules in private; that was what privacy was for. Roman fathers, who set the example of austerity which Victorians followed, and who in a part of their lives appeared to be prisoners of formality and legal niceties, sometimes had very close and tender relations with their daughters, who liked their fathers more than their husbands: in this they were inventing a form of friendship which neither the Etruscans nor the Athenians knew.

Fathers have not just followed tradition: they have also been inventors, and there is no reason to suppose that they will cease to invent new kinds of fatherhood, all the more so now because the meaning of heredity is no longer what it was.

'Idiots beget idiots': that was the opinion of St Thomas Aquinas, and of most people until quite recently. Nothing much could change if it was true. But when Francis Galton (1822–1911) looked carefully at the children of geniuses, he found they were usually not geniuses at all. In the last century and a half, ideas about inheritance have changed completely, but most people have not caught up with them in the way they think about their private lives. Parents become quite different if they are seen as a bag of genes and molecules.

It used to be believed that children resembled their parents, in one of three ways. Some said that the characteristics of the mother and father were blended together; others that the father's sperm contained a miniature version of him, so that the child was entirely a reproduction of himself, while a few claimed the miniature version was in the mother's egg. Why children, despite this, differed from their parents was not a mystery people wanted to solve, because they could not bear the thought that the rebellion of the young might be preordained by nature.

The new understanding of fathers can be dated from 1850, when a peasant's son failed his examination to qualify as a teacher. Humanity owes enormous debts to failures, heretics and people who give the wrong answers in examinations, and who, sometimes, change the direction of history. Gregor Mendel (1822–84) was failed by the eminent professors of Vienna University because he was self-taught. He was abnormal also, for his time, because, though a monk, he believed that the history of creation was not finished, meaning that the world would not always remain as it was. Indeed he did not take the world too seriously: continuing as an unqualified teacher, ignoring his failure, he gave his lessons jestingly, blushing and laughing at his own jokes, 'a roguish twinkle in his eye, never disposed to give anyone the cold shoulder', talking about sex in plain language, and if anyone tittered, he would say, 'Don't be stupid; these are natural things.' The only fear he had was of draughts, instantly putting his hat on the moment he heard the aeolian harp in his garden sounding its warning note. Little fears are useful if they protect against big ones. Mendel was not afraid that his ideas would be difficult to harmonise with those which were current in his day.

'Wherefore was man created?' he asked in an attempt at verse.
'Unceasing toil
The ennoblement and development of his energies.
These are man's lot here below.'

That optimism enabled Mendel to carry out experiments on different sorts of garden peas, which proved that there was no blending of inherited characteristics. By heretically combining mathematics with botany, he worked out the dance steps of dominant and recessive genes through the generations. Modern genetics is based on that one idea.

A century later, in 1953, Watson and Crick revealed the shape of what parents gave their children. Like Mendel, they had been anointed by failure: Watson was rejected by Harvard, Crick failed to get a First at University College London; Watson was a loner, Crick an extrovert. They both attributed their success in discovering the double helix, the molecular structure of the nucleic acids in genes, to their being so different, criticising, complementing and stimulating each other. The differences between people have traditionally been regarded as disturbing: their value is only just becoming apparent. The boundaries separating the different branches of science used to produce specialists who could not understand each other, but knowledge is increasingly being pushed forward by knocking holes in those boundaries. It was reading *What is Life?* by the physicist Schrödinger which convinced Watson and Crick that biology could be advanced by using the concepts of physics and chemistry, that at the border between the living and lifeless there was a bridge, not just an abyss.

Now it is clear that life never quite repeats itself: its irregularities and surprises are the charm it imprints on matter. The differences in each individual are reshuffled by each generation, and mutations are added for no apparent reason. Parents cannot pass into the bodies of their children the peculiarities they have acquired in their lifetime. They are much less awesome now when viewed close up, no longer as authorities who know what is right, but as a mass of DNA struggling rather blindly to produce yet more DNA.

The fourth fact of life is that 'family values' have changed so much over the centuries that their meaning has become increasingly uncertain. At times, children were scarcely distinguishable from servants, put to work as soon as they were able to walk, so that in the nineteenth century they might provide a family with as much as a third or a half of its income. When this was stopped, and wives went to work instead, another great emotional revolution occurred: the role of children became to spend their parents' money, instead of earning it. Today 70 per cent of the French say that parents have no right to demand sacrifices from their children; 62 per cent say they do not ask for any share of their children's wages.

To make love, instead of economic partnership, the basic family value, was not easy. In the USA the Reverend Dr H. J. Kerr, in his *Children's Story-Sermons* (1911), told of a child called Bradley presenting a bill to his mother: 'Mother owes Bradley, for running errands 25c., for being good, 10c., for taking music lessons, 15c., extras 5c. Total 55c.' His mother gave him 55 cents and presented him with her own bill: 'Bradley owes Mother: for being good, 0 c., for nursing him through his long illness with scarlet fever, 0 c.; for clothes, shoes and playthings, 0 c.; for all his meals and his beautiful room, 0 c. Total that Bradley owes his mother, 0 c.' Bradley's eyes filled with tears and he gave the money back, saying, 'Let me love you and do things for you.' This preacher's dream of unselfish, uncalculating families remained a dream so long as love was measured almost like money, though in unconvertible currency. Social workers complained, at around the same time, about immigrant parents: 'Although they love their children, they do not love them in the right way.' Not any kind of love was acceptable. Nevertheless, having someone to love became increasingly necessary; so though children became ever more expensive to raise, the value placed on them rose higher and higher. In the *Annals of the American Academy* (1908), it was laid down that 'the child is worthy of the parents' sacrifice'. Before that date there used to be people who would take an illegitimate child off your hands for ten dollars. By 1920 prospective parents were paying a thousand dollars to obtain a child to adopt, and Mrs Georgia Tann of Memphis was the first woman in the world to become a millionaire by running an adoption agency. An American survey in the 1970s showed children typically contributing three and a half hours a week to household tasks, compared to the mother's fifty hours.

However, when the family based itself entirely on love, the child was left uncertain as to what it was expected to do in return for the love it received. It could judge its own worth when it brought money into the home, but when it was economically useless, and a financial drain, it depended for its self-esteem on assurances of parental love and admiration, without any guarantee that the admiration was shared by the rest of the world. Sometimes it was required to pay the parents back by making fantasy into reality, by becoming the ideal person the parents would like to have been themselves. Sometimes it was urged to grow into an independent, 'happy' person, though the love might then be withdrawn if the result was too much of a shock. Then some parents decided that making their child happy could not be the whole reason for their own existence, that there were limits to the sacrifices they were willing to make, that though the child was king, it could be dethroned. Divorces proved that the child did not always come first. There was a big difference between the family value of love as a duty, and love as spontaneous, constantly needing to be renewed.

There was also always a conflict between the family value of continuity and the endless search for renewal. Apart from the pharaohs of ancient Egypt who married their sisters, marriage, for most families, has meant the introduction of new blood from strangers. Spouses have been an indispensable breath of fresh air, whether they caused havoc or not. It is not a sign of the collapse of the family when children do not fulfil their parents' expectations, nor is it a break with tradition when they do not receive the sole and undivided attention of their mothers. The family is the oldest of all human institutions because it is the most flexible. Its goals, over the centuries, have changed again and again. With one or two children it has little in common with the households which included retainers, lodgers, servants and illegitimate offspring as well as kin of every generation. A century ago, even in France, which pioneered small families, half of all children had at least two brothers or sisters.

It is a fact of life that the family is always changing its mind about what it is trying to be, and how to achieve its aims.

Its final about-turn has been to transform itself from an employment organisation into one concerned primarily with leisure. As a provider of employment, its record was poor. Parents used to play safe, having more children than they needed to ensure the continuity of their own trade and leaving their surplus offspring to solve their own problems in the cities or abroad. That is why orphans and quasi-orphans invented modernity: they had no other choice. Extending education was supposed to enable the young to do better than their elders, but, apart from rare periods of prosperity, there have never been enough jobs of the right kind or profitable outlets for those with free imaginations. In the 1820s and 1830s 'young people found a use for inactive strength in the affectation of despair', as Alfred de Musset put it: 'Scoffing at glory, religion, love, at everybody, is a great consolation for those who do not know what to do.' Adolescence, invented to be a paradise before real life began, turned out to be a breeding ground of neuroses and delinquency. Then adults became fascinated by this invention, by the mirage of being forever young and by the dangerous delights of cultivating the imagination. That is how it finally emerged, rather by accident, that the family was at least as good at stimulating adventure as at providing stability.

The most positive discovery parents made was how interesting children could be, how much more rewarding they were as human beings, full of unpredictable curiosities, than as obedient slave labour. Parenthood has increasingly become a gamble, as children have increasingly become like a fountain of unanswerable questions, which make the old answers look limp, forcing a rethinking of everything that used to be taken for granted. Don't hurry to understand the world, Tom Brown was told, you are not old enough to understand it. But the new attitude became: there may be many facts you cannot change, but there are others which may not be facts at all.

There has been so little progress over the centuries in making families more stable, and a more reliable source of virtue, that it is time to consider how all this uncertainty contained in their history can be put to better use. The whole of history, so far, has been an attempt to get rid of uncertainty. But it is not just that life would be dull without it. Security by itself has ceased to be an adequate ideal, because it has never been foolproof. Plans almost always go wrong. In the process much experience is allowed to go to waste, even though failures can be recycled as opportunities. The shape of hope is always uncertain, and uncertainty is indispensable to hope.

But it is not by itself enough to guide a life. So my next chapter will consider what new goals the family can set itself, even if they lead to unexpected results. One needs to know where one would like to go, but to know also that one is likely to arrive somewhere quite different.

Viviana A. Zelizer, *Pricing the Priceless Child: The Changing Social Value of Children*, Basic, NY, 1985 (quotes National Child Labor Committee of USA reporting in 1905, 'it is rare to find a family governed by affection); Michel Fize, *La Démocratie familiale: Évolution des relations parents–adolescents*, CNRS, Presses de la Renaissance, 1990 (42 per cent of French parents say they never discussed their personal difficulties with their children); Jean Delumeau and Daniel Roche, *Histoire des pères et de la paternité*, Larousse, 1990; Yvonne Knibiehler, *Les Pères aussi ont une histoire*, Hachette, 1987; R. A. LeVine, *Parental Behaviour in Diverse Societies*, Jossey Bass, 1988; Vitaly A. Rubin, *Individual and the State in Ancient China*, Columbia UP, 1976; Linda A. Pollock, *Forgotten Children: Parent–Child Relations 1500–1900*, Cambridge UP, 1983; Huxley J. Coale, *The Decline of Fertility in Europe*, Princeton UP, 1986; Pierre Guichard, *Structures sociales orientales et occidentales dans l'Espagne musulmane*, Mouton, 1977; Ursula Owen, *Fathers: Reflections by Daughters*, Virago, 1983; Morton H. Fried, *Fabric of Chinese Society*, Octagon, NY, 1974; Michael Mann, *The Sources of Social Power*, 1986; Eliezer Ben-Rafael, *Status, Power and Conflict in the Kibbutz*, Avebury, Aldershot, 1988; Paul Chao, *Chinese Kinship*, Kegan Paul, 1983; Caroline Blunden and Mark Elvin, *Cultural Atlas of China*, Phaidon, Oxford, 1983 (I am indebted also to Elvin's many articles on Chinese emotions); J. L. Domenach and H. Chang-Ming, *Le Mariage en Chine*, FNScPo, 1987; Colloque UNICEF, *L'Étranger vu par l'enfant*, Flammarion, 1986; David Cohen and Stephen A. MacKeith, *The Development of the Imagination: The Private Worlds of Children*, Routledge, 1991; Maurice Daumas, *Le Syndrome des Grieux: Les relations père/fils au 18e siècle*, Seuil, 1969; Judith Stacey, *Patriarchy and Socialist Revolution in China*, California UP, 1983; L. McKee and M. O'Brien, *The Father Figure*, Tavistock, 1982; Judith P. Hallett, *Fathers and Daughters in Roman Society*, Princeton UP, 1984; Pierre Duclos, *Les Enfants de l'oubli, du temps des orphelins à celui des DDASS*, Seuil, 1989; Maurice Godelier, *La Production des grands hommes*, Fayard, 1982; Yushio Markino, *When I was a Child* (in Japan), Constable, 1912; John Gillis, *Youth and History*, Academic Press, 1974.

22

Why the crisis in the family is only one stage in the evolution of generosity

'How were you brought up?'

'Badly.' He is eighteen, and has enjoyed all the privileges a successful professional family can give.

'I saw little of my parents, who used to leave for work at eight in the morning and return at eight in the evening. It's going to get worse, as women, and men too, work more frenetically. Children will no longer be cared for.'

His father is a distinguished barrister, defending human rights; his mother, Monique, has a top job in the world of music. In their youth, in the early seventies, they were rebels, pioneers of a new way of life: they met when both had scholarships to the USA, where all those who treated the future as an adventure went to be baptised in modernity. Monique is a very modern woman. The daughter of a radio journalist, encouraged to study rather than wash the dishes, often alone, left to think for herself, she graduated from France's top business school. (That was in 1968, before the explosion of feminism.) She spent a year teaching poor children in black Africa, and later helped organise Angela Davis's campaign in France. Her husband was a conscientious objector who instead of military service also taught in the Cameroons. That was where the son was born, who now says he is not French but a Cameroonian.

Monique did not want to be just a wife; and, in the French provinces, that was difficult to avoid when marrying into one of those prosperous provincial families where a woman could be intellectually powerful, but nevertheless allow her whole life to revolve around her husband's. Her mother-in-law, for all her conversational brilliance, did not know how to write a cheque. In a small city where every well-to-do family was under constant observation by every other, where people never tired of calculating who was the equal of whom, a new wife coming from outside was offered the same sensation as a young butterfly invited to have a pin stuck through it so that it could be admired by the neighbours. Monique sought distraction in Paris; and it was always in tears that she returned home. The duties expected of an important barrister's wife did not appeal; she preferred the conversation of artists on the margins of society. She refused to entertain her husband's colleagues, or to accompany him to routine receptions. 'I left him to manage his life himself.' Her work was her 'secret garden', ensuring her independence from him. He accepted this.

How to look after her two sons was her problem. Her husband could take care of himself. 'I always wanted to save the children, putting them before our life as a couple.' He did his share of the chores, and of the cooking for the children, and stayed home when she worked late. 'We led parallel lives.' It was easy getting domestic help. But now she feels guilty, or more exactly bemused, wondering whether she neglected her children. All her friends have the same problem, and it is the main topic of their conversations: when the children were small, they did what they were told; now, as they approach independence, they fill their parents with anxiety. 'We are rejected by the adolescents.' After all their brave rebellions, they find they have got little further than their mothers, who sacrificed themselves for their children.

Monique regrets that her elder son, now at university, is a stranger; there is tension in every conversation with him. One in five French adolescents, according to polls, complain that they are unable to discuss their problems with their parents. 'He doesn't tell me what he thinks; I'd prefer rows to his silence.' He refuses to eat at the same time as the rest of the family, treats home as a hotel, gets up at midday, comes back for a meal at four in the afternoon, never tells them where he is going to be. The boy moved out once, but returned, because his flat had no bath: he gives the impression of living at home only to use its conveniences. They never know where he is, just as he must have wondered, as a child, where his mother was.

Why cannot the mother accept the son's independence, since she has always been so attached to her own? 'Because he's not motivated. If he had a driving passion, we would approve.' But that is not so. She admits he shares many of her own values. He says he wants to devote his life to helping the poor, to live in the poorest country in the world, Burkina Faso, and bring it food, education, medicine. The rich countries have too many problems, there is no hope of solving them; so it is better to devote oneself to a humanitarian cause. 'Because my parents did not devote themselves to me, I shall devote myself to others.'

'So your arguments with your parents are your revenge?'

'Yes. When I have children, they will be my first priority.'

Perhaps some good has come from the neglect he has suffered: it has made him a person who cares about others.

'If your parents had brought you up properly, would you have been selfish?'

'Yes.'

He has no ambitions, he says, in the conventional sense. He does not mind getting bad marks. It is not that his parents do not understand him, but that they cannot accept him as he is, refusing the idea of him as a failure, his conscious decision not to be a brilliant student like them. The tension at home, he says, is very high.

The question is whether Monique's decision to limit her family life for the sake of her personal flowering has borne fruits to justify it. She probably could not have done otherwise. She is much admired as a model of what a modern professional woman can achieve, and she has certainly been a source of encouragement. Her son is equally impressive, and she is mistaken in believing that she has damaged him. But she does not consider that she has found the right way for a woman to lead a full life.

Despite her high qualifications, even in a supposedly modern advertising firm, she had to start as an assistant and to make her career in a 'feminine profession'. On the face of it, this career has been one which most women would envy: helping, as public relations officer, to set up a television station, organising cinema and music festivals, administering an orchestra under a charismatic conductor. What a wonderful succession of artists she has met over twenty years, talking with them into the early hours, back at her office in the early morning, travelling and welcoming performers from all over the world, raising funds from benefactors and people in power, exhilarating, exciting, exhausting work, requiring inexhaustible tact.

However, now in her early forties, she begins to ask herself questions. Should she spend her life always being number two? It is not that she wants power or more recognition; the men she works with have far more need for constant reassurance than she does; that is their basic weakness. The compliments she receives arouse her suspicions; she prefers to pay them to others, to charm people, to observe the process of charm at work, which is not just a sexual game; politicians live by charming, as her conductor charms his orchestra; women, she thinks, can have enormous power without ever being at the front of the stage. But a man would never be content to limit his ambitions as she has; she thinks she is too timid to want a top post; speaking in public frightens her still. Her hectic life, for all the pleasure it gives, impoverishes her because it does not leave enough time to read, to travel in a leisurely way, above all to develop her own creativity. All her life she has helped artists to create; is it not time to create something herself, all the more so since she can feel this job draining her energy? Is it advancing years? She feels more mature, capable, stable: has she used all up her talents? When she first went into television, her hope was that she would become a film director, and it would please her now to make documentaries, to do something new, the never-ending search for novelty. 'What I need is time for myself.' But the orchestra is constantly doing new things, giving concerts in schools, factories, prisons, followed by wonderful, emotional conversations: that makes her feel that what she is doing is justified; 'it could change the world'.

Why does a young man who dedicates his life to the service of the poor appear as a bundle of regrets, both on his parents' part and his own? Two historical figures mentioned in a previous chapter, who chose a life of poverty, would today have been sent to a therapist as problem children. St Francis of Assisi is by universal consent one of the most admirable men who ever lived, and yet he disappointed his wealthy father, first by spending his adolescence as the leader of a gang of disorderly, pleasure-seeking youths, and then by giving his possessions away to the poor, siding with 'the weak in spirit', 'calling animals by the name brother [and removing] from the road little worms, lest they be crushed under foot'. Though always uncomfortable in the cruel world, he remained a charming and lively personality, and yet his father would have preferred him to be a cunning cloth merchant like himself. Albert Schweitzer had a mother whom her daughter-in-law remembered as 'very hard, very severe', and whom he described as being frequently in tears because he was so backward at school: 'Youth's unqualified joie de vivre I never really knew. . . . I was by no means a merry character,' he said, recalling that being laughed at was the nightmare of his childhood. But far from turning bitter, he became a supreme example of generosity, as a doctor in Africa, making 'reverence for life' his creed. Uncomplaining about his upbringing, he insisted, 'My parents trained us for freedom.' Freedom has indeed often been found where there seemed to be none. Generosity is an expression of freedom.

The family has, ideally, been the place where people have been generous to one another, but at the same time, as a unit, it has usually placed its own interests before those of others and been merciless to those of its members who failed to fulfil their obligations to it. Its first great idea was to try to eliminate these tensions by forgetting about large families, having only a few children, concentrating its affections and shutting itself away from the world's meanness. This did not always produce the desired result. So since then the large family or clan has been reinvented in a new form, not as a group of blood relations, but as a circle of individuals united by affection and often closer emotionally than kin ever were. In their intimate lives, non-kin came to play as important a role as kin, and children learnt as much from their peers as from their parents. The internal politics of the family – concerned with how to avoid domestic rows and how to be less undemocratic – is only a part of its history; its foreign policy, how to treat outsiders, has been less talked about, but it has had a decisive influence on the growth of freedom and generosity.

Parents have seldom dared take the risk that their child might grow up to be a third-rate, unemployed saint. Their aim is to ensure that it is happy. That aim, however, has proved to be unattainable by the traditional method of opening all the doors of pleasure and desire, because it is impossible to be happy if others are unhappy. Of course many people have been happy in the limited sense of being content, closing their eyes to the horrors around them. Or they have tried to be happy by laughing instead of being angry at human folly. Or they have consoled themselves with moments of ecstasy, accepting that these are inevitably brief. But no happiness is complete if it is selfish. Being of no use to anybody implies self-contempt, which, even for those who cultivate it – because they regard humans as basically greedy and brutish – is a grim pleasure. To have happy children can only be the beginning of a dream. They can only be satisfied with their happiness if the world benefits from their passing through it. That gives child-rearing, and indeed life, more than a prosaic purpose.

The advice to children used to be to accept their lot in life, but some refused to listen. Then it became to find a sense of identity, defined by Erik Erikson as 'a feeling of being at home in one's body, a sense of knowing where one is going, and an inner assuredness of anticipated recognition from those who count'. History has few records of such perfectly coherent people, and those who have existed were too complacent to be anywhere near perfect. Erikson himself did not achieve what he preached; as an illegitimate child who did not know who his father was, his philosophy was inspired by a longing for a normality he never experienced. The idea of a sense of identity was invented for people who wanted the world to be less complicated. The alternative is to be a different person in different circumstances, to make the broadening of one's sympathies and understanding of others a higher priority than understanding oneself. Families wrapped up in their own cocoon have not generally prepared for that.

Freud, an eldest son who lived in the heyday of authoritarian government, not surprisingly made rebellion against fathers the focus of his diagnosis. Today, however, it is no longer the authority of parents that is the menace, but their aimlessness; it is not the struggle for power that matters most, but the search for a purpose in life. Parents no longer dominate the imaginations of their children. Two thousand Australian primary school children, questioned on their attitude to the world as they see it, reveal the change: they define a parent as one who cares for you, buys things for you, spends money on you, protects you: 'They love me because I'm theirs and I'm all they've got.' They add that a teacher helps to open up the world and explains it, though not enough, and 'she is not always fair'. But then a friend is of vital importance – a friend who understands you, shares both fun and things with you, and tells you not to worry. These children's main complaint is that they do not have enough time to do what they want: there are too many demands on them, too many objects of curiosity to choose from.

The desire to participate fully in what the world has to offer, to be human in the widest sense, has been obstructed by the suspicion and disdain that humans have always had for one another and by the difficulty of getting them to be generous to strangers. So far, experience in trying to cultivate generosity suggests that there is no necessary link between generosity within the family and its manifestation towards strangers. To concentrate on generosity and harmony inside the home, as modern thinking does, and to forget what goes on outside it, is like looking at a field and not noticing the horizon.

'Thou hast no sense. You French people love only your own children, but we love all the children of the tribe.' So said a Naskapi Native American (once known as 'Red Indian') to an eighteenth-century Jesuit who wanted to teach him what good behaviour was. Though, among the Amerindians in general, a child was not given systematic schooling, the exceptional amount of love it received, indiscriminately from parents, kinsmen and others unrelated, struck all European travellers. Any child was a welcome visitor wherever its fancy took it, and was always certain of getting love from somebody, though not always from the same person. Participating in adult as well as juvenile activities, it was seldom alone. Affection was given regardless of whether it was born in or out of wedlock; orphans were fostered in complete equality, as were prisoners of war, for those who were spared scalping were adopted and became kin: a substantial number of whites who became assimilated 'Red Indians' after capture refused to return to 'civilisation' when liberated. Among the Mohave, there was no word for punishment. A delinquent child was called wild, crazy or obnoxious, but was treated with only a slightly exasperated tolerance, for it was believed that misbehaviour was induced by supernatural as well as temperamental forces, which were beyond an individual's control: such a child was therefore recognised as a future shaman, who could communicate with the unseen, and often eventually did become a respected shaman. Only gross violence was beyond the limit.

Such complaisance towards children was possible because they were thought of as belonging to the community, rather than to their parents, but then that was the attitude of Amerindians to all property. Generosity was, to them, the supreme virtue. The chief was not the man who had the most possessions, but the least, for he was the one who gave most away, who had built up most gratitude. There was no buying and selling among them, only an exchange of gifts. Destroying everything belonging to the dead eliminated the temptation to build up family fortunes. Wealth had no prestige in most tribes, but dignity, wisdom and spirituality did. Anybody could walk into anybody else's home and expect to be given a meal; and could appropriate any object which no one else was using. The whites called the 'Red Indians' thieves, not understanding that possessions were recognised as private only so long as they were being actively used and needed. Though the men spent much of their time in the forest, with the women and children in the clearing, both sexes were supposed to know how to perform every task, the men to sew and cook, the women to shoot and follow tracks. In some seasons, families lived in separate wigwams, but in others they shared a longhouse.

However, the price of spreading their affections was that 'adults were fond of many persons, but seldom if ever deeply in love with or tied hand and foot to one person'. Marriages were fragile: divorce among the Apache was achieved simply by the wife putting her husband's clothes outside the home, which was the sign for him to go back to his mother; or he would say he was going hunting and never return. The price they paid for their easygoing attitudes, for their refusal to get angry, for their horror of face-to-face confrontation, was that they turned their anger against other tribes and destroyed themselves in war. War became their cure for grief; and they constantly needed new captives to replace the casualties of war. Worshipping the equilibrium of natural forces, and denying the existence of evil, they found a superficial peace by always saying yes – 'Thou art in the right, brother' – refusing to coerce each other, regarding a hostile feeling in the same way as a physical illness. But on the other hand they were tormented by the fear of trickery; their endless, inconclusive discussions, and their factiousness, undermined their powers of resistance. Theirs was a civilisation designed for wide open spaces, into which the dissatisfied could silently withdraw: 16,000 Cherokees had 100,000 square miles of Texas all to themselves.

The ancient Amerindian civilisations collapsed – among other reasons – because although they had evolved impressive domestic policies, they had no effective foreign policy. They were puzzled by strangers. They accepted conflict as part of the natural order but could not cope with aggressors who had a completely different idea of what was natural. It is possible to cultivate a little plot of generosity in the middle of prairies devoted to envy, but it doubles the difficulties. The Amerindians were ultimately demoralised because their sense of community had limits.

Today the descendants of the Amerindians, intermixed with descendants of Africans and Europeans, can be found still unable to deal effectively with the descendants of their invaders. In Brazil, for example, they are poor enough to recognise the infinite gradations between being 'humble' and being 'poor', between being 'truly wretched' and being a 'walking corpse'. In the Nordeste, where the minimum wage is barely enough to keep an individual alive, let alone a family, where women and children therefore accept even less to toil in the sun cutting sugar cane, where the main ambition of a good worker is to save enough money to pay his funeral expenses, where the rich nations have invested vast sums of money to make the rich richer, leaving half the population illiterate, destitution has nonetheless stimulated a certain kind of generosity. The poor worry that hunger has made them too greedy; they are conscious that no self-respecting person likes to ask for help; so they make it their task to foresee the needs of their neighbours, distributing little gifts – a few beans, perhaps, or bananas, wrapped in brown paper – while taking enormous trouble to ensure that in the distribution of charity, those who are worst off are given priority, even if it means that they receive nothing themselves and go home crying: their neighbours remember them next time. Thirty-eight per cent of women have fostered children who were not their own, but this is also the measure of the hordes of children who are abandoned or leave home because there is no food for them. 'Does your mother love you?' A nine-year-old beggar replies, 'She has to love me. I bring her money and food to eat.' A husband likewise is often simply the man bringing food into a home, though he may vanish at any time; formal marriage is, in such conditions, rare.

Generosity does not easily survive ambition. When migrants first move into the slums around cities, in search of work, it is each for himself; then they begin to co-operate; but once they start saving it is often each for himself again. The well-to-do believe that they express their generosity by taking in a poor child as a servant, and are content that the poor should continue to be fascinated by the rich, but the effect is to encourage the poor to view themselves as hopeless and worn out: I'm finished, they say, even in their youth. 'I am not afraid to die,' says a ten-year-old street urchin.

About seven million children live on the streets in Brazil today, side by side with the stray cats, a piece of cardboard as their bed, and the middle classes, who are immensely warm to their own children, often do not notice these vagabonds as they pass by, see them not as children but as threats to their property. How else can an abandoned child survive except by stealing? 'What is your ambition?' A nine-year-old girl arrested seven times for theft replies, 'To join the police.' Why? 'So that I can steal without being caught.' Shopkeepers employ murder gangs to eliminate these street children, in the interests of good order, and one death squad calls itself The Disciples of Jesus. A sixteen-year-old professional killer says, 'I have killed only people who were good for nothing.' This is in the only country in the world which has a Ministry for Children, with some of the most advanced legislation on children's rights.

Brazil is repeating the experience of nineteenth-century Britain, where the gentry used to visit Bedlam as a Sunday entertainment, but refused to visit orphanages, as Charles Dickens urged them to, for the orphans swarmed the streets 'like locusts' in the cities of the first Industrial Revolution (while the orphans of New York City used to be known as 'little Arabs'). The fairy-tale orphans Hansel and Gretel, who at least had each other, were invented as a counterpart to the dream of married couples living happily ever after, and they remain alive because prosperity has had the effect, at first, of making generosity more rather than less difficult, just as it does not increase the number of happy marriages. Thus the Argentinian mothers who united against the dictatorship which kidnapped its opponents, and who expressed concern for all children, not just their own, dissolved into their private family preoccupations once life went back, more or less, to normal. However, prosperity has not necessarily and regularly terminated in a blind alley of self-absorption. Enjoying the pleasures of home and family has encouraged the accumulation of possessions only as a first stage; then comes the time when the home, despite its cosiness, becomes too narrowly confining, not sufficiently stimulating. Collecting objects is replaced by collecting 'interesting people'. Ultimately curiosity becomes more important than comfort.

The role of curiosity as a stimulus to generosity has been crucial. However, parents and children have seldom felt able to reveal themselves completely to each other, and, in that relationship, there are limits to what curiosity can discover or achieve. Among the French, for example, though children on the whole give their parents good marks, seven out of ten saying that their parents are helping them 'to be themselves', parents remain doubtful: only four out of ten parents consider their children affectionate or are convinced that their children find them 'open-minded and young'; 26 per cent believe that they appear exasperated, 30 per cent demanding and authoritarian, 20 per cent worried; 44 per cent of children believe their parents think they are lazy, but only 12 per cent of parents believe that. In other words, they have a lot of trouble guessing correctly what is going on in each other's minds.

People have therefore often looked outside the family, where they have no obligations, to cultivate the pleasures of generosity. One can watch this process among the Swat Pukhtun of northern Pakistan, thanks to a terrifying picture of them which has been painted by an anthropologist, in just those violently contrasting colours, with just that touch of exaggeration, needed to illuminate a universal dilemma. He describes these people as farmers who dream of being self-reliant and strong, fighting against all others to maintain their honour, independent to the point of preferring to eat only the produce of their own soil, regarding it as demeaning to work for fellow-Pukhtuns, so that even doctors give their consultations free rather than be taken for servants. Individual pride and family cohesiveness are uneasy partners here, just as they are in the West. Competitiveness dominates the children, to the point that conversation is mainly a long disputation over possessions: This is mine, No it's mine. They are taught to be aggressive, to lie to avoid punishment, to fear only public humiliation, but to experience no guilt while devoting their energies to circumventing the unconcealed favouritism of their parents. Fathers are portrayed as proud of having sons, but liking them less with the years, becoming jealous of them, and the sons wait impatiently to inherit their land, because a man without land is nobody; their relations become starkly economic, a struggle for wealth and influence, and sometimes the rivalry culminates in murder. It is as though everything is organised to provide everybody with an enemy or a rival. The result is that everybody finds a friend, but always among those with whom they are not in competition.

Thus, although sisters are rivals, brothers and sisters are close, and theirs is the most powerful bond in this society. Sisters grieve most on their brothers' death, more than for their mothers; and a man goes to his elder sister first when he is in trouble. Fathers have 'warlike' relations with their wives – exchanging insults about each other's lineage – but a loving and even coquettish attitude to their daughters. Mothers, who show little affection to their daughters, care most for their sons, and protect them against fathers. Brothers are normally at war amongst themselves, cuckolding each other, and even the sons of brothers are enemies. However, these people have found a way out. They see Allah not as a father, but as a friend and lover: the Zoroastrian word for God is Friend. One of their prayers is 'O God, grant me a true friend who, without urging, will show me his love.' They need a friend whom they can trust, to whom they can show devotion and loyalty. The ideal friend comes from outside the tribe, demanding nothing, not seeking to dominate. They dream of a foreigner who will be their friend, on whom they can lavish hospitality, which is their greatest pleasure. In public they brag about their plans to win honour and wealth, but in private they complain that they feel imprisoned by family jealousies; so they migrate willingly, to lose themselves in another world. Then they complain that their new country shows less generosity than they knew back home, and much less hospitality. Again and again one finds that generosity flowers most freely in cruel soil, as though inhumanity is the best fertiliser for it. That is not a pessimistic conclusion, given that inhumanity seems to be available in limitless quantities.

All over the world, the disappointment individuals have had with their own brothers and sisters has led them to seek artificial siblings outside the family, to invent relationships from which envy would be excluded. Blood brotherhood created, with the aid of more or less bizarre ceremonies, the loyalty that ordinary brothers did not guarantee. In Timor, for example, the contracting parties used to slash their arms and fill a bamboo with their blood and with wine, as a symbol of their union. According to Tacitus, Armenian and Iberian princes tied their thumbs together, pierced them and each sucked the other's blood. According to the *Lay of Brunhilda*, Scandinavians made brotherhood by letting their blood mix in a footprint. In some countries, garments were exchanged, in others names, or weapons; in others still, the partners rubbed each other's saliva over themselves. Often, trees were planted to commemorate the event. Sometimes blood brothers shared everything, sometimes even their wives; but above all, they meant their bond of loyalty to be unbreakable, totally reliable, occasionally even thinking of themselves as having become two halves of the same person.

Travellers of the nineteenth century reported places where all youths initiated at the same time became brothers, and addressed each other's wife as wife and each other's children as child. In Montenegro they found three grades of voluntary brotherhoods, the little brotherhood being created by a kiss thrice repeated, while profounder ties involved the blessing of a priest and the taking of the Eucharist. Women, married and single, entered similar relationships by drinking wine together, kissing and exchanging gifts, their union being finally solemnised on the steps of the altar; and then, it is said, they addressed one another as 'little sister' (or 'my gold' or 'my little fawn'), often wearing identical clothes and ornaments. Ceremonies of this kind also occurred between persons of different sex, though they were reputed to have been more common in the distant past. In Serbia, Croatia and Bulgaria, there are records of bonds of brotherhood being renewed each year. Brotherhood could even be created by a girl going over the mountains alone, inviting the first man she met to be her brother; he was bound to guard her as if she were his own sister. A man in extreme danger could likewise appeal to another to become his brother. In Fiji, 'comrades in arms are spoken of as man and wife, to indicate the closeness of their military union'. But if brotherhoods limit themselves to two or a few individuals, if they have no foreign policy towards the rest of the world but war, they are in the long term self-destructive.

The numerous efforts to persuade everybody to be the brother of everybody, indiscriminately, on a mass scale, have not met with much success. Christianity, for example, while proclaiming the highest ideals of universal brotherhood, irrespective of race, sex, social status or character, in practice waged war on heretics and infidels, as communism did also. Every faith, once it tastes power, forgets why it wanted power; only recently, having lost their political status, have churches returned to their ideals. The French Revolution was more hesitant about fraternity than about liberty or equality, and did not get much beyond gestures, like planting symbolically welcoming, sheltering trees on frontiers: the lawyer who tried to implement fraternity by insisting that the jury of the tribunal of Paris should include a Catholic, a Protestant, a Jew and the only coloured man he knew, 'to consecrate the fraternity of colours', was called 'a snake crawling slimily between opposing parties'. That suspicion is reminiscent of the fear inspired, almost a thousand years before him, by the Islamic Brotherhood of Purity (founded in AD 951), which interested itself in all religions and found elements of truth in each one, only to have its *Encyclopedia of All Knowledge* condemned to be burned, by order of the calif. The modern welfare state, while giving every disadvantaged person a legal right to aid, has been unable to include human warmth in its offering; and now conservative theorists are burying the ideal of fraternity altogether, dismissing it as a threat to privacy and competition. The Black American leader Marcus Garvey's call for 'the universal fraternity of all races' remains a noble dream.

Meanwhile, however, small fraternities have been mushrooming discreetly, trying to do what big fraternities cannot. The girl on the farm near Cognac who pours out her private thoughts to her pen-friends in other continents, more than to her own family, is a sign of a worldwide search for soul-mates and confidants out of which may grow another kind of family, and attachments on a new level: families of the heart and the imagination, freely chosen, unable to impose punishing obligations. Humans can no longer be thought of as being nothing but residents of the town they inhabit, for they increasingly speak, write and listen to contacts miles, and sometimes thousands of miles, away, more than to their own neighbours. They are as much citizens of the republic of letters, or of science, or of business, or of football, or whatever their passion is, as of their own country. Every individual is slowly building up an international confederacy of personally chosen individuals. More and more humans are refusing to be carrots, stuck in the mud, inseparable from their one family root. Some still try to behave as such, to feed themselves entirely from their own sap, but it has become almost impossible; and they surreptitiously send out fibrils to distant parts in search of more varied nourishment. Now that the air contains not just oxygen, but radio and television signals too, no family, however close-knit, can seal itself off from the thoughts which fly in like bees through the windows, fertilising imaginations and moving pollen from one mind to another, making them relatives of people they have never met. This is a completely new kind of fraternity, more ephemeral, changeable, accidental, but less liable to be asphyxiating.

Close but not too close, each keeping a certain amount of independence; that is an ideal in accord with the changes in family arrangements which are called the crisis of the family: but in every crisis there is an opportunity. The ageing of the population also means that more grandparents, with greater serenity and detachment than parents, are available to act as mentors to the young, compensating in some respects for what elder brothers and sisters once did in large families: in the course of the last two centuries the proportion of people aged twenty who still had grandparents has doubled from a quarter to a half. The traumas of marital collapse have as their by-product a reshuffling of roles, so that children sometimes become the counsellors to their parents and parents ask indulgence of their children; and individuals who have no tie except that created by the erratic loves of their relatives have been brought together, in a new, looser version of the traditional large family; the rigid barriers which used to cut families off from strangers are coming down. Neither envy nor greed may be diminishing, but the vast increase in the number of people living alone has been like a mass migration away from direct confrontation: in several Western countries a quarter of households consist of one person. The growth of adoption, no longer as a means of passing on property but as a gratuitous act of affection, is very recent, and so too is open adoption, where no secrets are kept from the child, and adoption by single parents, and intercountry adoption: these are beginnings of yet another kind of fraternity, introducing choice into kin.

When traditional family obligations have lost their vigour, new relationships which appeal equally to the heart and the head have sometimes appeared to take their place. A new phase in the history of generosity has opened precisely when it has seemed most in danger, when competition has been honoured as never before. Generosity got tied up in knots in the past, because people wanted to keep the heart and the head separate, demanding of each other that they should do the right thing out of duty, whatever they felt, however grudgingly, convincing themselves that they would become virtuous eventually if they got into the habit of performing charitable acts; but those at the receiving end of charity demanded help as their due, and refused to adopt the values of their benefactors in return. To untie the knot, generosity has to go beyond giving and receiving, beyond tit for tat. With gift-giving of the traditional kind, the recipient is made a debtor, and a benefactor therefore can easily become an enemy. Generosity can avoid that only by being the joint effort of two people who have succeeded in putting themselves in the place of the other.

The smaller families have become, the greater has been their need for a foreign policy, but it is not a question of just deciding to have one. The favourite instrument for dealing with danger used to be magic, which controlled the meeting of the invisible with the visible, and it brought fear. Today, love is the magic most people believe in, when two strangers meet and discover they cannot live except in each other's arms, which also brings fear, the fear of losing love. The small family based itself on this magic. However, there is a third kind of magic, in which an individual can make a difference to the way the world revolves, by helping another individual without asking anything in return, without offending pride, without curtailing freedom, being purely and simply generous. The fear that goes with it is the fear of being made a fool of, or being taken for a ride. Scientists who studied such matters used to be scathing about those who believed this to be possible, insisting that people always want something in return, and that envy is one of the inevitable by-products of existence, like carbon dioxide.

Humans have the choice of having relationships which are either prosaic or magical, which are cold, economic and guaranteed, or which attempt something wonderful and amazing while also being more or less real. Now they are being encouraged to be bold by scientific discoveries showing that generosity is not mere fantasy. Thus it used to be believed that children were born selfish, but recent observation of the very young (from fourteen months) has revealed that they are capable of many different kinds of generosity, not haphazardly, but in ways appropriate to the needs of others. They can recognise other people's feelings and perspectives much earlier than was previously thought. Likewise, until the last century, scientists insisted that racial prejudice was natural, but now some studies of children have suggested that discrimination, far from being something they think up for themselves, is taught to them by adults, that it is most virulent when a small minority faces a large majority, but that it is slow to be accepted where black and white are equal in numbers.

It is not only the idea of human nature, what it means to be human, which is being modified, but the idea of animal nature too. The belief that animals care only for themselves in their struggle for survival has been challenged by the discovery, in some species, of intelligent co-operation. Even vampire bats, for example, contrary to their frightening reputation, returning successfully from a night's foray for food, give some of it to other bats who have had less luck. The white-fronted bee-eater does not just feed its own offspring, but also helps its parents to raise their new brood. However, the debate about whether humans and animals have a basic goodness, or nastiness, is not worth pursuing. It is the lessons they have learned, in the process of trying to be generous, that matter more.

Whenever families have widened their ambitions, they have had to invent new tools. For example, when they were no longer content with what their nearest neighbours could offer them, they used godparents to help them widen their horizons. In the Andes mountains, far from modernity, children may have as many as six godparents to make up for the inadequacies of their own begetters. People who are asked to be godparents cannot refuse. They are named at each of the child's major festivals – the first haircut at the age of four, at confirmation, at marriage – and having received a religious blessing, become real parents and behave as such. Fathers try to find godparents who will help the child in its career; peasants seek out merchants in the town, a mutually profitable relationship.

Now that understanding others has become the great ambition, the idea of godparents, and godsisters and godbrothers, takes on a new meaning. In the traditional system, each person has a different godparent, who does not have to be shared with a sibling and who can be an independent interpreter of what goes on in the natural family. The unexpected result of new technologies of communication is that it is possible to conceive a criss-cross of networks of people who expand that relationship, who care for each other, no longer just in the neighbouring city, but anywhere in the world, and who, because they are not too close and not competitors, can put generosity before envy. This is where the human rights movement inevitably leads, and it may turn out to be as momentous a development as the growth of nations, which were the result of much more primitive forms of communication.

There are a few schools in the West which try to teach children to understand other people by making them play the role of another, which in effect is an invitation to widen one's idea of what a home is. If home is where one feels comfortable and understood, but still retains one's privacy and mystery, if it is where one both takes care of others and is taken care of, while also having the right to be left alone, and if it is one of the great personal and collective works of art that all humans spend their lives attempting to raise up and to keep from falling down, then the art of creating homes, as distinct from building houses, still has a long way to go, and still remains within the province of magic. Instinct or imitation are not enough to make a home.

Ulla Bjornberg, *European Parents in the 1990s*, Transaction, New Brunswick, 1992; CEE, *Les Jeunes européens en 1987–1988* (half of the Irish and 40 per cent of Britons and Germans had no desire to learn any foreign language); Daniel Linhart and Anna Malan, *Voyage au pays de 18–25 ans*, Syros, 1990; GRIF, *Les Jeunes*, Issue no. 34, Winter 1986; Maureen Baker, *Quand je pense à demain: Étude sur les aspirations des adolescents*, Conseil consultatif canadien de la situation de la femme, 1985 (on low ambitions of the unqualified, and where they turn for advice); David M. Brodzinsky, *The Psychology of Adoption*, Oxford UP, NY, 1990; R. A. Hinde, *Relationships within Families*, Clarendon Press, Oxford, 1988; Robert A. LeVine, *Parental Behaviour in Diverse Societies*, Jossey Bass, San Francisco, 1988; B. Sutton-Smith, *The Sibling*, Henry Holt, NY, 1970; Glenda A. Hudson, *Sibling Love and Incest in Jane Austen's Fiction*, Macmillan, 1992 (on her preference for sibling affection); Christine Downing, *Psyche's Sisters: Reimagining the Meaning of Sisterhood*, Harper, San Francisco, 1988; No Addo and J. R. Goody, *Siblings in Ghana*, University of Ghana, Legon, 1957; Jacqueline Goodnow and Ailse Burns, *Home and School in a Child's Eye View*, Allen and Unwin, Sydney, 1985; Moncrieff Cochran, *Extending Families: The Social Networks of Parents and Children*, Cambridge UP, 1990; Claude Boisleme, *Une Génération inattendue: Les jeunes de 15 à 24 ans*, no publisher, B. N. Li 1 746(1), Montpellier, 1987 (good detail); Ségolène Royale, *Le Printemps des grand-parents*, Laffont, 1987; Judith Wallerstein and J. Kelly, *Surviving the Break-up: How Children and Parents Cope with Divorce*, Basic Books, NY, 1980; Carolyn J. Mathiasson, *Many Sisters: Women in Cross-Cultural Perspective*, Free Press, NY, 1974; Ira L. Reiss, *Family Systems in America*, 3rd edn., Henry Holt, NY, 1980; Sybil Wolfram, *In-Laws and Outlaws: Kinship and Marriage in England*, Croom Helm, 1987; Martine Segalen, *Historical Anthropology of the Family*, Cambridge UP, 1986; Mirra Komarovsky, *Blue Collar Marriage*, Yale, 1962; Sarah LeVine, *Mothers and Wives: Gusii Women of East Africa*, Chicago UP, 1979; Carole Klein, *Mothers and Sons*, Houghton Mifflin, Boston, 1984; Leigh Minturn and W. W. Lambert, *Mothers of Six Cultures: Kwang-Ching Liu, Orthodoxy in Late Imperial China*, California UP, 1990; Linda Rennie Farcey, *Mothers of Sons*, Praeger, NY, 1987; Turid Bergljot Sverre, *Mothers and Daughters as Portrayed by Norwegian Women Writers from 1984 to the Present*, University of Texas, Austin, Ph.D. thesis, in BM; Helena Hurme, *Child, Mother and Grandmother: Intergenerational Interaction in Finnish Families*, University of Jyvaskyla, 1988, in BM; Betsy Wearing, *The Ideology of Motherhood: A Study of Sydney Suburban Mothers*, Allen and Unwin, Sydney, 1984; Sandra Barry, *Elles aussi: La maternité*, PUF, 1988 (interviews with career women); Nichole Prieur, *Adolescents parents: Des rendez-vous manqués*, Casterman, 1981 (survey of opinions); Catherine M. Scholten, *Childbearing in American Society 1650–1850*, NY University Press, 1985; Stanley N. Kurtz, *All the Mothers are One: Hindu India and the Cultural Reshaping of Psychoanalysis*, Columbia UP, 1992; Ronald P. Rohner and M. Chalki-Sirkar, *Women and Children in a Bengali Village*, University Press of New England, Hanover, 1988; Charles Lindholm, *Generosity and Jealousy: The Swat Pukhtun of Northern Pakistan*, Columbia UP, 1982; Shanti Lal Nagar, *The Universal Mother*, Atma Ram, New Delhi, 1989; Suzanne Dixon, *The Roman Mother*, Routledge, 1988 (mother's influence on the adult child); G. Duby and M. Perrot, *Histoires des femmes en Occident*, 5 vols., Plon; Elaine Heffner, *Successful Mothering: The Challenge of Motherhood after Freud and Feminism*, Robson, London, 1980; Sara Ruddick, *Maternal*

Thinking: Towards a Policy of Peace, Women's Press, 1989; Yvonne Knibiehler, *L'Histoire des mères*, Montalba, 1980; Megan Vaughan, *The Story of an African Famine: Gender and Famine in Twentieth-Century Malawi*, Cambridge UP, 1987; Katherine Arnup, *Delivering Motherhood: Maternal Ideologies in the Nineteenth and Twentieth Centuries*, Routledge, 1990; Beverly Birns and D. L. Hay, *The Different Faces of Motherhood*, Plenum, NY, 1988 (international essays); Lee N. Robins and Michael Rutter, *Straight and Devious Pathways from Childhood to Adulthood*, Cambridge UP, 1990 (revisionist ideas on the influence of childhood experience); Eileen Simpson, *Orphans Real and Imaginary*, Weidenfeld, 1988; James Brabazon, *Albert Schweitzer*, Gollancz, 1976; E. G. Reynolds, *Life of St Francis of Assisi*, Anthony Clark, 1983; Miriam Lewin, *In the Shadow of the Past: Psychology Portrays the Sexes, a Social and Intellectual History*, Columbia UP, 1984; Janet Finch, *Family Obligations and Social Change*, Polity, Cambridge, 1989; Walter L. Williams, *The Spirit and the Flesh: Sexual Diversity in American Indian Culture*, Beacon Press, Boston, 1986; George Devereux, *Mohave Ethnopsychology*, Washington Smithsonian Institute, 1969; Diana Everett, *The Texas Cherokees*, Oklahoma UP, Norman, 1990; H. H. Stockel, *Women of the Apache Nation*, Nevada UP, 1991; Daniel K. Richter, *The Ordeal of the Longhouse: The Peoples of the Iroquois League in the Era of European Colonisation*, University of North Carolina Press, 1992; Fred Gearing, *Priests and Warriors: Social Structures for Cherokee Politics in the Eighteenth Century*, American Anthropological Association Memoir 93, vol. 64, no. 5, pt. 2, October 1962; Guy Lanoue, *Brothers: The Politics of Violence among the Sekani of Northern British Columbia*, Berg, NY, 1992; Gilberto Dimenstein, *Brazil: War on Children*, Latin American Bureau, London, 1991; C. Meyer, *Histoire de la femme chinoise*, Lattès, 1986; Wilson Carey McWilliams, *The Idea of Fraternity in America*, California UP, 1973; Ferdinand Mount, *The Subversive Family*, Unwin, 1982 (against fraternity); F. Furet and M. Ozouf, *Dictionnaire critique de la révolution française*, Flammarion, 1988 (on fraternity); Marcel David, *Fraternité et révolution française*, Aubier, 1987; Juliet B. Schor, *The Overworked American: The Unexpected Decline of Leisure*, Basic, NY, 1991; Alan Wolfe, *America at Century's End*, California UP, Berkeley, 1991 (interesting articles on family); Elizabeth Pybus, *Human Goodness, Generosity and Courage*, Harvester, 1991; H. Medick and D. W. Sabean, *Interest and Emotion*, Cambridge UP, 1984.

23

How people choose a way of life, and how it does not wholly satisfy them

I met these six women in six different towns. Each seemingly had a completely different story to tell. What have they in common?

Dominique Lepèze, the ballet dancer, does not express herself easily, or willingly, in words. Her statements come in the way she moves and holds herself; the wave of a hand, every slight gesture, is her substitute for a sentence; her walk is a paragraph. But there is no dictionary for her eloquence. When she is persuaded to try and explain what she feels, 'There seems to be a little person inside me who is amazed by the words I utter: "What on earth are you on about?" it says. As I talk to you, I say to myself, How badly I am talking. I am uncultured. Monsieur is very cultured.' The intimidation of the silent by chatterboxes is another unwritten history.

She began dancing at the age of twelve. Till then she had been unhappy, which she explains by saying, 'I have no roots.' Her father was a Lille worker who got a job in Morocco (where she was born) but who then disappeared; her mother found another man, whom she abandoned in her turn. Mother and daughter then settled in Toulouse, where Dominique won a place at the conservatoire. 'Dancing was my whole life, and it made me very

·396·

happy.' By seventeen, she had a part on stage; at eighteen, after seeing Felix Blaska dance, she was starstruck and swore to dance with him. Off to Paris she went, without a penny, presenting herself at an audition together with 300 other women. He chose three, and she was one of them. For several years she danced in a company which enchanted her – 'everyone in it is a soloist'. She had her own flat, a good salary and frequent travel.

However, something was wrong. Her dream had come true, but she was not happy. Felix Blaska could not understand why. Neither could she. Perhaps, she thinks now, it did not suit her to have a regular job like a civil servant, saving to buy a home. She resigned and went 'to search', for she did not know what, worked in the Italian theatre, in the National Ballet of Ecuador, in Los Angeles, tried to learn Indian dance ('but it requires a different set of muscles, I got cramps'), joined a flamenco group, experimented in a modern dance company. But she still was not satisfied. Her conclusion was, 'I liked dancing, but not the world of dancing', not its rivalry, its jostling to get to the front, its choreographers with their private obsessions: 'They drew out my dramatic, serious side, but I also have a comic side, and they never used it.' Dancing, for her, had to express the whole beauty of life. 'I was not a princess. I am not a person waiting for a prince. It is not technique that matters, but getting beyond yourself; it is what you communicate that counts.' The nearest she comes to finding a word for her aim is 'purity'. But that was not what her fellow dancers wanted to talk about, preferring endless discussion of technique, and gossip about the critics.

Finally, one day she got a glimpse of that indefinable 'purity' she was groping for: she read Krishnamurty. 'He spoke to me in that book.' By chance, one of her colleagues, a Dutchman by origin but fluent in French – whom she later married – was also finding an answer to his own problems in Eastern mysticism. He lent her his books on yoga, Tantric and various modern gurus. A vaguely Indian community, run by a German couple, situated idyllically in the Pyrenees, attracted her, but it proved too authoritarian. A move to another guru, the Bhagwan, did not answer her needs either, because he was the exact opposite, asking nothing of his disciples except that they should wear red clothes, otherwise he left them to their own devices, saying, for example, that if you wanted to stop smoking, just wait for the cigarette to fall from your mouth.

Eventually, she found a guru with whom she could communicate without words. He was an American and she could speak no English. He wore very few clothes (Indian style), unlike the Bhagwan who was always beautifully dressed and who was beautiful, but whose fault was that he always said the same thing. This new guru, by contrast, looked both vulnerable and immensely strong. Franklin Jones, a graduate of Columbia and Stanford, had set up the Free Daist Community with headquarters in Fiji and ashrams in Holland and California. Her husband became a dedicated follower, but she hesitated – 'He is always one step ahead of me' – until she met the guru in person. She was the first of his students not to know English, attending his lectures without understanding a word, and she was captivated, 'by intuition'. 'He was a man like no other.'

Her husband began translating the works of the master, but things were more difficult for her. 'I doubt all the time. I say to myself, "You are amusing yourself, Dominique. What are you up to?"' She was attracted but also annoyed, because now she had a child, and this guru was monopolising her husband's whole attention. In the Dutch ashram, the common language was English, which she could not speak; worse, the guru insisted that they live in dormitories, so as to become liberated from an excessive concern with their self. Nor was the food to her taste – she was French, after all. And the converts had group discussions in which they talked openly about sex. 'A Frenchwoman can't do that; it is very interesting but don't ask me to speak. It gives me a stomach ache. It is super, but it isn't me. I get very uncomfortable.' For all her glamour, she was a mother from a provincial town.

The marriage became strained. Last year, however, they went to visit the guru in his Fiji retreat. 'The lavatories were awful. Mud everywhere, and frogs clamber up your legs. I hated that. But I was happy. Everything I feared, I lived through laughing. Seeing the guru is like being in love, and I awoke in the morning saying, "Where is he?" He gave audiences to his visitors, but in silence, and her whole body jumped, as though electrified by the experience. 'With him, my attention was total. He accepted me, even though I did not speak his language. Going to see him, I feel the same stage-fright as in a ballet performance, and I leave as though I have wings, completely free. I was no longer afraid. I said, "I am not afraid of death, I would be happy to die here." Everything made me happy. I stayed a month in Fiji, not seeing him every day, but feeling his strength, feeling that he was working with us. Dancing in Fiji gave me the greatest happiness; it was the best show I ever put on, much better than at the Palais des Congrès. It was really love, an escape from the body. I now take my lack of confidence less seriously. Being famous is not what makes you happy. Now, I have no illusions. Because I cannot speak, I am taking up dancing again. I can give people something with my dance. One day, I should like to dance for the Master, to see his look. I danced only for the women who serve him.'

But Dominique's problems are far from solved. She does not feel liberated. She still panics, and wishes she could behave with more detachment when she does. Married life, half in public, half in private, leaves her distraught. She no longer feels jealous of her husband, who might have had a more brilliant dancing career than she – he was invited to join Béjart's company, but refused, perhaps so as not to outdo her, settling down instead as a translator. But she cannot see her future clearly. 'I should like to be reassured about it.'

Catherine teaches people how to be successful. By that she means success in earning a living, making a profit, attaining prosperity, which always involves money. In a small room in which a dozen people are a crowd, in a modest office block, she offers a course to people who want to set up their own business, and Catherine is the director and lecturer in one. Most of the students are unemployed; only a third of them, at most, will turn their dream into reality, and of these, quite a few will fail. And Catherine herself, though as a human being she is a very impressive person, though she has extraordinary warmth, which kindles enthusiasm even in the most timid student, though she can talk with amazing lucidity, fluency and speed, is not, in terms of profit and money and the results she achieves, a total success. She is not rich, she is divorced, she never finished the thesis she meant to write, she never became the business manager she once aspired to be. How does one cope with the refusal of reality to resemble one's dreams?

Her business career began full of promise: a job in the prestigious National Bank of Greece, six years' experience in international commerce. Then 'I discovered I had talents of which I was not aware': psychological tests showed she was just what a specialist in industrial training should be. The bank paid for her to go to France to study personnel management. But then it would not give her the promotion she felt entitled to: she did not have, she says, the patronage without which nothing can be obtained in Greece.

Her studies came to a halt. Though educated by French nuns in Salonika, and equipped with a French as well as a Greek baccalaureate (and able to speak English and Italian too), she was a foreigner in France. The Sorbonne had no room for her, so she moved to the provinces. But her director of studies was not interested in her, nor in business either, nor in her proposal to do research to discover whether the productivity of banks is increased by training courses. Though she obtained a diploma, the important book she planned was never finished. Her grant ran out. Her husband left her. She needed a job. 'I considered my studies as an investment, and I wanted the profit.' But her bank did not give it to her, and she left. Alone with a small son, she had to start again, quickly.

Her present job is one she has invented, and persuaded the authorities was needed. That is what being an entrepreneur means, meeting an unsatisfied demand. Catherine is proud of the achievement, because it is a job she enjoys. Moreover, she has introduced a new method of training, originating in French Canada, which emphasises not experience but self-knowledge. What students discover on her courses is the answer to what she calls the 'Shakespearean problematic' – to be or not to be an entrepreneur. Not every one can be one. That is not because she takes an elitist view, she insists: people just have to be honest with themselves and ask whether they feel they can take risks. Business is all about risk. Unfortunately, she says, women are more wary of risk than men; it is a question of education; so in France they start less than a third of all new firms. Nonetheless, the majority of her students are women.

Women, she laments, tend to want to go into commerce, marketing, counselling, rather than industry, even though they would have a better long-term chance of succeeding if they undertook production; there are not enough women engineers; and not one of her women students has had an original idea for a business. But does she tell them that? No, she is very tactful, she has to be prudent. The purpose of her course is to encourage. (There are limits to self-knowledge.) What matters, she likes to repeat, is ambition, will power, audacity. 'Everything is easy, it is we who make things complicated.' She says that even while recounting her own difficulties.

Of course, Catherine has not gone into industry herself, nor even into a job which yields profits in the conventional sense; and her work has not made her rich. 'One does not create a firm to become rich,' she retorts. If you want to make money quickly, go into the oil business. But if you want to combine business with intellectual pleasure . . . her own mixture suits her well. It is true she is a salaried employee, which is not quite in keeping with her ideal of independence, nor her family tradition. Her father, a Greek from Asia Minor, but partly French, educated by the Franciscans in Aleppo, expelled by the Turks, established a textile firm in Salonika, which lasted forty years. Her mother, born in Bulgaria, the daughter of a grocer, set up her own little dressmaking business. But Catherine did at least create her own job out of nothing, even if it is salaried.

France is her adopted country now; she feels perfectly integrated into it; she has chosen it freely; she likes French logic; perhaps the failure of her Greek employers to appreciate her has left a bad taste, but she could have chosen to return to Greece. Her son cannot speak Greek, preferring to make English his second language – it is more useful for business. 'He takes after me in being ambitious.' Her husband has been content to be a secondary school teacher all his life; that is why they broke up: he could not understand ambition, let alone the desire to make a profit. 'A pure hard-line communist', he has remarried, and his new wife does not work; perhaps he is frightened of ambitious women. Catherine is still friends with him, and admires his intellect; they shared a passion for history and philosophy, but he 'lacked a practical side'. The best he can do to earn extra income is to give private lessons. They should have realised that they had different ambitions, but Greeks were not in the habit of living together before marriage; they did not really know each other until long after their wedding; they underestimated the class difference between them, he being a son of workers, who disapproved of her capitalist father. Her ex-husband is quite happy as he is. Catherine says she is satisfied as she is. But in addition, she dreams. 'I dream for myself and for my son.'

At the same time, however, she says that the way to uncomplicate life is to plan, to be organised. She lives close to work, so she can find time to read, to cook ('not from obligation, but in the same way as I read a novel'), to entertain friends. 'I believe a lot in friendship.' What friends can give most of all is sincerity; they are no use if they are hypocrites, flatterers. 'I am very hard on myself; I love to criticise myself; it is a superb thing to do, and to induce others to criticise themselves, because that way you will have no grudges against people; you won't say you failed because the bankers were nasty; you will understand what you are not capable of.' That is how she reasons: will power is what counts; but on the other hand there are things that you cannot do. There is no conflict if you know yourself. But does she know herself? Catherine's art of life is both to see and to avoid the complexity of life. She believes in planning, but she will not do anything to find another husband, she would like a meeting to happen by chance. She believes in friendship, but has few close friends. 'If you expect friends to be faithful unto death, you won't find them.'

Love is even more double-faced, a stage young people have to go through: 'It is important to make mistakes, not to start philosophising too young; you won't regret your mistakes if you draw lessons from them. There is a positive side in every experience.'

So what is the aim of the new modern hero, the entrepreneur? As far as she is concerned, she does not want 'to remake the world', but only to adapt to the world as it is. 'The world is what it is, and it isn't all that bad.' On the one hand she says that women do not suffer from inequality: 'It is all in the head; it is a matter of will power, which one either has or not.' On the other hand will power has not worked any miracles for her. Yet she thinks of herself as having got her own way. 'The women I admire are those who have succeeded in getting their aspirations and their ideas accepted.'

Is Catherine French or is she Greek? 'European,' she answers. She leaves one speculating whether Europeans may be those who do not fit completely into existing nations, who see everything from two sides at the same time, as she does. She leaves one conjuring up an image of a Europe of people who have had many disappointments, but are determined to transform these into sources of strength.

Everyone in Victorine's family was dark haired, but she was born blonde. The family's secret was in danger of being revealed. Her mother – though none of the neighbours in Fontainebleau knew – was the daughter of a German soldier, passing by in the Second World War. To conceal that stigma, Victorine was given away when she was two weeks old. The Alsatian farming couple who adopted her were kind and loving, but when she was twelve they died in a car crash. Her adoptive brothers said they could not look after her. Her real parents were traced and she went to live with them, but her happiness was gone. She left as soon as she could. Recently, on the birth of her daughter Mélodie, her parents came to visit her and asked to be forgiven. 'I told them I forgave them, but inside me I do not. Perhaps I shall never be able to. I would do anything for my daughter, whether she was a blonde or a redhead, I would love her all the same.'

As a teenager, Victorine spent five years living in hostels, moving repeatedly from one to another, disqualified by this or that regulation, begging to be made an exception. Sometimes she was allowed to clean or cook without wages, in return for a bed, but 'every time I changed hostels, I lost my friends. . . . I was all alone. It was very hard. I did talk to the social workers, but they were only doing their job, there was no bond of affection. My morale fell very low, especially at Christmas, when I saw other girls holding hands with their mother, going into shops and buying things. I became sullen, irascible: everything made me angry, and when people got angry with me, I did not care. I was alone. If you had met me when I was eighteen, I would have refused to speak to you, or I would have denied that I had problems. I invented a family. I lied.' Her training for the profession of office clerk did not lead to any offer of a job.

Surviving with no money – or with the occasional allowances of the social security – means one is under constant threat of being sent to prison. When Victorine travelled by train, she did not pay. When once she did not pay in a café, she was put in a police cell for forty-eight hours. Other youths, in the same predicament as herself, 'tried to get me into trouble; I was tempted, but I said no. I was afraid of going to prison for twenty years, or more. The hostels were tough enough. Prison would be harder than hard. It wasn't that I was timid. I was at rock bottom and I didn't want to be squashed still further. I did not have the pill, but I was too scared to get pregnant. I did not want to have an unhappy baby.'

One day, on a walk, her bag was stolen and she had nothing left in the world. Her social worker had just refused her any more food coupons and the hostel had asked her to leave. She decided there was no point in being alive. She was about to jump from a bridge into the Loire to end it all, when a tall, slender young man with a nice smile passed by. He stopped to watch her. 'Can you swim?' he asked.

He too lived in a hostel; he too was unemployed. 'Come with me,' he said. He smuggled her into his hostel and shared his food coupons with her. They talked; he was gentle and sensitive, and he had problems as overwhelming as her own. 'It was a friendship. We became mates.'

Born in the slums of Paris's nineteenth arrondissement, Antoine had never known his Breton father. His Martiniquaise mother had breast cancer and asthma, and was obliged to send him away to a wet-nurse, who 'did not want me, because I was black'. His Creole-speaking grandmother, a priest's servant, then looked after him, but 'she died of fatigue'. Moved from one convent school to another – six in all – he was harassed by complaints that he was refusing to learn his lessons. 'I wanted to learn,' he says, but he never succeeded. He has no qualifications of any kind; and he has failed his driving test four times: 'I cannot remember the signs.'

He has a wonderfully warm personality, laughs a lot, talks fast with great energy and gives an impression of dynamism and curiosity, bubbling over with plans for the future. He would love to be a vet. Or a lorry driver, as a step to becoming a racing driver. He tried to get a job with Renault, but was turned down 'because I have no qualifications'. Or he would breed horses and enter riding competitions and become famous and represent France, so as to become independent, and have a nice house. . . . Or perhaps he will go to the West Indies to reclaim his ancestral plot of land.

A racing stud was sufficiently impressed by his enthusiasm and joviality to give him a job, but it emerged that his hopes were mere fantasies; his claim that he knew about horses was exaggerated, and he was sacked. He has been lucky to get a job in a supermarket: he comes home tired, 'particularly since I do not get paid much'.

Now Victorine and Antoine are married and they have a beautiful daughter, so they have been given a two-roomed council flat in a dreary block on the outskirts of a small town. There is so much damp and fungus on the walls that they have to be repapered every year. The bare furniture is of the sort that others discard: the webbing of the chairs is broken, their covers torn, the plastic tablecloth has holes. Though the children in the neighbourhood say good morning to them, the adults on the whole do not; sometimes parents speak to them, about children, but when they are not with their child, they are ignored. Last summer they decided to go to a discothèque and were refused admission. She is afraid to walk with him in the streets of nearby Tours, which she finds 'suffocating'. People say their daughter (who is white) is sweet, and ask, 'Where is your papa?' When Antoine replies that he is the father, they turn their backs. He is 'deeply hurt' by racism and the diatribes of Le Pen.

'I don't mind others having a more luxurious life,' says Victorine, 'because I am in charge of my home. I can do what I like here.' Above all else, Mélodie is happy. Mélodie is the centre of their world and the source of their joy. 'Sometimes I still feel alone,' she says, 'but rarely with the baby.' And she feels that her husband has felt 'fulfilled' by the birth of the child, even though he dislikes his work.

However, they do not plan to have another child. It is very hard managing on his single, minimum wage. 'We tighten our belts for Mélodie. She comes first.'

In England the collared dove (*Streptopelia decaocto*) is a much-loved native bird. But for centuries it lived only in Asia. In the early twentieth century examples of the species crossed the Bosphorus and settled in the Balkans. Between 1930 and 1945 it colonised central Germany. By 1970 it had become a common bird in Britain, northern France and southern Scandinavia. Then it was seen setting off westwards over the Atlantic: perhaps it will become American soon. The experts say that a genetic mutation may explain this sudden expansion over three continents.

Parwin Mahoney also belongs to three continents. Her ancestors originally lived in India, and then moved to East Africa, where she was born. At the age of ten she was sent to boarding school in England and has seen her parents, who stayed in Africa, only once every other year. She married an Irish barrister settled in England, who got a job in Strasbourg at the European Court of Human Rights. Their children went to a French school; the eldest took German as his first foreign language and now spends the week in Freiburg at a bilingual German-French school.

There was a time when humans had no more need of a passport than do birds when they want a change of scene. But with the multiplication of government officials, objection was taken to their moving even from town to town without papers to prove that they were who they said they were. The French Revolution abolished passports, deeming them inconsistent with the freedom of the individual, but they crept back. It was in vain that in the nineteenth century the British simply refused to carry passports when they visited France, even though the law said they could be arrested if they did not. For a time the law was ignored in their favour, and indeed repealed after the revolution of 1830, which had faith in liberty. However, the fear of spies returns every time there is a war; passports were revived. In 1872 the British were once again freed from the need to carry passports and were required simply to sign their name when they crossed the frontier, but world wars brought passports back yet again. The reason they survive today is of course that rich countries are afraid of poor people invading, and dictatorial countries of their enslaved subjects leaving. However, Parwin is one of those rare ones who have managed to wander around the world using a passport like the wings of a bird, with grace.

'When I lived in England, I felt English and I was accepted as English. I think of my children as English. We speak only English at home.' In Strasbourg, the children have picked up French habits, like wiping their plates with bread, or dunking it in their coffee, or shaking hands; they play in French. But the rule is that they must not speak French at home. The plan is that they will go to an English university.

How do others think of Parwin? I was told that she is a 'Strasbourgeoise'. She says, 'My home is where I am.' They originally thought they might spend only a few years in Strasbourg, but her husband's work has been very interesting, while she has made a name for herself and become a useful citizen. In the first week after her arrival, she told everyone she met at the Council of Europe that she intended to find a job. They laughed. Some wives were lucky if they picked up occasional typing or proofreading. They laughed when she said she hoped to make French friends. She went home and cried. 'Is this city darker than darkest Africa?' The Alsatians seemingly had no use for her. All her efforts to be accepted were in vain. She determined to take the initiative, 'No one is going to say, "You look foreign, I'd like to get to know you."'

At first Parwin and her husband tried to behave like the natives, whose main preoccupation seemed to be their food, talking about food when not eating it: 'The English tell jokes more, discuss politics less.' Parwin began cooking like the ladies of Strasbourg, because she had nothing better to do, 'but it only made us fat and poor, so we went back to English ways'. At the time, in the mid-1970s, the women of Strasbourg seemed to her to be second-class citizens: estate agents, for example, would not take her seriously. Refusing to give up hope, she went to the university to learn French. There she met a woman who invited her to teach English, for which there was a big demand. 'Don't be silly,' said Parwin. 'I have never taught.' We'll help you, they said, and so they did, and that was how the Alsatian world opened up to her.

Through teaching she made friends. But after the birth of her third child, she decided to give up. Her pupils would not let her: they would come to her home for private lessons. So she created what has become a sort of school of English at home, open during the hours when her children are at school. Her home has become a meeting place for people of all professions, lawyers, doctors, women who say they need English for their cocktail parties, men who say they want to learn English in a month and are willing to pay any price. A tax consultant, after consulting a speech therapist who told him that he 'had no ear for the English language', came to her for lessons, and has mastered the language 'in 210 hours': tax consultants have to be precise. Parwin chooses whom she will teach, because they come to her home; it is like organising little parties for four or six or eight. They read texts to suit their interests.

The basic experience round which everything turns is Parwin's affection for her family, and teaching is fitted into the hours when the family does not need her. A lot of preparation is involved, however; the demand has grown, so people now say to her, 'You look tired', or 'Is your marriage not working?' Her husband has in fact made it all possible: if he was like other men she knows in the city, wanting his slippers and supper, she could not have done it. 'If I want to go out in the evening, I say to Paul, "Please feed the children", and he says, "OK, have a nice time." A French husband would say, "What, without me?" He would want his meal. You couldn't tell him to have sandwiches.' (French people do not boast about their sandwiches but if there was a world cup in sandwiches they and the Americans would be in constant battle for the championship.)

Now the problem is whether Parwin will be destroyed by her work, even though she is not ambitious. Her only ambition is for her family's success. 'I don't live off my work. I could not give 100 per cent of myself to it. I want to be able to forget things, to see my family; I'm not willing to sacrifice too much.' But the trap of technology is threatening to close upon her. Perhaps, she thinks, she should invest in visual aids, slide projectors, copying machines, expand, make her teaching into a proper business, get help, become a consultancy. Will she have more or less freedom then? She thinks about expansion even though the only way she would like to change herself is to be able to worry less.

At present she worries about being able to talk intelligently with her husband and his colleagues: that is more difficult than answering her children's scientific questions, for she can read their magazines before them to do that. She is forcing herself to read Martin Amis's *Money*: 'I can't get into it. But I'll read it. I always do what I've determined to do. And I'll read it also because I like to have an argument.' Before that she had read *The Bonfire of the Vanities*: once she found what the plot was, it gave her no problems – and no surprises either, which was a disappointment: 'I like to get away from real life. The work is hard, so the evening must not involve too much effort.'

Parwin has created a world to suit herself: 'I always have,' she says.

One great difference between the present century and all previous centuries is that there are now many more young women whose preoccupation is not to find a husband but to find themselves. Corinne, with her long, glossy, loose hair, is not an advertisement for a luxury shampoo, but a mind with two faces. Sometimes she is thoughtful and her face is an elegant veil, so that one can only guess that furious mental acrobatics are going on behind it. Sometimes she smiles, not with Mona Lisa's half smile, but with the equivalent of a curtsey, which says that your thoughts and her thoughts are ready for a dance together. And naturally, since she is so intent on knowing what she thinks, she has made it her profession to discover what others think.

It is what children think that interests her most. She began her career as a 'psychomotrician', helping infants who would not speak or would not walk. 'To walk, you need to want to go somewhere,' she says, and the phrase does not apply only to babies. Gradually, she has discovered why children mean so much to her: when she was one herself, she felt that she had to listen to others, but nobody listened to her. The adult world kept on saying to her, you must not do this and you must not do that; everything was dangerous. Then, much later, she realised that she had been trained to model herself on adults who, now that she was an adult, did not seem to be worthy models, who were too selfish to be admirable. Adults used to frighten her with their power and their knowledge, until it became clear that she was idealising them. Around the age of twenty or twenty-one, it dawned on her that the unassailable truth adults had preached to her did not exist, was like an empty box, that the world was not as simple as they had made it out to be. She regrets that she was never allowed to live her childhood fully. She is searching for the pieces of it that got broken. And in her work, she tries to enable others to do the same. 'I want children to be listened to.'

But what they have to say is not obvious. 'A child is an enigma.' The world, in her view, is full of enigmas. 'One never knows much about other people.' She is an enigma to herself, also: 'I feel there is something that I cannot understand when I look into myself, and no one can tell me. Conformity is what frightens me most of all: to be captured in a mould, to be passive, dependent – as though we could all be the same, which I do not believe. My aim is to discover something in myself, and if I conform, I shall be failing to do this. So being an enigma energises me. I believe I am building foundations for myself. I am confident that I am on the way to creating something new. I used to be afraid of what I was, of disappearing under the disapproval of others. It is important to create a self others cannot touch.'

Working with infants who are afraid of failing, or who have been damaged by pressure, she does not attempt to offer a miracle cure, or to prove that she can do what parents have not done. 'I am not attacking parents; without parents a child would be an animal, and it needs their protection.' She does not claim to know what is making the child suffer, and only says to it, 'Let us do things together, to discover what is difficult, we can always find another path.' The child will find the path for itself. There is no need for it to say in words what is unbearable; it can do so while playing; she will play with it; if it plays at being mother, for example, it will show all that needs to be revealed about its mother. What matters is to try to do things differently: 'Let us try together. I succeed when a child is able to be alone and I can leave.' A child cannot live independently, but it is full of potentialities, and can teach adults more than they realise.

The fear of failing preoccupies Corinne in her personal life, too: she is feeling her way to an attitude that will drain failure of its poison. She had a relationship with a colleague. He tried to create a sense of rivalry between them, to see who could get first to each objective. She found that unbearable. He protested that she was refusing to do things with him. She admired some things he did, but 'they were foreign to me, they were not me, and I did not want to go by the same road. I did not recognise myself in him.' One day he accepted an invitation for them both to speak at a conference, without asking her. He said it was important that she should have the same experiences. 'That was the break between us. I haven't got a voice that carries far; I like writing, but not publishing. He enjoyed speaking in public and was good at it. I am different. I have different potentialities. I wish to avoid rivalry. If I lose, I can still go on working, knowing why I have lost. A winner always has the losers against him, he will have to use his energy not to improve himself but against others, to keep his place.'

Not everyone, she says, can or wants to change, because it means abandoning the familiar. Each individual must be left to decide whether they want to change: those who do are people who realise that they are not perfect and have humility, who know they cannot change alone, and need help. Her experience is that highly educated people are the most difficult to help. Working with illiterates has been most rewarding: it gives a sense of affection exchanged.

'I am always afraid of repeating myself, of thinking I am changing when I am not. It is in periods of crisis that I do change, when I lose my confidence, and so my illusions.' The first change in her life occurred when she started working, the second when she separated from her friend, losing faith in her ability both to love and to do her job. Her cure was to train for a year in a new career, as a psychotherapist. That brought the confidence back, but did not solve everything. 'I feel I have something to achieve, but do not know what.' Every modern person sings that song. 'We always have a choice,' she insists, but there are so many doors between which to choose.

In former times, people were able to acquire a sense of community through fighting wars. Corinne's parents' generation sustained itself with the myths of 1968. Her own generation, she says, has no obvious goal. However, what she likes about schoolchildren today is that they are trying to find something in common, but on the basis of accepting that they are all different. Accepting differences, being accepted as different, is where she starts. So far she has been able to apply that more easily to children than to adults. Adults have still to learn to give more weight to the child that is inside them.

Every week, *Glamour* magazine, in its French version, amid hundreds of photographs of models and clothes and fashionable things to do, has one article that seems out of place, written by a philosopher, a historian or an anthropologist, trying to explain what life is really about. Anne Porot finds the great minds and tells them what to explain, with a mixture of thoughtfulness and humour. How does she know what her readers are interested in? Is she there to give them what they want to hear, or does she hope to put ideas into their heads? What ideas has she which she would like to pass on? Those are not the questions to ask. Anne Porot has never bought a woman's magazine in her life. 'I have a poor opinion of people who buy women's magazines.' *Glamour* has never tried to use polls to discover its readers' tastes. It is a work of art, and it is up to the readers to make sense of it. She lives with Jean-Pierre Mougin, editor of the comic magazine *À Suivre*, whose friend Martin Veyron, an equally famous strip cartoonist, is the husband of *Glamour*'s editor, Anne Chabrol. The connection with strip cartoons is not fortuitous. The Parisian edition of *Glamour* is in some ways an extension of France's peculiar comic magazines, which became enormously popular because they combined relaxed intelligence and wit with a sense of initiation into a detached view of the world. Anne Chabrol, who used to be assistant editor of *Elle*, has metamorphosed *Glamour* into a series of studies of the art of not taking things too seriously. The fashion photographs are the equivalent of cartoons. The people who write for *Glamour* never take anything for granted.

Anne Chabrol's own regular feature, *Un homme mis à nu* (A

Man Undressed), interviews a famous personality who, though photographed as naked as he will allow, is not expected to offer a total revelation of his soul. It is a verbal game she plays with him, with irony and puns, witty but deliberately superficial, a game recognisable as a form of Parisianism, the art of transforming relationships into epigrams. The magazine's slang is so up to the minute that many of its own staff cannot understand it; the effect is to create a feeling that there are wonderful private worlds, private cliques, private languages to penetrate, which distinguish those who know from the mob.

Anne Porot is hard on her readers, because she has a lot in common with them, because she is very hard on herself too. Her problem is that she suffers from being educated, intelligent, observant, self-conscious, while at the same time having no real confidence in herself. If she is to believe Freud, her parents are to blame. But she does not believe Freud, not any more; her lack of confidence is, she says, nobody's fault, to be accepted as one might accept having red hair, while trying to limit the disasters it causes, and consoling oneself with the belief that confident people do not seem to be better off than oneself. 'I live with fear most of the time. I kick myself twenty-four hours a day. Till the age of thirty, life was a permanent challenge, I was in painful conflict with myself. Now things are improving a little.'

She has tried several solutions. Coming from a highly educated family which never oppressed her, she had no difficulty getting into a university, but at eighteen she had no idea what she wanted from life, even less than she does now. 'I was in such mental disarray that feminism came at the right moment for me.' Deliberately playing down her good looks she fought for the cause, and the positive result, she says, is that women now have privileges which they have acquired not by birth or divine right, but by their own efforts, through merit, which is a satisfaction men do not have; so women can enjoy being alive with more intensity, with more sense of adventuring into new experiences. 'We have turned what used to be the mark of our oppression, our interest in cooking or flowers for example, into a delicious pleasure.' She is sorry for men who lack such sensitivities.

On the other hand, feminism has, for her, confirmed that there are two species of humans, male and female. 'All the better, say I. I do not mean that no communication is possible, but there is a lack of understanding between the two. However strong the bonds of love, each remains a stranger to the other. There are things we can say only to other women. There is nothing wrong with that; it is enriching, exciting, that there should be a difference. . . . When I have doubts, I calm myself by saying, "If I love him, then he is lovable." ' However, this means that it is not easy for men to help women to become confident, for women now have to solve their own problems; and she cannot think what women should do next. She is distressed that she can no longer formulate what she is trying to do, where she is going, even while arguing that women have 'basic attitudes, but never reflect on them, being content with common assumptions, enriched with anecdotes and little signs, which they recount to one another'. When she considers what the solution for today's women's problems might be, she replies, as though by automatic reflex, 'We must fight.' But then she quickly adds, 'I would fight only if I had the energy, but I don't really want to, because we would not succeed.'

Anne Porot then sought to solve her lack of confidence by trying to win the respect of people she respected. 'What I suffered from in my work was lack of recognition as a professional.' A short cut suggested itself, a job which automatically gives prestige and a sense of having arrived: as an editor of art books she became part of the whirl of fashionable people exchanging mutual admiration. Now she thinks she made a mistake: it was a superficial solution. The snag was that one was constantly watched by others, constantly judged, one had to be witty at all times, avoid saying anything stupid: it was a struggle against a new form of oppression, the lack of time to do all that one was supposed to do. It suits only people who cannot live without constant movement, and who, if they had their telephone cut off, would feel they had been put into a horror movie.

'But I am different. I really need solitude.' She has now moved to a rural cottage an hour and a half away from her office, and comes in only a few days a week. Surrounded by flowers and birds, she can work ten hours a day without getting tired, or at least not in the same way: there is none of that permanent fret created by having people around one. She can gaze at her garden and try to be nice to her family. All the social stimulation she needs she gets in those few Parisian days, and then it is back to the cocoon and the trees.

These pastoral pleasures trouble her a little, though: they seem selfish. Sometimes she thinks that all you can do when you see that things are not as they should be is to close your eyes, but that is seldom totally possible. The idea of rethinking everything, doing a spring clean of prevailing opinions, requires, she says, the kind of broad intelligence she does not have. 'If the chance to do this came up, I would like it, but I would not provoke it. Asking questions, putting oneself in the dock, is tiring. So too is coping with the unexpected. With people one knows, one is still in the dock, but less frantically so.' When she was twenty, she could live in a world which was unpredictable, but now getting to know new people is exhausting. Work involves this, readjusting every day. What a relief to come home to people with whom one has things in common. 'My work provides me with enough novelty to assuage my taste for it. At home I can return to the deeper side of life, away from the struggle for perfection, back to easy friendships, to spending a whole afternoon preparing a meal in order to show affection.' Creativity, she claims, is beyond her ambitions; it would be enough if she avoided annoying those around her, and if she got on well with her man.

So at home she is doing the opposite of what she does at work. In Paris she cracks Parisian fireworks, and she rushes back to the country to avoid the fallout of ashes. Modern living is about living in two worlds.

That is true of the editor of *Glamour* also. The interest in culture, self-improvement and achievement, all expressions of a forceful personality, is balanced by uncertainty. Anne Chabrol is stern with the famous people she interviews for her magazine, who are never allowed to see in advance what it decides to say about them. The only exception to this rule has been her clairvoyante, for whom she has an almost religious respect. The clairvoyante used to be a professor of philosophy, but times change.

Humans have so far distilled six lessons from their attempts to find the best way of surviving with the minimum of pain. They seem to have concluded that there are six ways of travelling through life, six forms of transport. What these six women have in common is that they have each opted predominantly for one of these ways, as though they had decided that, for them, it was best to travel by bus, or train, or plane, and to stick to their choice. The most popular moral philosophies of the world, which give advice on how to live, are of one of six kinds, but since each believes it alone has the right answer, there has never been an equivalent of a tourist office to give visitors to life on earth a full selection of these possibilities. It may seem that there are thousands of alternatives, that history is a giant garbage dump of different sorts of advice that have been tried and discarded, but in fact most advice has been of one of these six kinds.

The first way is to obey, to defer to the wisdom of others, to accept life as it is. In the past, probably the majority of humans travelled by this method, often because they were forced to, but no less because it promises peace of mind and the reassurance of being in harmony with one's neighbours. To convince those who doubted it, experiments have been done on rats showing that those avoiding confrontations are healthier and less stressed than dominating ones, which are rigid with anxiety, as though worried that the exploited might cease to obey them. Humans also have tried to limit the number of decisions they make for themselves, in the hope of sleeping more soundly. There is no census showing how many still opt for this strategy, beyond the statistic I have quoted (confirmed over a three-year period) that one-third of Britons say they would rather be told what to do than take responsibility. The reason why obedience has survived so tenaciously is that it is impossible to make up one's mind for oneself on every issue, so most people have always remained conformists in more or less large segments of their behaviour.

It has never been easy to obey, and it is becoming harder all the time, as the range of choices increases. Many have believed that religion is simply a matter of following rules, but that is only a beginning: the spirit in which they are followed counts for more, and the most deeply religious have been those who have been most conscious of the difficulties; there is no limit to how perfectionist they can attempt to be. It is not surprising that every religion has quarrelled within itself about details, which sometimes seem trivial or academic, but in religion every detail counts. Islam is the religion of obedience: the word means 'surrender' to the will of God, but it also means reconciliation, and that requires constant effort. It is not just wrong-headedness that explains why Christianity is now split between over 300 independent churches, each obeying differently.

Ultimately, the particular religion or ideology individuals profess says less about them than how they practise it. Those who accept a faith and cease to question it have a lot in common, whatever that faith might be. On the other hand, religion can also mean disobedience, a rejection of the world, a search for something better, perpetual self-questioning. When Dominique Lepèze, the ballet dancer, turned away from her Western heritage to Eastern philosophy, she was both rebelling and hoping that obedience would give her its opposite, freedom, but she still has not found freedom from the uncertainties which plague her.

The religion of obedience to the rules of fashion or etiquette, searching for other people's approval, doing as others do, has as many dupes as followers. It has become increasingly difficult to discover exactly what others do, what lies are concealed behind the façade of conformity, or to know what to conform to. People have wanted to believe that they could settle their problems by making one big decision, but that has not eliminated the vast number of little decisions which still have to be taken every day. When they get on the bus of obedience, it has always been a journey with many stops, and many temptations to get off.

The second method of travelling is as a negotiator, bargaining to get the best possible deal out of life. Pagans, ancient and modern, practise it in preference to all other methods. People alive today still behave like the Romans, for whom the world was full of powers who might harm or might help them, but whose favours could be bought. The skill is to know the lowest price that needs to be paid, the minimum sacrifices that need to be made, to obtain one's desires or, in modern language, to be successful. Bargaining could be enjoyable, a game, as some modern competitors still find it. The Romans followed their sacrifices to the gods with gastronomic feasts, and made friends and influenced people in fraternities which honoured individual gods; Bacchus was almost a friend. They had no inquisition into private beliefs; their religion consisted simply in the performance of rituals in exchange for benefits. Their originality was to insist that humans could win success as a result of their negotiating skills (with a little help from the gods), whereas other religions and philosophies asked awkward questions about morals and justice, or else were convinced that everything is decided by uncontrollable forces (and many still hold that view).

If you choose to be a negotiator, you eliminate worry about whether you deserve to be successful. Personal success became the aim of democracies because everybody has an equal right to it; and it can be had in this life, no need to wait for the next. The pagan system for gratifying personal desires is ancient Rome's most influential legacy to Europe and America. Those who govern no longer claim to be divinities, but they still grant favours in the same way as pagan gods used to, in return for sacrifices, relieving fears, promising security, breaking their promises. The snag is that it is hard to know what one's desires are, and to feel satisfied when they are fulfilled. Security is difficult to buy.

The trouble about bargaining, once one abandons the belief that there are gods in the sky willing to do deals, is that when one loses in a particular competitive negotiation, one's chances of winning the next negotiation are frequently diminished; one despairs of winning; then one no longer knows what to offer others in exchange for what one wants, and too many people never get a chance to put in a bid. If one chooses to drive one's own car, the likelihood that someone will eventually crash into it is not negligible. Catherine, the expert on business enterprise, has not found all aspects of the negotiator's life to her taste, and has had to seek consolations in other worlds, where there is no need to win.

The third option is to cultivate one's garden, to shut leaders, rivals and prying neighbours out of one's world and to concentrate on private life. In the beginning, there was no such thing as private life, no refuge from the public gaze and its ceaseless criticism. Then the middle classes began cultivating secrets. The more you kept to yourself, they thought, the more likely you were to avoid envy of your wealth, or of your tastes, and to enjoy esteem, provided you presented a carefully polished public image. A room of your own – a great rarity before this century – became a declaration of independence. Couples put off their public wedding until they privately decided whom they loved. In private, you could not only think your own thoughts, but make mistakes without being scolded or despised.

Those who cultivate their garden leave the big world to stew in its own ever more complicated sauces: they treat democracy as the right to be left in peace, in return for paying their taxes, like protection money. Political debates no longer have meaning and politicians become, for them, puppets on a stage, inebriated by fantasies of power. Even in the communist world, where adultery, drinking habits and bickering between neighbours used to be discussed publicly in Komsomol meetings, privacy gradually made headway against government pressure, against the inquisitiveness of colleagues, against bullying in the family. In Moscow and St Petersburg, only one-tenth of the inhabitants now see their neighbours regularly. Privacy means seeing only people whom one chooses to see. The rest do not exist, except as ghosts or gods on television, the great protector of privacy.

Cultivating one's garden is how those to whom the world has been cruel, like Victorine and Antoine, can take a first step to freedom. But the exploration of privacy, particularly by women, has revealed some problems. When they could afford to stop working and withdraw into domestic seclusion, freeing themselves for what they really wanted to do, they sometimes found that privacy could be a prison too. Cultivating one's garden simply for one's own benefit is like growing plants and not knowing what to do with them when they are ripe. Victorine and Antoine have obtained much satisfaction from the discovery of a tiny bit of the world which they can, however temporarily, call their own, but it is not for themselves that they live, but for their daughter.

One may imagine that getting on a boat which takes one away from it all means that one can opt to drift and forget about all purpose. But that is a purpose too, and there have been surprisingly few people who have not tried to steer or drop anchor from time to time. One may think that one can devote one's life to minding one's own business, but to do this well ultimately means becoming a saint or a sage, whose ambition includes being in some way useful to others. One can never be interesting enough to be the sole object of one's own attention.

The fourth way is to search for knowledge. The idea that it is something anyone can acquire is recent: for most of history knowledge has been rare and secret, and this esoteric heritage, with its dream of mastery and mystery, survives in the jargon with which every profession protects itself. Knowledge is still a serpent eating its own tail.

India, between about the fifth and eleventh centuries, comprised almost half of humanity and accumulated enough knowledge to enjoy the best food and clothing in the world: all who sought a better standard of living envied it and tried to obtain a share of its cotton, rice and sugar. India produced perhaps the most important scientist of all time, the anonymous mathematician who invented the system of counting with nine digits and a zero. And yet Hinduism taught that the only knowledge really worth having was that which eliminated desire and showed that an individual was a mass of illusions: knowledge, it insisted, did not abolish suffering. Likewise China, the first technological civilisation, though it amassed an army of a million men under the Sung Dynasty (960–1279) and developed the world's largest iron industry, capable of producing 16 million iron arrowheads a year, found that many of its ablest citizens acquired knowledge only to pass examinations and to memorise government regulations. Learning has repeatedly decayed into repetition, duplication, numbness of the mind. And the Arabs, whose science made possible the discovery of America, who were the first to appreciate that knowledge is an essentially international pursuit, who established the world's first academy of translation in the ninth century under the much-travelled doctor Hounain, and who made the Baghdad of that time one of the world's centres of intelligent discussion, nevertheless became exhausted and irritated by the conflicts of knowledge, and for many centuries put a lid on curiosity.

It may seem that the West took so long to discover the joys of knowledge because, after the intellectual fireworks of the Greeks, Christianity put charity above all virtues: though a person might have the knowledge to move mountains, said St Paul, without love 'I am nothing'. Luther called reason 'the devil's whore'. But others have adopted the same attitude, and even the Chinese who worshipped scholarship had Taoists who said that the acquisition of knowledge leads to the loss of happiness.

It is possible to have a most enjoyable journey in search of knowledge, but it is probable that one's train will be shunted into a siding, refusing to move further, forgetting its destination. Seeking knowledge for its own sake is another way of avoiding having to decide what one wants it for. Parwin Mahoney has no difficulty in finding pupils who want to learn languages, and what they do with that knowledge is their own business, but she herself, like all professional merchants of knowledge, knows that knowledge is not enough.

The fifth way is to talk, to pour out one's opinions, to reveal oneself to others, to get rid of one's gloom by bringing out all one's secrets, memories, fantasies, conscious and unconscious, advancing by smashing hypocrisy and decorum. It is like travelling by bicycle, in full view, waving to everybody one meets. The faith in talk had first to overcome the belief that to use words is dangerous, that words, as the Sumerians said, were part of the breath of the gods which 'caused the earth to tremble'.

In Antigua, talk is sometimes treated as a form of music, and people are said to 'make noise' (in Shakespeare's day noise could mean a quarrel, or a band of musicians, or an agreeable sound with each instrument making its particular statement) when they boast, curse and argue, each of which has a separate tune. There is no need to wait for a listener, because talk comes from within the speaker, so several can speak at the same time, and when another joins the group, he will begin talking when he is ready, but not necessarily on the same subject, and eyes may or may not turn to him; but if he is not heard, he will repeat himself, again and again, often with the same remark, until eventually he is heard or he gives up. People boast just because they feel good, and it makes them feel better, as singing does, as the boxer Muhammad Ali used to, as Beowulf and the Vikings did, to vent their feelings.

The present century, in proclaiming the advent of a new age of communication and information, and inventing recording machines to give talk immortality, forgot to deal with the great problem of talk, which is how to find someone to listen. For more and more people, merely to talk, mainly about oneself, as birds sing from tree-tops, is not enough. Humanity's pride in being able to communicate better than any other creature is belied by most talk being greeted with silence or incomprehension. The frustrations of sex are nothing compared to the frigidity of listeners. Corinne is far from being alone in being dominated by the search for someone who will both listen and understand.

These five methods of transport through life retain their attractions, despite their disappointments. There remains a sixth which has been tried much less, called 'being creative', which is like travel by rocket. Originally only God was a creator, the Creator. Only in the 1870s did the word 'creative', applied to ordinary mortals, enter French slang, used by artists when the crowds in the salons could not understand them. Only when originality came to be valued – which it is only in some circles – did it become possible to conceive of individuals as being creative. The geniuses of the past who today excite admiration because they proved everyone else wrong invariably suffered in their own lifetime. But the ideal of creativity is spreading fast.

In a French enquiry among adolescents of both sexes about the kind of work they would ideally like, almost a half wanted jobs which would allow their artistic, creative side to flourish; money, leisure and security were not everything. In another enquiry, among French women aged between eighteen and twenty-four, 32 per cent wanted to be journalists, 30 per cent artists, 29 per cent barristers, 26 per cent heads of firms, 25 per cent heads of an advertising agency, 25 per cent doctors, 19 per cent scientific researchers, 18 per cent housewives, 13 per cent cinema actresses, 12 per cent executives in banks or finance, 8 per cent engineers, 7 per cent television stars, and 5 per cent politicians. Today's adolescents do not pine for a lonely martyr's life. Nor do they say, as Mozart did, that death is 'the true goal of our existence . . . the key to our true happiness' (4 April 1767).

However, a few years ago the Palestinian publisher Naim Attallah thought he would like to understand Women with a capital W, by interviewing 300 successful ones in several countries, which resulted in his coming 'to think that perhaps women were more interesting than men'. He had not expected it; but each woman seemed to have a different opinion. However, they agreed on one point: they did not believe that women were creative, or at least they said women were rarely geniuses, because they dissipated their energies, were too sensual, were satisfied with the world as it was, or were not sufficiently brave, or ruthless, or as unbalanced as men; creativity was the male's substitute for childbearing, which was enough to absorb all women's creative urges. One of the women Attallah questioned, an Egyptian educated in Malaysia, Nigeria and England, said women were more creative than men socially, as hostesses, mothers, friends; however, no credit was taken for this, because it was not seen as a central feature of their lives, but as a complementary quality which enriched them and which it was unwise to stress, for women had to yield to the pressure to behave like women.

Doubtless most men, if they were honest, would answer with similar reticence about their own creativity, that it was stunted by pressures to conform, that they lacked the single-mindedness which was needed, that they were too absorbed by their pastimes, their families or their work. Too many people (like Anne Porot) worry that they cannot reach the level they wish. But the great problem about creativity remains, that there is no guarantee that it will produce results that anybody will value, or understand, or indeed that are not imitation in disguise. So it is necessary to go one stage further.

The world has so far treated its method of transport in the same way as it has sought a spouse: wait for the magic click, then settle down, hoping it lasts a lifetime. That was not silly when life was short, when, as in ancient India, married life lasted on average seven years, or when in France's prosperous Belle Époque fifteen years together was as much as the majority of couples could expect; but when a lifetime lasts nearly a century, it is time to rethink whether one wishes to spend it all travelling in the same bus, whether trying out all six methods is necessary for anyone wanting a full life, or whether even that is not enough.

For, when one has decided about one's method of transport, one still needs to know where to go.

Obedience and ethics

James Hastings, *Encyclopedia of Religion and Ethics*, 13 vols., Scribner, NY, 1908–26; John Carman and Mark Juergensmeyer, *A Bibliographic Guide to the Comparative Study of Ethics*, Cambridge UP, 1991; British statistics of obedience in Henley Centre, *Leisure*, July 1992 (cf. *Nouvel Observateur*, 22 Feb. 1990, p. 86); Barrington Moore, *Injustice: The Social Bases of Obedience and Revolt*, Macmillan, 1978; Rudolf L. Tokes, *Dissent in the USSR*, Johns Hopkins UP, 1975.

On the search for new and old beliefs

Thomas Robbins, *Cults, Converts and Charisma*, Sage, 1988; Joseph H. Fichter, *Autobiographies of Conversion*, Edwin Mellen, 1987; Eileen Barker, *New Religious Movements*, HMSO, 1989; Eryl Davies, *Truth under Attack: Cults and Contemporary Religion*, Evangelical Press, Durham, 1990 (suggests new cults have won 96 million converts); Marc Galanter, *Cults and New Religious Movements*, American Psychiatric Association, Washington DC, 1989; Marc Galanter, *Cults, Faith-Healing and Coercion*, Oxford UP, NY, 1989; Stuart A. Wright, *Leaving Cults: The Dynamics of Defection*, Society for the Scientific Study of Religions, Washington DC, 1987 (78 per cent of defectors move to another religion); Nathan O. Hatch, *The Democratisation of American Christianity 1790–1840*, Yale UP, 1989; Sharon Keely Heyob, *The Cult of Isis among Women in the Graeco-Roman World*, Leiden Brill, 1975; Helen Ralstron, *Christian Ashrams: A New Religious Movement in Contemporary India*, Edwin Mellen, NY, 1987; Frans Bakker, *Da Love-Ananda*, Free Daist Community, 1983; R. I. Moore, *The Birth of Popular Heresy*, Arnold, 1975; Michael Cole, *What is New Age?*, Hodder, 1990; Rupert Sheldrake, *The Presence of the Past*, Collins, 1988; Sorcerer's Apprentice Press, *The Occult Census*, 1989; Wilfried Floeck, *Esthétique de la divers-*

ité, Biblio, 1989; Daniel Offer, *The Teenage World: Adolescents' Self-Image in Ten Countries*, Plenum, NY, 1988; William M. Johnston, *The Austrian Mind*, University of California Press, 1972.

24

How humans become
hospitable to each other

'I can take hard knocks, even though I've had a difficult childhood.' For Francine, a Frenchwoman of African descent, not quite eighteen, that certainty is the foundation on which she is building her self-confidence, deliberately, as though building the foundations of a house. Her method is to reflect on her experiences coolly and at length, never taking it for granted that a misfortune must necessarily mark her for life.

Her parents divorced; Francine was placed in a hostel, but she has no grudges. Her mother, she says, had problems. Instead of resenting the hard times she has suffered, and 'quite a few things happened', she marvels that she has survived them, and that makes her feel that she will be able to cope with any more hostility she might meet in the future. She did not have such self-confidence before, but she has been thinking a lot about herself, now that she is alone, and she has reached her own conclusions.

'I have achieved it all on my own.' She has no friends. 'Out of every ten people I meet, only one is likely to interest me, just a little. Since reaching the age at which I could choose my friends, I have found only three people who interest me, but they are not friends.' One of them she might almost call a friend, but that word is too strong, she prefers to say mate. 'It is not that I am difficult,' she insists. It is just that it is not enough for her to find a person agreeable. 'It's natural that a person should be agreeable if he or she has done nothing bad. I start from the premise that everybody is nice.' But then she searches for those who are more than that, more than just amusing. 'Amusement is only part of my programme.'

So what is her programme? 'I enjoy discussion.' Her first criterion for a friend, if she ever found one, would be that he or she had a taste for reflection, a fondness for talk about a great variety of subjects. Then there must be 'a certain complicity', which means neither resembling each other, nor being complete opposites, but something in between, so that they can 'click'. Their values must be similar, 'because it is values which make an individual, and I think my values are broad enough for anyone'. Most of all, the friend must be demanding, push her to surpass herself, and that should be mutual. She will not accept a friend who has self-interested motives: the meeting must happen 'just like that'. She sets out these requirements fluently, as though she has pondered them carefully.

No boy has approached her – 'I see boys from quite a distance' – for which she has two explanations. Her own shortcomings, first of all: she is not a sufficiently warm person, and she thinks that may be inherited. All the girls at school kiss when they meet, but she cannot: 'It's completely beyond me.' At most she holds out her cheek. But that could change: 'It is a habit, warmth is something I lack, I know why, so I can remedy it.' More difficult to deal with is another weakness: she is slow, both mentally and physically. She eats too slowly, everyone finishes before her (but she is making progress in that). Worse, she thinks slowly: 'My school reports say so, people tell me so, I have to believe it.' Despite admitting this, she believes another defect is that she is not sufficiently modest. 'I was recently told I was pretentious.' It is no easy matter building up self-confidence if you are then attacked for it. 'I like myself as I am. I have a good deal of narcissism.' It used not to be like that, but Francine has been privately giving herself good marks for survival and has slowly been raising her opinion of herself, as though she were another person. And yet she wonders what other faults she is unaware of.

Her second explanation of her friendless state is that her school almost discourages the making of new friendships. Children who move from school to school and from class to class tend to stick to friends they already know. Her class is exceptionally cliquish, she believes; there are plenty of pupils who have never spoken to each other; the academically successful do not help the weaker ones; the timid keep their distance from those who follow the fashions; there is little real conversation, except when one is alone with someone else. 'We put up with each other without getting to know each other.' She herself sits at a double desk with the same girl who sat next to her in another class a year ago, simply because they had done so, not because they were close; and that coupling cuts her off from other couples.

All the same, Francine has been elected prefect of her class. The reason is not that she is popular, but that she protested loudest when the headmaster, whom she refers to ironically as 'the charming gentleman', decided to shut down some courses against the wishes of the pupils. 'I got noticed without realising it. I accepted, not because I wanted to be prefect, but for the pleasure of being elected', which she is somewhat ashamed of. Her class was not mistaken about her. When a teacher gave lessons which everyone found incomprehensible, she was brave enough to tell the teacher, who did not take the criticism in good part. Francine's marks immediately fell ('I know what marks I deserve'); the teacher kept turning to her asking whether she understood, 'as though I am stupider than the others, ignoring the fact that I had spoken on behalf of the whole class'. However, her classmates have never told her that they appreciate her efforts; they say very little to her. 'Some may not be satisfied because they expect me to do things for them which they can do themselves; and I put them in their place.' But she believes the teachers she meets in the class council think that she is a good delegate, 'because I defend those pupils who deserve to be defended'. Her conclusion is, 'There are things in which I know I am good.'

On whom does she model her behaviour? She has no models. No character in literature attracts her sufficiently: Madame Bovary, she says, going through the fictional characters she knows, is no model, Mérimée's Columba has qualities but she is no model either. Nor are politicians heroes to her. But surely there must be someone she admires? Yes, when pressed, she approves of Jack Lang. 'He appears to be serious, honest, to have a strong character, to be interested in hearing what everybody has to say, and, on top of that, he's a man of the left.' Has he any faults? Not that she can see. Are there any politicians that she does not admire? All the rest, they are feeble, and get lost in explanations. Lang, by contrast is dynamic, he seems to want to fight for worthwhile causes, to achieve beautiful deeds; his merit is that he is not wholly a politician.

Jack Lang is always denouncing discrimination of every kind. But Francine has worked out her own answer to racism, again by solitary reflection. 'I have my ideas on the subject.' Le Pen exasperates her, but she considers that those who support him are not so much racist as stupid. They do not reflect enough; if they cannot find work, they assume it must be because there are too many immigrants. Or they are people who, because they have had an unpleasant experience in the Métro, draw false conclusions. Or they are rich people who are concerned only with their own physical comfort. If only they met good people, they would appreciate them. Francine looks on racism as part of daily living; it has to be coped with; as a child she had some fights, mainly with boys, because boys are more racist than girls; but she feels she can cope.

Though she has done so much reflecting, Francine often answers questions by saying, 'I have never asked myself that.' Apart from a vague nostalgia for a simpler way of life, for more contact with nature (but she likes towns nonetheless) and for a slower pace – she is willing to give up washing machines and to make her own clothes – the Third World appeals to her, but 'without its problems'. She has no recipe for eliminating those problems. She wants to be a psychotherapist, not in order to know herself better, she feels she knows herself enough, nor to understand her mother, the same goes for her, but just for the pleasure of studying, and later to help others. However, her goal is very specific. She wants to be happy. That does not mean being happy as other people are. What they call happiness she calls only well-being. The experience she is interested in is closer to ecstasy, 'a peak which comes from time to time, which lasts briefly, which one cannot induce'. But it is possible to 'provoke it', and she hopes to achieve this by having a family.

A family, for her, means a husband and children and a nice house, and that everyone should blossom. But are there not many families, including her own, which are unhappy? Yes, but again she believes intelligence can solve that problem. 'There must be families, though I've never come across any, where children speak about their problems. Parents should be able to deal with their children's problems.' In her own house, there was never any discussion, her mother was always right. And she has never talked about families with her mate, who is not interested in the subject, only in social success. But Francine insists that she is not putting all her hopes in a family. She says people do not realise that what they want is happiness. That is what she wants. The family is only a means to it. If it does not work, she will seek happiness by another path.

She is ready for failure, too. Philosophy is one of the subjects she has most enjoyed at school, and it made her think about things she had not thought about before. The most important lesson she has drawn from it is that there is no need to be afraid of death. If ever she had to cope with death in her family, she would be able to avoid taking it tragically. That is the measure of Francine's march to independence. She can conceive of being even more completely alone in the world.

If the way to cope with life is to think harder about it, does one need to be a genius to do it well? From an early age Olga had good reason to believe she might be a genius. By the time she was twelve she was touring the USSR as a chess prodigy. It looked as though she might be a future world master. But when she met other prodigies, she thought they were cleverer than herself and decided that she would never reach the very top. That is the process of discouragement which hundreds of millions of children have gone through, whether geniuses or not, but even genuine, recognised geniuses are frustrated by not being clever enough. Indeed, the cleverer one is, the more goals there are just outside one's grasp.

Deciding also that being a chess champion was not a 'real job', Olga studied mathematics instead, and became a near-genius at that, eventually obtaining one of the best jobs in the country, in the Academy of Sciences. Now, she says, 'the most important thing in my life is my work, which comes before my family and my children': she says it openly in their presence, while conceding that her attitude 'is not admirable'. Being a mathematician is, for her, like being an artist, and not a routine job. It requires her to 'have intuitions', like visions, which then have to be proved; mere technical skill cannot make a good mathematician. The reward is the pleasure of thinking. 'I am drugged by it. And it is the best way of putting all else out of one's mind.' Geniuses do not find it easy to decide what is worth thinking about and what is not. They have the reputation of being totally absorbed in their speciality, but creative thinking is in fact the very opposite, a wandering in unknown territory, a search for connections where there seem to be none. What distinguishes geniuses is the conviction that they will one day find the clue, and emerge from the jungle; they are not frightened of being lost.

Before Glasnost, the most important thing in Olga's life was fighting the regime, as the editor of an underground journal. All her friends had been arrested. There was no one else to continue its publication. 'I had to do it.' Politics did not interest her, and she was sceptical about what dissidents could achieve; but she could not bear it that the journal should disappear, that anything good should disappear; and besides people have a right to information, to as much of it as possible: 'I am crazy about information.' She knew she would end up in prison, too, but was not frightened, waiting almost fatalistically for it to happen, thinking simply: 'Prison would be part of my career.' What could frighten her then? 'I am not afraid of anything at all.'

Once the KGB had marked her as a dissident, 'I had more freedom than the man in the street, who is terrified of all contact with it.' The KGB's ways become familiar to those it hunts: she learnt how to recognise what was safe and what was not; the enemy is frightening only when you know nothing about him. Just when she was about to be arrested, she became pregnant, deliberately, saving herself, knowing that the KGB, for all its ruthlessness, did not jail pregnant women. Of course, she felt physical fear sometimes, as when she saw her friends being arrested; but that would not make her hesitate, if they were arrested again, to become a dissident once more. 'But now life is better, I can travel abroad, I can make money.' This distinction between the fear of the body and the fear of the mind has been one of the bases of heroism through the ages.

The most important thing in Olga's life, at other times, was being 'madly in love'. This used to be a problem, because until the age of twenty-five she was sure that she could control herself completely. By 'self-education', begun at seventeen, she had tried to 'master myself at every moment'. Being shy, she had set out to transform herself from an introvert into an extrovert. When she met her first husband, her belief that she was in control was not challenged, because she chose him 'rationally': 'I think I tricked him into marrying me.' The marriage meant she would be able to stay in Moscow. 'Of course I told myself I was in love.' However, when she was twenty-five, she had a nervous breakdown, not because of thinking too hard about mathematics, nor as a result of her divorce, but because she could not choose between two men. The agonising convinced her she was 'not totally rational'. The next time she fell in love, with her present husband, it was 'true passion, not chosen, physical, love at first sight'. Looking back on it, she thinks he represents her attempt to get closer to what being human involves. 'It is not enough to be self-sufficient, nor to have a happy marriage. I need to be with other people.' Getting married, whether one is in love or not, is a useful experience; getting divorced is no bad thing; 'both cure your complexes'.

Friends are also very important to her. Again, her choices are not totally rational: she likes them as people, more than for their qualities. But what would disqualify someone from being her friend? If he was a traitor. Her own father was the first to lose her love for this reason. He had brought her up to believe that certain liberties could never be sacrificed, that it was, for example, dishonourable to be a member of the Communist Party. But then he joined the party: being a biochemist doing experimental research, and passionate about his work, there was no other way he could get the equipment he needed. She felt he had betrayed himself, and was never close to him again. That happened when she was eighteen. Was she not too hard on him? 'Yes, I am a hard person.'

Self-sufficiency is not enough, but she values it, and claims she possesses it, as well as self-discipline. 'I can decide what I don't want to think about, and put it out of my mind. I like to do one thing at a time.' She can get rid of 'unpleasant thoughts' by doing mathematics, or reading a novel. It was while living in a hostel that she acquired this skill: having five other people in the same bedroom taught her to blot out their conversation; sharing with one other person was much harder: the only escape was to learn to concentrate. Now 'when my family speaks to me, I don't hear what they say'. To have a room of her own is her dream, because though she tries to solve mathematical problems in her sleep, they often keep her awake and she likes to walk about at night, thinking. Individualism has flourished most in places where people can afford to have a room of their own, and the complexes which go with it.

At any rate, traditional household arrangements do not suit Olga. Her husband is her wife. While she works, he looks after the children. 'Our roles are reversed not only in this, but in my being a scientist, while he is an art historian.' Hiding behind his beard, he speaks in a quiet voice, moves softly, smiles gently. Always ill as a child, physically weak, unable to fight back, finding school a torment, the first conviction of his life was that 'nobody liked me'. Though he got excellent marks, he left school as soon as he could, to become an anonymous assistant in a library, putting books on shelves. Starting at eight in the morning, he was able to do all that was required of him by 10.30, and was then free to read, safe with his imagination. At the age of nineteen, he got married in order to escape his parents, who were always lecturing him on what he ought to be doing: all of them lived in a single room. 'I thought I was in love, but the important thing was to get married.'

As soon as he got his job, he had started drinking, and when he divorced at twenty-one, he dedicated himself to drink. 'I was not happy with what I knew about myself.' Added to which, an attack of gastritis took him to hospital, where doctors promised him that he would have cancer in fifteen years' time. He believed it, and waited for it. Waiting for death dominated everything. At first he thought he could stop drinking if he wanted to; then he realised he was an addict, and that he did not want to stop, because he thought his life was over. At the age of thirty, he decided that drinking was the best way to kill himself.

The great consolation of drinking was that it could be done in company. Loneliness haunted him; since the age of five, he had been afraid of being alone, and nothing had ever changed that cold feeling. He cannot explain it: 'It is irrational.' But there is some use in loneliness, it makes people come together. 'For me, the good thing is to be together.' When he married Olga, he was an alcoholic. Her love brought him off the bottle. 'She did not persuade me,' he says. Now he is sober, but he admits it with hesitation, there is always a danger that he will relapse, particularly when he is left alone. Olga has abstained from all social drinking so as to keep temptation away. However, when the Soviet government rationed vodka, giving everyone an entitlement to some, they could not resist taking their allowance, which was hoarded in the kitchen like an armoury of bombs.

Another refuge from loneliness is his art collection. 'It does not matter what you collect,' he insists. His particular interest is postcards, which are a special form of Russian art, since many great artists produced work specifically for postcards. He has become a great expert and has written a book. But it is only, he says, 'a way of existing', partly 'a retreat from reality', and most of all an opportunity to be with people who share his interests.

He earns his living buying and selling these works of art, which was a brave thing to do. Giving up regular employment was a crime in the communist era, making him officially a 'parasite'. But he does not like regular work. Looking after the house suits him and her. He has invented his own kind of courage.

When perestroika came, Olga also took courage into her hands and resigned her privileged state employment, joining with a dozen others to start a private business, selling statistics and public opinion polls. Her salary has doubled; but above all she is freer than ever to meditate on abstract mathematical problems and to apply her thinking to medical and political computer software. But she at once began to think about going abroad. 'I am a cosmopolitan, a citizen of the world, and have been since childhood. Patriotism is for fools.' One of her poems, about the Motherland, says: 'My country is not a mother, but the first love of my life, who wants to continue to be that for ever, a jealous love.' She is too much of 'an adventurer` to be content with the land of her birth; her interest is always for 'what I have never experienced before'. She loves the unknown, or more exactly the almost-unknown, visions that have been partly glimpsed.

Olga used to be a poet, until she gave herself to mathematics, which she sees as verse in numbers. Literature remains another of the most important things in her life. Proust may be dead, but he has friends everywhere, more numerous every year, and she is one of those who have been touched by his spell. Paris is holy to her, because it was the capital of the world for so many great Russian writers, who remain, irrespective of who is in the Kremlin, the true rulers of Russia, the guardians of its imagination. But she has read Proust only in translation; her French has fallen a long way behind her fluent English; language limits her. So though she is a cosmopolitan, she has as yet found no room in her heart for the civilisations of the East.

That reveals at last the only thing she fears, which is the passing of time. There is no time to do what needs to be done. It is not advancing years that are worrying, because her fear of time has been with her since she was twenty-five, when the complexity of life first overwhelmed her. It is obvious that the task of the genius today is to fold time in two, or three, to make it last longer.

Meanwhile Olga tries to look beautiful and to dress well, loving clothes also because 'it is nice to be admired'. But she does not need admiration from many people. In her work, too, it is enough for one or two people to understand her learned articles. An idea lives, even if it is ignored. 'When work is done, it is enough to have done it.'

Far from having a single purpose, being clever multiplies one's aims. She sums up her purpose in life as 'to live decently, to have a reason for self-respect. To succeed without being a hypocrite. There are certain things I will never do. For example, I will never lick anyone's boots.' The mark of geniuses is that they do not compromise, they have to think that other people are wrong, they have to believe in themselves. Olga says, 'I feel I am the equal of anybody in the world of science, even of Academicians.'

It is just as well that she does not believe she can have much influence on the world, because she has had very little on her teenage son. He is a fan of the gloomiest Soviet rock music, a devotee of the group Cinema, whose song he quotes with approval: 'If you have a packet of cigarettes in your pocket, it means things aren't too bad today, and if you have a ticket for a plane which leaves its shadow on the earth: we knew it was always like that, that fate loves best those who live by other people's laws, and who die young.' He hums another song: 'Money, money, money, the rest is of no importance.' His ambition, he says, is to buy fashionable sports shoes and a leather jacket. The people he is jealous of are the Georgian businessmen who can afford to wear such luxuries, thanks to their profits from selling fruit and vegetables, and who can afford to smoke and eat at McDonald's. His hobby is karate, which makes him feel strong, 'physically and morally able to defeat other people'. At school, however, he defeats nobody, getting consistently bad marks. He defends himself saying that is because he pays no attention, because what is taught is not interesting – except for history and information technology (but he does not get good marks in them either, says his mother). What would he do if he could go to America? 'Wash cars,' he replies.

This boy has just found his first girlfriend, with whom he has nothing in common. 'We don't have the same attitude to life, I'm an optimist,' she says, meaning that she has recently started going to church, having been baptised at the age of ten, introduced to religion by her grandmother. The world for her is divided into the honest and the dishonest – all those who go to church are honest, and she is convinced they will eventually be in the majority. The boy says he sometimes goes to church with her, 'to clean my morality', as though his soul was a car needing occasional washing. She is always afraid that she will get bad marks at school. He says he is not afraid, because he knows for sure that he will get bad marks. Her parents, she complains, do not understand her, because 'everybody is different', but she hopes that with age she will acquire new talents and 'perhaps I will see everything from their point of view'. Despite his studied coolness, he is worried that her answers might be thought better than his. He sees her home, and the telephone rings: he has got into a fight and is hurt. 'Perhaps I wasn't a sufficiently attentive mother,' says Olga.

With how many people do you feel you have a lot in common, or a little, or nothing at all? Francine is an immensely impressive person, but she has such a sharp mind that she can barely recognise anyone in the whole world with whom she can agree, because she directs her thinking to the differences between people more than to their similarities: she is imprisoned in her uniqueness. Olga can make the most ingenious connections between mathematical signs, but hardly any between the thought processes of the members of her own family, who remain mysteries to her, as though they each inhabit a different planet. Individuals can hardly call themselves fully free if encounters with others are so difficult and if they feel even more disconnected from their neighbours because they belong to religions, nations, classes, sexes which have all, throughout history, gone through phases of incomprehension toward those who are different.

Do people find it more or less easy to speak to strangers than they did in the past? The answer can be found in the history of hospitality. Today, in the rich countries, hospitality means, above all, entertaining friends or acquaintances in one's home; but once upon a time it meant opening one's house to total strangers, giving a meal to anyone who chose to come, allowing them to stay the night, indeed imploring them to stay, though one knew nothing about them. This kind of open hospitality has been admired and practised in virtually every civilisation that has existed, as though it fulfils a basic human need.

When in 1568 a European missionary arrived out of the blue at Nagasaki, he was given a Buddhist temple to stay in, and banquets three nights running: the hospitality continued until he abandoned the role of stranger and began interfering in local politics. The Lithuanian word for guest is clansman (*svetjas*), because by eating and sleeping in another's house, a guest became a member of his host's clan. In Albania, a host who gave hospitality to a stranger was obliged to take revenge on anyone who harmed him before he reached his next destination. In seventh-century Ireland, King

Guaire of Connaught, of whom it was said that 'he was so constantly giving away that his right hand grew longer than his left', understood hospitality to mean that when visited by a crowd of 150 poets 'and as many pupils, as many valets and as many women', he felt bound to erect a building specially for them and to give them all they asked, even though they tested his generosity with the most outrageous demands for the rarest foods: it was remembered as a mark of his virtue that he put up with them for a year and a day before hinting that they should go. China's Great Plan, in the third millennium BC, lists 'the entertainment of guests' as one of the eight objects of government. Ancient Indian texts required every person to perform five sacrifices every day: the worship of the World Spirit, of ancestors, of gods, of all things living and finally 'the worship of men, by showing them hospitality'. Faint, very faint, echoes of such traditions linger in the phrase 'Hello, stranger'.

The decline of hospitality was first noticed in sixteenth-century England, when bishops were accused of limiting it to friends and relatives. As soon as the rich appointed almoners to do their charitable work for them, they lost direct touch with their visitors; as soon as distress was dealt with impersonally by officials, hospitality was never the same again. By the eighteenth century, Smollett was writing that English hospitality was a sham. The French *Encyclopedia* explained the decline by saying that there were now too many people travelling and too many thinking in commercial terms. Free hospitality was superseded by the hospitality industry, and survived only in remote and poor regions: in twentieth-century Andalusia total strangers were still being invited to share a meal, even a restaurant meal; rural Greece and Bedouin Arabia continued to amaze the traveller. But there were few places left where all comers were entitled to take fruit from orchards, as they once could in colonial Virginia, where it was an honour to give, a pleasure to see a new face. The pedlar who sold unusual goods, the wanderer who told amazing stories, the stranger who brought interesting news, were no longer needed in the age of television and supermarkets.

A new phase in history begins when this ancient and simple hospitality is succeeded by a deeper hospitality, which alters the direction of human ambition. That happens when people become hospitable to strange ideas, to opinions they have never heard before, to traditions that seem totally alien to them, and when encounters with the unknown modify their view of themselves. When foreign travel becomes a necessity, and no longer an exception, when television news is about distant parts more than about one's own city, when one's emotions are roused by the misfortunes of total strangers, what goes on elsewhere becomes a crucial ingredient in the shaping of one's life. It becomes impossible to decide what to do unless one knows about everyone else's experience. This is a deeper hospitality because it is not just politeness, but involves admitting new ideas and emotions temporarily into one's mind. For this to happen, the mind has to work in inhabitual ways.

Though it has been discovered that the brain has ten billion cells, each capable of making 5,000 connections, many connections are never made, and messages, feelings, visions and thoughts never register, simply bump blindly into each other without any result. I went to Russia to watch what people could do with their brains when they were suddenly free to think and to say what they pleased, and I saw how political freedom is always only a first stage: the mind has to struggle to break away from its habits; political freedom does not automatically liberate it. Cognitive scientists have shown that in the connections the mind makes, in the categories it uses, in what it considers relevant and what it ignores, it tends, if left to its own devices, to follow well-established patterns. That is why history is so full of opportunities that have gone unnoticed, why so many thoughts and feelings have been sterile, like sperms and eggs that never meet. Not everyone immediately recognises a piece of flotsam as a possible bridge to a shore that seems unreachable; most people who have tried to start a new life have found themselves reliving the same life. So feeding more new information into the brain has only worsened the traffic jams in it: people do not hear much of what is said to them, they cannot be hospitable when they do not notice what is staring them in the face. Only when they deliberately change the way they think, the way they perceive, remember and imagine, are they able to stop being afraid of strange ideas as though they were monsters.

Most thinking has always happened automatically, unconsciously, but there are parts of the process which it has been possible to steer. Thus though the perceptions which nourish thought are normally classified into conventional categories – so that strangers are automatically put into the pigeon-hole of the dangerous or the ridiculous or the bizarre – occasionally they are not. Artists have picked out – from the millions of bits of information received from the world – what others have not noticed, like the shape of empty space; perfumers have sensed thirty ingredients in a single smell; what was most obviously noticeable about humans for my mother, who was a dentist, was the decay of their teeth and the quality of the repairs they had undergone. Perceptions have been modified by interrogating them with different questions, or by giving them another goal to aim for, or by discouraging them from reinforcing stereotypes, which they do because that costs them the least effort. Information has never arrived in convenient chunks, let alone with all its implications spelled out or its contents clearly labelled; so the lorries full of facts driving round our minds have seldom been properly unloaded; only if they break up their cargoes into little packets can they deliver them to a destination where attention will be paid to them. This is the first reason why there are so many traffic jams in the brain: the highways are blocked by convoys of miscellaneous containers all wanting to drop off their goods in the same few places. Picking up perceptions of the right size, in infinitely small detail or in broad panoramas, is an art, the basis of all art and of all achievement. This book has tried to show how great a difference to the conduct of daily life the ability to alter the focus of one's perceptions can make. To be hospitable to the nuances of life, it is no use treating the mind as an automatic camera; only by composing one's picture and playing with light and shadow can one hope to see something interesting.

The sense that has been made of perceptions has depended on the assumptions with which they are surrounded, and the most important of these come from memory, which has been the second major source of blockage in the mind. Memory has traditionally been lazy, preferring to remember the same things: some memories have established a supremacy like tyrants, and most information has been used to reinforce them, confirming old beliefs, instead of being dissected to reveal new facts, which are not easily recognisable. However, there have been people who have succeeded in turning their memory into a source of energy, rather than being controlled or constricted by it. They have done this either by drawing new implications from old memories, stimulating them with new questions, or else by expanding their memories, incorporating the experiences of others into their own experience. Many Russians I met could not stop talking about the memories of their own country, which had been traumatic; even though they were fascinated by everything outside it, they had difficulty in looking at their own memories through the eyes of an outsider. The mind has to be trained to do that. Throughout this book, I have tried to show how memory has been misused, underused, overused, and how many memories, belonging to others but capable of being borrowed, have been neglected. Believers in progress have generally assumed that tradition needs to be expunged from memory to create a better world, but memory is indelible, though it may be temporarily mislaid, which is why reformed institutions have so often ended up behaving like the ones they replaced. Handling memory is also an art that has to be learned. It is not enough simply to memorise, a ritual which too easily becomes an obsession.

The enhancement of memory by the imagination – the third factor – has sometimes helped people through the traffic jams of the brain, but sometimes it has done the opposite, making it impossible for them ever to get out. Imagination has long been looked upon as dangerous. The Bible condemns it as evil (Genesis 6:5), because it implies disobedience. Even those who have wanted to liberate humanity from tyranny have feared imagination as a threat to reason. The philosopher John Locke (1632–1704), for example, who was an enemy of all dogmatism, warned parents who discovered a 'fanciful vein' in their children to 'stifle and suppress it as much as may be'. Sensitive to human frailty though Locke was, consumptive, asthmatic, preaching toleration for all, he nevertheless believed there were things the imagination should not be allowed to do, like imagine that there was no God.

On the other hand, the Romantics, who put their faith in imagination, expected too much of it. Hating the prosaic ordinariness of existence, they saw imagination as a magic carpet which could whisk them to unknown destinations, making them godlike, enabling them to penetrate the mysteries of the universe. With its help, they did have remarkable visions of beauty and sometimes felt they had discovered themselves, but more frequently they were only made conscious of what they could not achieve. Imagination carried to the point of heroism has produced tragic figures, who have typically died young or gone mad, experiencing only a few moments of unutterable bliss.

Imagination has only been truly liberating when it has been constructive, when it has arranged fertile marriages between images and sensations, when it has not only dissolved the obstacles confronting it, but also recombined them so that they become useful, when it has spotted both what is unique and what is universal in them. But there are risks in seeing beyond superficial appearances, in guessing at what people or situations might have in common, in giving meaning to apparently meaningless events, or in investing emotion in unpredictable encounters. It is impossible to feel fully alive unless one takes risks, so those who have refused to take risks, assuming that one either possesses imagination or does not, and that there is nothing one can do about it, have hesitated to live fully.

The direction of thinking has been decided by intuitions, which are sometimes hypotheses and sometimes judgements reached so fast that the reasons·for them are not noticed. Women's intuition has been neither magic nor genius but the result of close attention to minute signs and an interest in unspoken emotions: it is as rational, and elusive, as medical diagnosis, using past experience in the face of uncertainty; but it is never easy to learn from experience, because two experiences are seldom exactly alike; an imaginative jump is needed to spot the similarities. That means being hospitable to facts which are usually ignored. Unfortunately, though humans ruminate, cogitate, brood, play with ideas, dream and make inspired guesses about the thoughts of other people all the time, there has been no *Kamasutra* of the mind to reveal the sensuous pleasures of thinking, to show how ideas can flirt with each other and learn to embrace.

'He thinks too much. Such men are dangerous,' said Shake-

speare, and too many people have issued warnings against thinking, as they have about sex. Freud was in that line when he claimed
that women were not interested in thinking because they were
forbidden to think about sex, which is what mattered most to
them, so there was a 'physiological imbecility of women' (3 May
1911, speech to the Psychoanalytical Society of Vienna); but in his
view men were not much better, because they engaged in thinking
to satisfy the erotic desire for possession. Now that sexist and
militaristic ideals have ceased to be credible, it is possible to escape
from the vicious circle of reason and emotion being alternately
fashionable. It has been said that for those who 'feel', life is a
tragedy and for those who 'think', it is a comedy. There is no need
to live only half a life. For those who both think and feel, life is an
adventure. Doing both is the way to be hospitable towards all that
is alive.

How, I shall now illustrate by some examples. In the preceding
chapters, I have tried to show how contemporary preoccupations
can be elucidated with the help of the memories of other civilisations. Now I turn from the particular to the general, and to the
barriers between civilisations as a whole. From a distance, each
looks like a fortified castle, forbiddingly alien, surrounded by a
moat. But each has many windows, too, and people looking out of
them with whom it is possible to communicate, however profound
the differences that separate these civilisations may be.

This is not the first time in history that humans have thought
that everything is becoming too complicated. In Ming China, for
example, they complained that 'the sky was falling' when they
became unhappy with their government, frustrated with their
bureaucracy, incensed by crime, made desperate by famine and
helpless before the apparent collapse of all values. Out of the
infinite richness of their reactions, I pick one, that of Lu K'un
(1536–1628), who has no place in most history books. A magis-
trate by profession, he slowly rose to be assistant censor in chief.
He drew the emperor's attention to what his censorship revealed,
that 'the people were harbouring rebellious thoughts' and were
indignant that taxes were used for wasteful luxuries. The emperor

paid no attention. So Lu K'un retired from government service and spent the rest of his life seeing what he could do as an individual. 'Working for fame,' he concluded, was not what interested him; he ceased to be impressed by his titles: 'I am just me,' he said. To solve all problems was not his ambition: he compared himself rather to a doctor, who must not write out his prescription in advance, for every individual is different. So whereas others who shared his dissatisfaction went round the villages preaching a return to traditional values, he tried to do something more practical, helping people to meet and work together on practical projects in their own locality.

His 'Song of Good People', written to be chanted by the illiterate, explains what he meant. The 'separation' between the upper and lower classes, and between individuals, seemed to him to be 'most disastrous for human feelings'. What people needed to learn was how to put themselves in the place of others, but without illusions, because every individual was different. 'Regarding others like the self, while realising that others are not necessarily like the self, is comprehension.' Education was not the whole answer, for 'educated children are rebellious'. Expect nothing from the government bureaucracy, who merely waste paper and vermilion ink. Do not try to end poverty simply through charity, because that perpetuates dependence. Recognise first that 'all good people are sick', that there is something wrong with everyone: it is dangerous to believe one is right and others wrong. The only cure is to 'share personal experiences'. But 'only men with bitter minds can share illnesses and be sympathetic'. Shared suffering could be the origin of a sense of community. The educated should not pride themselves on their polish, but 'share everyday difficulties with ordinary people': only the degree of insight they had into their own ailments, rather than their wealth or examination results, made them deserving of respect. Individual self-interest, he conceded, would always be the basis of action, but he thought it could lead to co-operation and community, if the preoccupation with hierarchy was replaced by the exchange of information. The main cause of social discord was not selfishness and greed, but the inability to put oneself in other people's shoes.

It was customary to promote officials after their death, with honorific posthumous titles: Lu K'un was named Minister of the Board of Punishments. The government could not put itself in his shoes; it did not understand that he was not interested in punishment. But he was part of a movement which tried to do things governments could not do. Many voluntary associations were established by people like himself at this time, 'The Society for Sharing Goodness', for example, or 'The Society for Spreading Humaneness', which improved daily life in small ways that slowly added up, building roads and schools, while also offering opportunities for enjoyment as clubs for drinking and conversation.

History is full of people who talked as though they were alive today, and yet it is assumed that the experience of the Celestial Empire was so exotic that it has no relevance to the West, which sees China simply as a developing country waiting to follow the Western path to prosperity. But of course China had its own Industrial Revolution, and experimented with mass production, while Europe was still in the Dark Ages. China went through a financial and communications revolution a thousand years ago, when it invented paper money and printing and a cheap system of water-canal transport, creating both a vast national market and an export industry so enormous that it was the world's principal source of luxury products. China probably profited from the discovery of the Americas more than any other nation, for half the silver mined there before 1800 ended up in its coffers, in payment for silk, ceramics and tea, which were the refrigerators, televisions and computers of the day. By then, Kwantung, which has recently caused the West surprise by becoming the fastest growing economy in the world, was already a precursor of the service-sector economy, living off craftsmanship and commerce, and importing its food. Agriculture was so efficient that wheat yields were 50 per cent higher than in France. By AD 1108 the Chinese already had treatises listing 1,749 basic medical drugs. The empire survived so long because, despite war and corruption, taxation was reduced over the centuries from 20 to 5 per cent of GNP. The army, instead of being a burdensome expense, farmed and fed itself. The civil service, recruited by examinations requiring candidates to be poets as well as scholars, used to be the envy of all bright and ambitious Europeans whose careers were impeded by the privileges of birth.

In the Taiping Rebellion of 1851–64 absolute equality of the sexes was the declared aim; women were given an equal share of the land and formed their own armies. Though the Chinese wreaked havoc on the environment like any other nation, especially in deforestation, their awareness of the claims of nature, their sense of being part of it, was particularly deep, for whereas in monotheistic countries nature was believed to exist to serve mankind, the Chinese attitude was that humans were part of nature. The Mahayana Buddhist attitude was that plants and trees could be Buddhas, and were all part of the same moral community. The hydrangeas, forsythias, rhododendrons, magnolias, wisterias and tea roses in Western gardens, all originating from China, should be a reminder of the many other arts of life which the Chinese pioneered.

China also experienced the inconveniences of prosperity: it was so successful, so comfortable, so full of cultured people, that a time came when there seemed to be no point in inventing anything new. Ideals which were once liberating became rigid. Prosperity came so easily that the serious inequalities which accompanied it were ignored by the rich, and the country then almost destroyed itself fighting over what justice meant, enabling foreign competitors to enter like vultures to pick at the spoils.

There certainly have been profound divergences in outlook between China and Europe. After the initial politeness, Christian missionaries were often greeted with bafflement. Confucians could not understand how it could be fair that a wicked, antisocial person could obtain forgiveness simply by repenting, nor why the pleasures to be had on this earth should be held in such suspicion; they insisted that humans did not need threats of hell to make them behave decently and that an action could only be moral if it was disinterested, without hope of reward. They could not understand the Christian distinction between fleeting appearances and reality, because for them reality was in constant flux, ever changing, a vision quite close to that of today's science. But Sun-Yat-Sen, China's first president, was a Christian.

The Chinese have never been prisoners of a single philosophy. Different sides of the personality were catered for by Confucianism, Buddhism, Taoism, and they knew how to mix these. For example, they put aside their disagreements in the Tao temple, which was not just an abode for the gods – many different gods – but also a meeting place where every sort of person could come for the pleasure of conversation, where musical, theatrical, charitable, chess, reading, boxing and medical clubs could meet, each placing its own patron god in the temple. The main attraction was the 'sharing of incense', obtaining a sense of community, or rather the sense of belonging to a few among many hundreds of different communities, which, loosely interlinked, formed a popular, unofficial cultural network, existing side by side with the state, but independent of it. Here humans appeared as complex as the universe, having many souls each, needing life to value both feminine and masculine qualities in order to be complete. Women enjoyed equal status inside these temples, even if they did not outside them; they could hold the rank of Celestial Master, a dignity which each Master shared with his or her spouse; initiation could be obtained only in couples; and there were as many female as male Masters. Sex meant more than a meeting of the genitals; they considered it possible to unite all the senses, 'the eyes, the nostrils, the breasts, the hands', beyond mechanical orgasm, aiming for refreshment and transformation rather than sexual exhaustion. Their ideal was not to be a god – they were sorry for gods, whom they thought of as wandering souls seeking rest – but a mountain, standing above the world's cruelties. But as happens with most religions, magic came to be used as a short cut to bliss, and charlatanism often became a substitute for wisdom.

China has of course, like every country, suffered from cruelty, violence, oppression, callousness and every vice that humans have perpetrated on each other throughout the globe. The way to be hospitable towards what happened in China, nonetheless, without forgetting this, is to approach it first through small crevices, through the intimate experiences of individuals, through details, through emotions, not through the massive laws and doctrines which are barriers to keep foreigners out. One would need many lifetimes to meet even a small fraction of the Chinese sages, scientists, wits and poets who have something interesting to say to the present day; it is frustrating not to be able to write endlessly about them, there is so much pleasure to be had from them. But this book is not a summary of history: it has deliberately limited itself to finding locks that look as though they will not open, and to showing how they can be opened.

One example must suffice to show how a Westerner can become hospitable to things Chinese. The biochemist Joseph Needham (born 1900), author of the many-volumed *Science and Civilisation in China* and one of the most important contributors to Western understanding of China, was the son of an Aberdeen anaesthetist, who in his spare time composed songs still sung today, but who quarrelled almost unceasingly with his wife. Needham said he grew up in the midst of a battlefield. Historically, marital disputes have had a positive side too: in Needham's case they encouraged his interest in reconciliation. He suffered all his life from 'anxiety neurosis symptoms', but these only stimulated his Protestant faith, to which he always remained attached, saying that he did so because he happened to have been born into it, but he interpreted it in his own way, and worked actively to improve its attitudes on sex, race and social justice. To his Anglicanism, he added a sympathy for many other religions and philosophies, maintaining that the essence of religion was not dogma but poetry and ethics, an awareness of what could not be proved scientifically. The teachings of Confucius, he said, though they had no place for a Creator, should present no difficulties to those brought up to admire the ancient Greeks, and he appreciated the master's instruction: 'Behave to every man as one receiving a great guest.' Taoism pleased him because of its emphasis on the victory of weakness over strength, its belief in spontaneity and naturalness and its mystical love of nature, and he stressed the leading role played by Taoists in the growth of natural science and technology. Buddhism worried him at first because of its denial of the possibility of redeeming the world, but later contact with Sinhalese Buddhists convinced him that its compassion was more significant than its idea of the emptiness of everything.

Confucius wrote, 'For him who respects the dignity of man, and practices what love and courtesy require, for him all men within the four seas are brothers.' Needham's interest in China was more than courtesy, more than pleasure at discovering the unity of the world: he found in China also much that was different from what

he believed in, and he valued that as essential to the full appreciation of his own principles.

The most challenging of all locks is perhaps the one that keeps
Islam separate from unbelievers. And yet Arabic is the only language in which the word 'man' comes from the root 'sympathy': to
be a man, etymologically, means to be polite or affable. In their
proverbs, Arabs define themselves as 'people who like to be liked'.
Is it right to believe that there can be no end to the conflict of
Muslims and their neighbours, since holy war (*jihad*) against infidels is an Islamic duty? The Islamic ideal of the good life is sociability, not war. There is almost no mention of war in the early,
Mecca chapters of the Koran. The Prophet, on returning from one
of his military campaigns, expressed pleasure that he could now
turn from the lesser war to the greater war, which takes place
inside each individual's soul; and over the centuries the spiritual
side of Islam has become increasingly dominant in the private lives
of believers. It is true that after Islam's rapid military victories,
the 'sword' verses of the Koran were held to have superseded the
peaceful ones; but theologians, as usual, disagreed. Sayyid Ahmad
Khan (1817–98), for example, argued that holy war was a duty for
Muslims only if they were positively prevented from practising
their religion. It was the sense of not receiving due respect, and of
being humiliated by colonisation, that brought the 'sword' verses
into prominence once more.

'Beware of innovations,' said the Prophet. Superficially, it might
seem that Islam is hostile to all modernisation, but during its first
250 years it allowed great scope for individual reason. *Jihad*
means not only war but effort. There was another kind of effort
which was also encouraged – *itjihad*, which meant that the faithful
were required to work out individually how they should behave in
matters not directly covered by the holy texts. Those who studied
the Koran and did their best to form their own opinions were
assured of a reward from the Prophet, even if they were wrong.
Divergences (*ikhtilaf*) were declared to be allowed by Allah by
Abu Hanifah (700–67), one of Islam's greatest jurists, founder of
the school of law which has the largest following of all; and three

other schools of law, though different, were considered to be all equally legitimate.

A time came when some theologians demanded that the age of personal judgement (*itjihad*) should be closed, because their accumulated judgements had settled all possible uncertainties. But others insisted that it should remain open. Ibn Taymiyya (1263–1328), for example, though he inspired conservative forms of Islam, pointed out that a Muslim was obliged to obey only God and his Prophet, not ordinary mortals; each had a right to give his opinion 'within the limits of his competence'. Both he and Abu Hanifah spent time in prison because other Muslims rejected their views. Schism and disputation have been a permanent part of Muslim history. Outsiders who have perceived it as a monolithic, unchanging totality miss completely the enormous richness of its traditions, the complex emotions that it shares with the history of other religions, and the significance of the Prophet's statement that internal beliefs may be judged only by God. Though submission to God has been greatly emphasised by Muslims, the Qadariya sect insisted that humans had total freedom of will; the Kharajist sect even argued that it was legitimate to have a woman as a prayer leader (*imam*), and in one of its rebellions, led by Shabib b. Yazid, it used an army of women. The Azraqite sect made it a duty to rebel against an unjust government. The Shiites have alternated between quietism and challenge to the legitimacy of the secular state in their insatiable search for moral perfection.

There are obviously limits to how far Christians and Muslims can agree; the monotheistic religions have been like brothers who have each gone their own way; and centuries of insult and caricature are not easily forgotten. The fact that Jesus is mentioned 93 times in the Koran, and that St Thomas Aquinas refers to Ibn Sina (Avicenna) 251 times cannot alter much. It is not through theological disputation that people come to appreciate one another; but it is possible to find in Islam as broad a range of attitudes as in any other civilisation.

Today, many Muslims have as their hero the poet Al-Mutanabbi (915–65), the symbol of the defiant person who refuses to be cowed, who is not ashamed to boast about his own abilities – he claimed his verse was equal in beauty to the Koran – and who is at once quixotic, independent and generous. But equally admired is the blind vegetarian Al-Maari (973–1057), author of the *Epistle of Forgiveness*, who ridiculed dogmatism, insisting that the only truly religious people were those who were helpful to their fellow creatures, regardless of their beliefs; and he meant all creatures, for he was one of the first great defenders of animals. The most celebrated of Persian physicians, Ar-Razi ('Rhases', 850–925), is a precursor of Voltaire in his fierce anticlericalism, a model of a sceptical scientist; he was an expert on psychosomatic illness, the placebo effect and quacks.

Muslims are as much heirs to the legacy of ancient Greece as Christians are. In the romance about a child on a desert island written by Abu Bakr Ibn Tufayl (died 1185) of Granada, which inspired *Robinson Crusoe*, the point of the story is that it is possible to arrive at truth by thinking for oneself. And thinking has nurtured doubt as much among Muslims as among others: the most celebrated example is Al-Ghazzali (1058–1111), who resigned his professorship to spend many years travelling alone in an attempt to resolve his hesitations, and who finally showed how the many sides of Islam, rational, mystical, legal, political, could be combined to satisfy sophisticated minds.

The aim of many of the most pious Muslims – the mystic Sufis – has been to be 'a friend of God'. To be human, said the Sufi poet Rumi (1207–83, of Persian origin, a resident of Konya, Turkey), is to be confused, distraught, in pain, in love, unable to decide what is right and what is wrong – but there was no need to wallow in suffering. Through music and dance it was possible to discover what really mattered.

Oh hear the flute, how it does complain
And how it tells of separation's pain.

He founded the order of the Whirling Dervishes to enable people to dance themselves out of pain, out of uncertainty and separation, into ecstasy. 'I don't know how to identify myself,' he wrote. 'I am neither a Christian nor Jewish, neither Pagan nor Muslim. I don't come from the East or from the West. I am neither from the land nor the sea. I am not a creature of this world.' Of course, among the many congregations offering guidance and companionship in the search for a timeless bliss, some were diverted from their spiritual purpose and became refuges of anger, even hatred. Mystical experience has been sought in virtually all religions, and yet they have seldom seen how it unites them, even if it is not the same God that they find.

The position of women in most Muslim countries is probably the most serious, seemingly insuperable, obstacle for Westerners, but, despite the apparently uncompromising holy texts quoted to support their subordination, Zainab the granddaughter of the Prophet has become the symbol of the autonomous and assertive woman, in contrast to Fatimah, who is the model of obedience and submission. The position of women in Islam has varied considerably over time, in different regions and in different social classes; in subsistence economies, where everyone works, they have sometimes enjoyed considerable power; in some countries, educated women have been able to lead independent lives, in spite of restrictions. The fundamentalist attempt to put women back in the home has been partly an urban reaction, a response to crisis: it is not peculiar to Islam, and is to be found, in varying degrees, wherever fundamentalism is growing, in every continent and in every religion. Nor have even the most conservative Muslim societies been totally unyielding: Ayatollah Khomeini himself, saying that it was absurd to want to destroy modern civilisation and to make people 'live in shackles or in the desert forever', gave permission for women to appear on television and he authorised contraception; but of course he balanced that by other controls. There have been Muslim leaders who have tried to advance the women's cause, as Mustafa Kemal Atatürk (1881–1938) did when he turned Turkey into a secular state; the results may be inconclusive now, but may eventually be surprising if that country is admitted into Europe. The Mufti of Cairo, Muhammad Abduh (1849–1905), who spent many years as a political exile in Paris, envisaged a feminist movement within Islam, and his influence is not dead.

His name is best remembered in Indonesia, the most populous Muslim country today. Here Islam, for many centuries, did not attempt to obliterate the Hindu and Buddhist traditions, nor the firmly established conviction that an individual's relationship with the divine was a private matter and that pressure for orthodoxy was unacceptable. It began as a religion of merchants, accepted by the native inhabitants as an additional source of supernatural protection, incorporated into pagan and mystical traditions, so that the goddess of the Southern Ocean continued to remain a spiritual force for many. Here Muslim cities were founded five or six centuries later than the ones in the Middle East, and they developed different values. The Muslim Renaissance in South-East Asia – which happened before Europe's and which has only recently been rediscovered from hoards of Malaysian manuscripts that are just beginning to be studied – gave these merchants a vision not of an unchanging, fixed order but of the very opposite, a world full of uncertainty, with the individual passing through as a stranger; anything was possible, the poor could become rich, and faithful observance of ritual was not enough to ensure survival for the rich; the need for generosity and almsgiving was emphasised, ostentation was condemned and an idea of equality was held up. These city states were comparable to those of Renaissance Italy and Flanders: the idea that Islam has always been linked with Oriental despotism is a myth. Trade being inevitably precarious, the merchants favoured multinational diversification and created widespread international commercial networks; they were cosmopolitans, co-operating with Chinese, Indian, Armenian and Arab merchants, reaching the height of their prosperity in the nineteenth century, when their trade – and that of the Chinese – expanded faster than the trade of the invading European merchants. Though they were ultimately defeated by Western efficiency, having such a long history of success, they could not consider that reverse as more than temporary. The Indonesian archipelago's most esteemed poet, Hamzah Fansuri (another of Shakespeare's contemporaries), author of *The Beverage of Lovers*, insisted that God could be seen in every person and every thing, and that the superficial differences between humans should not mislead:

> *The sea is eternal: when it heaves*
> *One speaks of waves but in reality they are the sea.*

However, it may seem that fundamentalism has now drawn a curtain between Islam and the West, closing all possibilities of mutual hospitality. The bitterness and violence of many fundamentalist movements has indeed created uncompromising confrontation. The word 'fundamentalism' was first used in the 1920s by Protestant sects in the USA, where about one-quarter of the inhabitants now share their outlook, almost exactly the same proportion as are members of new religions in Japan, different though those are. As an attitude, fundamentalism has appeared again and again in history. When there is a massive increase in population and old institutions are unable to cope, and families can no longer provide work, and children have to go off to fend for themselves in cities where there is no security or moral support, and where the rich and powerful are corrupt, flaunting their pleasures before those who cannot afford them, then some kind of consolation or religion has to come to the rescue.

The fundamentalists in Egypt, for example, are reacting particularly to the lack of hospitality in cities. All cities are inhospitable, the old residents always resent the new ones, though they depend on them to do the unpleasant jobs. But when cities both attract and reject, when they do not offer enough jobs to immigrants, the underprivileged have to help themselves. So the fundamentalists include highly talented scientists and students, as well as the illiterate and the poor, who share, in different ways, a feeling that there is no welcome for them in the world as it exists, no proper place for them to exercise their talents, and no success of the kind they can be morally proud of. They want to restore 'family values' often because their families have failed to provide them with security or harmony or the kind of help and influence which families were once able to give. They are uncompromisingly nationalist, because they dream of an ideal nation which looks after all its citizens and guarantees them self-esteem. They organise themselves in brotherhoods and sisterhoods, which draw the young away from their parents, creating a substitute family. Here many women, by wearing traditional dress, find security against censure as well as friends, a sense of belonging, and often the possibility of playing an active role in the organisation, with the satisfaction that they are helping to shape the next generation. By accepting that each sex and age group must perform only certain roles, they limit competition; with advancing age, an individual can look forward to a new role. Far from believing that they are returning to the Middle Ages, they are convinced that they are providing a workable solution to the crisis of modernity, that they are enabling the young to triumph over destitution, creating a new morality, supported by their peers. There is a great deal of discussion about how one ought to live in a hostile world, even if blind conformism is the ideal of many. In the leading Cairo newspaper, *Al-Ahram*, Abdul Wahhab, who answers readers' letters, quotes Shakespeare, Balzac, St Francis of Assis, Dante, Einstein and Helen Keller to support his moral advice. As always, a big gap separates the sophisticated arguments of philosophers – like the Iranian Abdolkarim Soroush (born 1945), who is determined that Islam should not repeat the persecution of Galileo and that it should not be an impediment to science – and the unemployed and illiterate, for whom despair is the only alternative to anger.

There are Christian fundamentalists in the USA who have said that they do not want their children's horizons widened, that they are against the extension of hospitality towards strange ideas, that they are determined to control their children and to protect them from the unpredictable behaviour of other children, and so they prefer to withdraw into a safer world from which the abominations of modernity are excluded. The Pentecostal variety of fundamentalism in Latin America, in Venezuela for example, also stresses male supremacy in the family, but it is largely the creation of women, who have found in it an answer to absent, unemployed fathers: by giving prestige to men in the family, they have restored their pride, and their own acceptance of a submissive role in fact conceals the creation of a new kind of family, with the wives having a more active part and the men being more domestic and affectionate. In Ecuador personal conversion, or re-birth, has given a new sense of status to both men and women who had been unable to cope with competitive individualism, and who have recreated reciprocal social relations and a welfare system around their churches.

Fundamentalism cannot be dismissed simply as extremism: it is a force as powerful as communism once was, and it represents a not dissimilar reply to injustice and bewilderment. The more fundamentalists are repressed, the more they feel excluded. Their rejection of dialogue is based on the conviction that they are not understood. It is impossible to be hospitable to violence, but only those with short memories can believe that a new Cold War is the solution. It should not be forgotten that when Islam was at the height of its prosperity in the tenth century, it had no difficulty in being generous and tolerant to outsiders: the Brothers of Purity of Basra wrote then, 'The ideal and perfect man should be of East Persian origin, Arabic in faith, of Iraqi i.e. Babylonian education, a Hebrew in astuteness, a disciple of Christ in conduct, as pious as a Syrian monk, a Greek in individual sciences, an Indian in the interpretation of all mysteries, but lastly and especially a Sufi in his whole spiritual life.'

In a previous chapter I discussed the vigorous flowering of hospitality of the mind in India, and its fragility. I add a few notes about other terrains where further varieties are to be found. East Asia has the advantage that it has passed through a phase of what the Chinese call *chi*, humiliation, what the Koreans call *halm*, remorse or bitterness, and what the Japanese call *nin*, endurance and patience waiting for better times. Both the Germans and the French felt something very similar in 1945, and discovered what benefits humiliation could yield, as do Americans who remember their immigrant traditions. The strength of many Eastern countries is that they have been equally hospitable to a whole variety of Western and Eastern ideas.

Korea, for example, has been extracting, over several centuries, a unique 'practical learning' (*Sirhak*) out of Confucianism, Buddhism, shamanism and Christianity. The Sirhak movement has a history as long and remarkable as socialism; it borrowed with discrimination from every useful source, seizing quickly on Catholicism for its emphasis on equality, and evolving comprehensive programmes for an ideal society, as in the works of Yi Ik (1681–1763), author of the *Record of Concern for the Underprivileged*, and Chong Yag-yong (1762–1836), author of the *Design for Good Government*. It is not so much the fact that between 25 and 30 per cent of Koreans are now Christian which has shaped their originality, but how they have conducted their endless discussions on being Korean and 'modern' at the same time.

The Japanese, to take another example, have made hospitality of the mind into a refined art, following their tradition of turning almost everything into an art. The process of absorbing the influence of China, and then of Europe, and finally of the USA, has been more subtle than imitation. To cushion the impact of an import, they have put on protective layers of clothing which prevent their traditions getting hurt. While absorbing foreign technology, they half laugh at themselves for doing so, and revive, refurbish, re-create old religions to balance their imported creeds. The result is that they live in two or more worlds at the same time. This is a particularly interesting variety of hospitality, in which one is not swamped by the guest, and the guest is not allowed to become an intruder.

There can be misunderstanding between guest and host, but the Japanese have thought hard about this and about the difficulties of communicating, producing an ideal of direct communication from mind to mind (*isin-denishin*) – without words, which never say enough – an ideal of perfect understanding, almost akin to telepathy, or even religious revelation. Sometimes they believe that the only person one can communicate with effectively is oneself, which explains the national habit of keeping a diary, or the widespread custom of complaining that they are not understood, and the despair of young people who run away from home. But unlike Greek tragedy, which complains about Fate, Japanese tragedy is concerned more with not being understood. Etiquette requires that they should not reveal their thoughts too directly, for that would imply disrespect for the sensibility of the person they are addressing, and to argue one's point of view with the artillery fire of logic is a sign of immaturity.

Japan's imperialist and militarist impulses have thus been balanced by the very opposite, a passion for warmth. Its endless investigations of what makes it special and incomprehensible to others (*nihonjinron*), and its image of inscrutable silence, are balanced by the feeling, as Suzuki Takao wrote, that 'we Japanese unceasingly require that some suitable person understand what we are thinking and what our true feelings are. The wish to gain the assent and approbation of others, to savour their sympathy in one way or another, appears in every kind of our behaviour in contexts of interpersonal relationships.' To imagine such sentiments are peculiar to themselves is a sign of the difficulty of putting oneself in other people's shoes.

The Japanese, by analysing the problems of hospitality, have revealed some of its problems. They are still a long way from knowing how to interest every part of the world in their culture and their history. It is not by defining oneself that one achieves that: the novelist Natsume Soseki was already mocking that idea in 1905 when he wrote, 'Is the Japanese spirit triangular or is it quadrangular? As the name indicates, the Japanese spirit is a spirit. And a spirit is always blurry and fuzzy. . . . Everyone has heard about it but no one has yet encountered it.' But it is difficult to break old habits. The art of encounter, despite its long history, is still far from having been perfected.

That has been shown too by churches all over the world which have been trying to talk to one another. Negotiations to reconcile points of divergence in their theological positions have seldom drawn them much closer, for that involves defining their identity, which has usually led them to draw more rigid frontiers than existed in practice, ignoring the fuzzy edges. The first World Parliament of Churches, which met in Chicago in 1893, and the World Council of Churches (established 1945) have been almost as fruitless as disarmament conferences, even if there have been some reconciliations, even if different sects collaborate in such tasks as the translation of the Bible. There is still no sign of an end to wars of religion. Churches, as institutions, have been as reluctant as nations to compromise their sovereignty. The great surge in hospitality between religions has been due mainly to individuals and informal groups interested in reviving the primitive spirit of religion, ignoring the caution of those who hold power, and expressing their faith by devoting themselves to the disadvantaged.

But humanitarian organisations quickly discovered that they were not immune to sectarianism, and have also found themselves fragmenting into rival sects, just like the churches. 'When one tries to start a fraternity, one raises up hatred . . . the wars between philanthropists are the worst of all,' says the founder of Médecins sans Frontières, Bernard Kouchner. 'The organisers quarrel among themselves for control of the victims of disasters and fight to the death amongst themselves after having risked their lives together.' All the experience of history confirms that sharing the same beliefs has been a preliminary to quarrelling about their interpretation. Co-operation has worked best between those who have only a few aims in common, who are not rivals, who are untroubled by thoughts about who should control whom: when they come from different origins and intend to return once their mission is accomplished, there may be friction, but there have been fewer cancerous animosities. Close encounters must, in part, remain distant too.

That is why hospitality between believers and unbelievers holds out particular promise, now that the war is almost over between them, now that churches are no longer attempting to dominate government and, having become poor once more, are turning their attention to the poor, emphasising compassion more than dogma, and human relations more than ritual. In practice, believers and unbelievers often find themselves on the same side when faced by violence or injustice, and in such circumstances, the crucial distinction between them – belief in God and a sacred text – becomes a stimulus to the imagination of both, when the believers discuss their doubts and the unbelievers reflect on their own search for values. They are made to be intermediaries in each other's worlds.

A final insight into how people who, after being deaf to each other's words for centuries, begin to hear what the other is saying, and to understand, is provided by the history of music. If religions and nations have souls, music is the air they breathe. And yet no nation has ever had difficulty in learning foreign musics. The sound of Mozart and Rossini used to echo through the forests of Madagascar, played by a national orchestra formed by King Radama. The Sultan of Turkey chose Donizetti to be musical director of his court. It is perfectly true that music has been used as an affirmation of ethnic or national difference; and some patriotic Africans have denounced Nietzsche's statement that music is an international language as an insult, insisting that its function is to unite them to their own gods and spirits. But the musics of different continents have met and married and had beautiful children nonetheless. The way this has happened provides a highly instructive lesson in the art of hospitality.

The most successful encounter between musics has been that of Africa and Europe and America. African music is now considered by Western musicologists to have more variety and complexity than that of Europe; and such is its diversity, that there is no fundamental feature which unites it. However, it does contain some elements which have fitted into certain elements of European hymns and folk songs, and the union of these elements has produced a new music which is compatible with both traditions. This is an example of what I understand by soul-mates, not a pair which matches perfectly and becomes a single whole, but two different beings with some qualities, like some molecules, that are able to combine to make something new and interesting.

It is not proximity which makes an alliance possible, because native North American music has not combined with the music of the settlers. Nor is it similarity: the music of India has a considerable amount in common with that of Europe, but there has been almost no interpenetration, apart from film music, because Indian classical music is an essential part of Hindu religious events and temple worship, and is firmly based on the memory of great composers and flourishing musical schools. By 1900, there were already 4,000 recordings of its music available in the catalogue of just one gramophone company. So India has borrowed Western musical technology, but not its sounds. There was no gap for Europe to fill. Iran, by contrast, in the mid-twentieth century, had only about 2,000 connoisseurs of its classical music left, and its popular composers therefore found more inspiration from abroad. Japan and Korea, meanwhile, taught their musicians both Western and native music, and cultivated a double musical ear, two separate musics side by side.

Those who fear that hospitality must damage native traditions have not only this counter-example, but also many more in Africa. It used to be thought that its music was something that went back unchanged to the beginning of time, but research has revealed it to have evolved a great deal, independently of imperialist pressure. Thus the juju music of the Yoruba, which has its origins in an older 'palm wine music' (music performed while palm wine was drunk), has in the course of the past century been expanded with Afro-Brazilian borrowings, and then dipped back into traditional poetry, adding a wider range of traditional talking drums, then soul, reggae, country and western and Indian film music. The 'father of juju' was an African Christian preacher, while one of its leading modernisers called his band The International Brothers. Historians have shown likewise that 'Benni' music, which Westerners imagined was an imitation of themselves, because the participants wore naval uniforms, has ancient origins, and was revitalised by a rebellion of young Africans against their elders, incorporating new ideas with careful and changing discrimination. The major influence on modern African music has probably been the 1977 Festival of Black Arts, which created new contacts and new forms of hospitality among Africans who had barely known each other before.

In the Middle Ages, the music of Muslim Spain was taken into Europe by the troubadours, through whom Arab ideas about love penetrated into the Western world, eventually transforming the behaviour of people who could not speak a word of Arabic. In the twentieth century, African music has helped to strengthen the revulsion against racial prejudice. But music needs intermediaries to help it across frontiers. The majority of whites in the USA segregated themselves from black music until it was reinterpreted for them by people with whom they could identify: that was what Elvis Presley and the Beatles achieved. The modern intermediaries between Arab and Western music have still to make their breakthrough: the Tunisian singer Amina Annabi, though selected to represent France in the 1991 Eurovision concert, has had more success in Japan. The tension between Muslims and Europeans is still too great. Hospitality cannot be forced through memories that have gone rigid.

Thinking

Sophie de Mijolla-Mellor, *Le Plaisir de pensée*, PUF, 1992; Mary War-nock, *Imagination*, Faber, 1976; David Cohen and S. A. MacKeith, *The Development of Imagination: The Private Worlds of Children*, Rout-ledge, 1991; Senko Kurmiya Maynard, *Japanese Conversation*, Ablex, Norwood, NJ, 1989; Lee N. Robins and Michael Rutter, *Straight and Devious Pathways from Childhood to Adulthood*, Cambridge UP, 1990; Douglas Robinson, *American Apocalypses*, Johns Hopkins UP, Balti-more, 1988; Frank L. Borchardt, *Doomsday Speculation as a Strategy of Persuasion*, Edwin Mellen, Lampeter, 1990; Doris B. Wallace and H. E. Gruber, *Creative People at Work*, Oxford UP, NY, 1989; Barbara Rogoff, *Apprenticeship in Thinking*, Oxford UP, 1991; Hannah Arendt, *The Life of the Mind*, Secker & Warburg, 1978; Linda Silka, *Intuitive Judgements of Change*, Springer, NY, 1989; Jurgen T. Rehm, *Intuitive Predictions and Professional Forecasts*, Pergamon, 1990; Weston H. Agar, *Intuition in Organisations*, Sage, 1989; D. Kahneman, P. Slovic and A. Tversky, *Judgement under Uncertainty: Heuristics and Biases*, Cambridge UP, 1982; David Le Breton, *Passions du risque*, Metaile, 1991; Margaret Donaldson, *Human Minds*, Allen Lane, 1992; Dan Sperber and Deirdre Wilson, *Relevance: Communication and Cognition*, Blackwell, Oxford, 1986; Dan Sperber, *Rethinking Symbolism*, Cambridge UP, 1975; Dan Sperber, 'The Epidemiology of Beliefs', in C. Fraser and G. Gaskell, *The Social Psychological Study of Widespread Beliefs*, Oxford UP, 1990.

Relations between East and West

M. E. Marty and R. S. Appleby, *The Fundamentalism Project*, 3 vols., Chicago UP, 1993; Denis MacEoin and Ahmed El Shahy, *Islam in the Modern World*, Croom Helm, 1983; Henri Corbin, *Histoire de la philo-sophie islamique*, Gallimard, 1964; Gilles Keppel, *Les Banlieues de l'Islam*, Seuil, 1987; Henri Laoust, *Les Schismes de l'islam*, Payot, 1965; Anne Marie Schimmel, *I am Wind, You are Fire: The Life and Work of Rumi*, Shambhala, Boston, 1992; Patricia Crone, *Meccan Trade and the Rise of Islam*, Blackwell, Oxford, 1987; Richard W. Bulliet, *Conversion to Islam in the Medieval Period*, Harvard UP, 1979; Henri La Bastide, *Les Quatres voyages*, Rocher, 1985; G. E. von Grunebaum, *Unity and Variety in Muslim Civilisation*, Chicago UP, 1955; R. M. Savory, *Intro-duction to Islamic Civilisation*, Cambridge UP, 1976; Bernard Lewis, *The Muslim Discovery of Europe*, Weidenfeld, 1982; J. Berque and J. P.

Charnay, *Normes et valeurs dans l'Islam contemporain*, Payot, 1966; J. Berque and J. P. Charnay, *L'Ambivalence dans la culture arabe*, Anthropos, 1967; M. R. Woodward, *Islam in Java*, Arizona UP, 1989; Denys Lombard, *Le Carrefour javanais*, 3 vols., EHESS, 1990; Malcolm Wagstaff, *Aspects of Religion in Secular Turkey*, Durham University, 1990; Bernard Lewis, *The Emergence of Modern Turkey*, Oxford UP, 1968; Nehemiah Levtzion, *Conversion to Islam*, Holmes and Meyer, 1979; Sigrid Hunke, *Le Soleil d'Allah brille sur l'Occident*, Albin Michel, 1963; Jacques Gernet, *Le Monde chinois*, Armand Colin, 1972; Jacques Gernet, *Chine et christianisme*, Gallimard, 1982; Joanna F. Handlin, *Action in Late Ming Thought: The Reorientation of Lu K'un and Other Scholar Officials*, California UP, 1983; Hung Ying-ming, *The Roots of Wisdom: Saikuntan*, Kodansha, Tokyo, 1985; Mark Elvin, *The Pattern of the Chinese Past*, Eyre Methuen, 1973; Mark Elvin and G. W. Skinner, *The Chinese City Between Two Worlds*, Stanford UP, 1974; J. Baird Callicott and Roger T. Ames, *Nature in Asian Traditions of Thought*, State University of NY Press, 1989; K. Schipper, *Le Corps taoiste*, Fayard, 1982; S. A. M. Adshead, *China in World History*, Macmillan, 1988; S. A. M. Adshead, *Central Asia in World History*, Macmillan, 1993; Tu Wei-Ming, *Confucian Thought: Selfhood as Creative Transformation*, State University of NY Press, 1985; H. Nakamura, *Ways of Thinking of Eastern Peoples*, Hawaii UP, 1984; Joseph Needham, *Science and Civilisation in China*, Cambridge UP, 1954–78; Henry Holorenshaw, 'The Making of an Honorary Taoist', in M. Teich (ed.), *Changing Perspectives in the History of Science*, Heinemann, 1973; K. G. Temple, *China: Land of Discovery*, Multimedia, 1986; Marcel Granet, *La Pensée chinoise*, Albin Michel, 1988; L. Carrington Goodrich, *Dictionary of Ming Biography*, Columbia UP, 1976; Harvey Cox, *Turning East: The Promise and Peril of the New Orientalism*, Allen Lane, 1977; W. D. O'Flaherty, *The Origin of Evil in Hindu Mythology*, California UP, 1976; W. D. O'Flaherty, *Dreams, Illusions and Other Realities*, Chicago UP, 1984; Daniel Gold, *The Lord as Guru*, Oxford UP, NY, 1987; Abbé Pierre and Bernard Kouchner, *Dieu et les hommes*, Laffont, 1993; John B. Cobb, jnr., *The Emptying God: A Buddhist-Jewish-Christian Conversation*, Orbis, NY, 1990; Peter K. H. Lee, *Confucian–Christian Encounters*, Edwin Mellen Press, Lampeter, 1991; Rodney L. Taylor, *The Religious Dimensions of Confucianism*, State University of NY Press, 1990; Albert H. Friedlander, *A Thread of Gold: Journeys Towards Reconciliation*, SCM Press and Trinity Philadelphia, 1990; Leonard Swidler, *Muslims in Dialogue*, Edwin Mellen, Lampeter/Lewiston, 1992; Robert J. Sternberg, *Wisdom: Its Nature, Origin and Development*, Cambridge UP, 1990; John Hick and Hasan Ascari, *The Experience of Religious Diversity*, Avebury,

Aldershot, 1985; Unni Wikkan, *Managing Turbulent Hearts: A Balinese Formula for Living*, Chicago UP, 1990; L. Swidler and P. Mojzes, *Attitudes of Religions and Ideologies Towards the Outsider*, Edwin Mullen, Lampeter/Lewiston, 1990; Kenneth Cragg, *To Meet and To Greet: Faith with Faith*, Epworth, 1992; Peter N. Dale, *The Myth of Japanese Uniqueness*, Croom Helm, 1986; J. Eckert, *Korea Old and New*, Ilchokak, Seoul, 1990; Ki-baik Lee, *A New History of Korea*, Harvard UP, 1984; Martina Deuchler, *The Confucian Transformation of Korea*, Harvard UP, 1992; In-sob Zong, *A Guide to Korean Literature*, Hollym, Seoul, 1982.

World music

Irene V. Jackson, *More than Drumming: Essays on African and Afro-Latin American Music and Musicians*, Greenwood, Westport, 1985; Bruno Nettl, *The Western Impact on World Music*, Schirmer, NY, 1986; Deanna Campbell Robinson, *Music at the Margins: Popular Music and Global Cultural Diversity*, Sage, 1991; Marcia Herndon and Susanne Ziegler, *Music, Gender and Culture*, Florian Noetzel, Wilhelmshaven, 1990; Jean-Pierre Arnaud, *Freud, Wittgenstein et la musique*, PUF, 1990; Christopher Page, *The Owl and the Nightingale: Musical Life and Ideas in France 1110–1300*, California UP, 1990; William P. Malm, *Japanese Music and Musical Instruments*, Tuttle, Tokyo, 1959; Daniel M. Neuman, *The Life of Music in North India*, Chicago UP, 1980; Judith Hanna, *To Dance is Human*, Texas UP, 1979; John A. Sloboda, *The Musical Mind*, Oxford UP, 1985; Philip Sweeney, *The Virgin Directory of World Music*, Virgin Books, 1991.

25

What becomes possible
when soul-mates meet

Humanity's most long-lasting purpose has been to produce more humanity. Once that meant having as many children as possible, but the amount of kindness given to children has come to matter more than their number. Today humanity is above all an ideal of caring and kindness extending to every age and to every living being. The first rumblings of this historic shift were heard many centuries ago, but now large parts of the world are being shaken by it.

Imagining the form that a renaissance built on such a foundation might take has been difficult, because people have never been able to have a new vision of the future without first revising their idea of the past. The spectacles I have offered my readers are designed to facilitate such a revision, to show that history did not have to happen the way it did, and that what exists today is not its logical conclusion. There is no freedom where history is a straitjacket. I have written this book to present human experience as a source from which a sense of purpose may be derived, without any implication of inevitability or necessity, for it also opens up a vast array of options.

I see humanity as a family that has hardly met. I see the meeting of people, bodies, thoughts, emotions or actions as the start of most change. Each link created by a meeting is like a filament, which, if they were all visible, would make the world look as though it is covered with gossamer. Every individual is connected to others, loosely or closely, by a unique combination of filaments, which stretch across the frontiers of space and time. Every individual assembles past loyalties, present needs and visions of the future in a web of different contours, with the help of heterogeneous elements borrowed from other individuals; and this constant give-and-take has been the main stimulus of humanity's energy. Once people see themselves as influencing one another, they cannot be merely victims: anyone, however modest, then becomes a person capable of making a difference, minute though it might be, to the shape of reality. New attitudes are not promulgated by law, but spread, almost like an infection, from one person to another.

The dispute about how to achieve a better life, whether it should be by individual effort or by collective action, has no point any more, because they are two sides of the same coin. It is difficult to do anything without help or inspiration from outside oneself. Individual struggles have simultaneously been collective ones. All the great movements of protest against contempt, segregation and exclusion involve an infinite number of personal acts by individuals, making a small change in the whole by what they learn from each other, and by the way they treat others. To feel isolated is to be unaware of the filaments which link one to the past and to parts of the globe one may never have seen.

The age of discovery has barely begun. So far individuals have spent more time trying to understand themselves than discovering others. But now curiosity is expanding as never before. Even those who have never set foot outside the land of their birth are, in their imaginations, perpetual migrants. To know someone in every country in the world, and someone in every walk of life, may soon be the minimum demand of people who want to experience fully what it means to be alive. The gossamer world of intimate relations is in varying degrees separate from the territorial world in which people are identified by where they live and work, by whom they have to obey, by their passports or bank balances. The rise of Christianity and other religious movements in the Roman Empire is an example of a new gossamer spreading over a rotting civilisation; though outwardly emperors and armies continued to give orders as though nothing had changed, individuals, feeling that official institutions were ceasing to be relevant to their needs, sought their consolations from each other. Today, a similar switch in attention is happening: the earth is in the early stages of being criss-crossed afresh by invisible threads uniting individuals who differ by all conventional criteria, but who are finding that they have aspirations in common. When nations were formed, all the threads were designed to meet at a central point; now there is no centre any more; people are free to meet whomever they wish.

However, if everyone met the people they dream about, it would not mean that all discontents would evaporate. Soul-mates have often had tragic histories. Meeting God has not prevented pious people from being cruel in His name. Friendships have frequently deteriorated into sterile routine. Most lives have come to a halt in early adulthood, after which new encounters have yielded nothing new. Those who have suffered most from this kind of arrested development have dedicated themselves to crime, becoming interested in nobody but themselves, which is the final bankruptcy of the imagination.

Imaginations, nevertheless, are not doomed to fossilise. Mutual discovery has often led people to care for each other as much as for themselves. Usefulness to a fellow human has from time to time been recognised as a more deeply satisfying pleasure than the pursuit of self-interest, though it is an increasingly delicate one, hampered by ever more complicated sensitivities. Some relationships have been established which go beyond the belief that humans are basically animals, or machines, or lifelong invalids in need of permanent medical attention. But the art of encounter is in its infancy.

A new age always involves a new kind of hero. In the past, humans have admired heroes because they had a low opinion of themselves, with very few believing they were personally capable of being heroes; but they have also repeatedly unmasked their heroes as fakes, and most efforts to invent new kinds of heroes have ended in disappointment. Machiavelli's hero was too callous. Gracian's involved too much pretence. The romantic idol had charm but carried sensitivity to the point of self-torture. The workaholic Hero of the Soviet Union eventually felt duped. Heroes were once conquerors, but to subjugate is no longer admirable, and those who command are now valued less than those who encourage. Success in a career is no longer enough to make a person a hero, because private life is now valued as much as public achievement. Religion can still inspire some enthusiasts to be martyrs, but very few choose to be saints. The charismatic orator and the revolutionary leader are increasingly viewed with suspicion, for the world is tiring of broken promises and prefers someone who listens.

'Happy is the country that needs no heroes,' said Brecht. No, they will be missed if they vanish. But too many have mistaken themselves for gods: there has been a shortage of modest heroes. That is why antiheroes were invented, who could never disappoint. Today a hero is not so much one who sets an example for others to follow, because the ideal relationship involves each partner being made more alive by the other; heroes must be able to receive as well as to give, because influence which travels only one way can become dispiriting or corrupting. To benefit from a hero, one must be a bit of a hero oneself; one must have courage; the heroic relationship is an exchange of courage. Heroes need to be intermediaries, who open the world up to one another. To be an intermediary who does not cheat is within everyone's grasp.

However, it is not enough to focus only on the minute synapses of personal encounters. It has become possible, as never before, to pay attention to what is happening in every corner of the globe. Humans each have a personal horizon, beyond which they normally dare not look. But occasionally they do venture further, and then their habitual way of thinking becomes inadequate. Today they are becoming increasingly aware of the existence of other civilisations. In such circumstances, old problems take on a new appearance, because they are revealed as being parts of larger problems. The shift in interest away from national squabbles to broad humanitarian and environmental concerns is a sign of the urge to escape from ancient obsessions, to keep in view all the different dimensions of reality, and to focus simultaneously on the personal, the local and the universal.

Justice – humanity's oldest dream – has remained elusive

because the art of doing this is only gradually being learned. In ancient times, justice was blind, unable to recognise the humanity that is in everybody. In modern times it has been one-eyed, narrowly focused on the principle of impersonality, imposing the same rules on everybody so as to avoid nepotism and favouritism, but unable to notice what people feel when they are treated impersonally and coldly, however justly or efficiently. The impersonal monetary compensations of the welfare state have not been able to heal the wounds of unfairness, because nothing can compensate adequately for a wasted life, least of all when even in the USA, which has studied efficiency to its limits, it takes seven tax dollars to get one additional dollar of income into the hands of a poor person. Only with both eyes open is it possible to see that humans have always needed not just food and shelter, health and education, but also work that is not soul-destroying and relationships that do more than keep loneliness out; humans need to be recognised as persons. This book is a history of persons.

Humanity can only have a satisfying sense of direction when it can calculate its achievements with the help of an economics which is about persons as they really are, which incorporates irrational and altruistic behaviour in its computations, which does not assume that people are always and fundamentally selfish and which understands that success even in the material world is not obtained by the exclusive pursuit of self-interest. This two-eyed economics is in the process of being born, as is a two-eyed politics, concerning itself not simply with giving victory to the majority, but with offering mutually acceptable alternative victories to the losers, encouraging, without jealousy, the cultivation of multiple loyalties.

Religion has always been two-eyed, essentially universal, encompassing both the material and the spiritual, balancing personal salvation and regard for others, though many believers have chosen to be one-eyed, and to see nothing but their own truth. In the twelfth century Maimonides said that all human beings could expect to go to heaven, whatever their theological beliefs, provided they behaved with decency, accepting the 'Seven Laws of Noah', who was everybody's father when 'the whole earth was of one language and one speech', laws which demanded no more than respect for others; Jews agreed that a Muslim or a Christian could be 'a righteous person'. The universal essence of religion is rediscovered whenever it is remembered that doctrine divides while action unites; or that *ying* and *yang* are not opposed, but interacting; or that Hindu devotion (*bhakti*) involves learning the art of listening and befriending; or, as the Warsaw rabbi declared before his community was annihilated, that nobody is alone – *yahid* (solitary) and *yahad* (together) are separated by only one letter. Discovering compatibilities across the frontiers of dogma is the next item on the agenda for believers and unbelievers alike, who do not wish to be confused by the different metaphors which each system of beliefs adopts. The ghosts of the past can be put to useful work, and do not have to cause havoc.

However, so long as each section of humanity, forgetting that the search for respect is a universal preoccupation, demands it only for itself, the results will be mediocre, as they have been in the past. The traditional methods of agitation, legislation and slow infiltration into positions of power have never been enough to change mentalities. Women winning their way into previously closed professions have generally had to accept the rules of those already in power, who make concessions on the understanding that the newcomers play the game more or less as it has always been played. Moreover, economic independence, the right to work and equal pay are not ends in themselves, but a means towards a more complete life, which most jobs were not designed to encourage. Beyond the struggle for power lies the possibility of seeking self-respect by helping others to respect one another.

I have tried to provide a basis on which to build not a withdrawal from public affairs into private self-obsession, but an awareness of what is most genuinely public, what humans share. What is unique about the present time is that humanity has never been so conscious of the primacy of its intimate concerns, nor expressed them so openly, in almost every part of the globe. Searching for what we have in common, despite our differences, provides us with a new starting point.

'My life is a failure.' Those were the words with which I began this book, and I finish it with the story of a murderer who repeated that phrase many times, until one day . . .

Half a minute is enough to transform an apparently ordinary person into an object of hatred, an enemy of humanity. He committed a murder and was sentenced to life imprisonment. Then in his desolate jail, half a minute was enough to transform him again, into a hero. He saved a man's life and was pardoned. But when he got home he found his wife living with someone else and his daughter knew nothing of him. He was unwanted, so he decided that he might as well be dead.

His attempt at suicide was also a failure. A monk summoned to his bedside said to him, 'Your story is terrifying, but I can do nothing for you. My own family is wealthy, but I gave up my inheritance and I have nothing but debts. I spend everything I have finding homes for the homeless. I can give you nothing. You want to die, and there is nothing to stop you. But before you kill yourself, come and give me a hand. Afterwards, you can do what you like.'

Those words changed the murderer's world. Somebody needed him: at last he was no longer superfluous and disposable. He agreed to help. And the world was never the same again for the monk, who had been feeling overwhelmed by the amount of suffering around him, to which all his efforts were making only a minute difference. The chance encounter with the murderer gave him the idea which was to shape his whole future: faced by a person in distress, he had given him nothing, but asked something from him instead. The murderer later said to the monk: 'If you had given me money, or a room, or a job, I would have restarted my life of crime and killed someone else. But you needed me.' That was how Abbé Pierre's Emmaus movement for the very poor was born, from an encounter of two totally different individuals who lit up a light in each other's heart. These two men were not soul-mates in the ordinary, romantic meaning of that word, but each owes the other the sense of direction which guides their life today.

It is in the power of everybody, with a little courage, to hold out a hand to someone different, to listen, and to attempt to increase, even by a tiny amount, the quantity of kindness and humanity in the world. But it is careless to do so without remembering how previous efforts have failed, and how it has never been possible to predict for certain how a human being will behave. History, with its endless procession of passers-by, most of whose encounters have been missed opportunities, has so far been largely a chronicle of ability gone to waste. But next time two people meet, the result could be different. That is the origin of anxiety, but also of hope, and hope is the origin of humanity.

Acknowledgements

This book is the result of help, stimulation and encouragement from many people. I should like to thank all those who have discussed their experience of life with me; each one has, in a different way, increased my respect for the human species. My colleagues at St Antony's College, Oxford, have shown me great kindness and freely shared their erudition with me. Journalists in different countries, who know far more than they publicly say or write, have generously given me valuable pointers in my local enquiries. Having spent a vast portion of my life reading books, my debt to other authors is immeasurable, as it is to the librarians and booksellers who helped me find them. I owe much to the business people, voluntary associations, politicians and public officials who, having invited me to discuss their aims or strategies with them, have enabled me to enter worlds normally closed to outsiders and to observe at first hand the practical problems which book learning cannot wholly grasp. Publishing has been made a pleasure for me by Christopher Sinclair-Stevenson, Claude Durand, Hugh Van Dusen, Christopher MacLehose, Jean-Bernard Blandinier, Eric Diacon, Roger Cazalet, Andrew Nurnberg, Robin Straus, as well as by Edith McMorran and Louise Allen. I have benefitted greatly from discussions with Christina Hardyment. And, as always, my wife Deirdre Wilson has shown inexhaustible generosity with ideas and encouragement. I do not know why people say that writing is a solitary occupation.

Index

Index